Praise for *E[illegible]*

"[*Empires of Light*] provides a wealth of colorful anecdotes and fascinating detail." —*The Washington Post Book World*

"In *Empires of Light*, Jill Jonnes shares a rollicking story of competitive zeal.... [The book] delivers richly on its promise: chronicling a vital stage of American progress as seen through the lives of three mavericks." —*The Wall Street Journal*

"Entertaining and informative ... a lively account of how personal ambitions and hostilities fueled the interaction between science and business during the long War of the Electric Currents." —*Los Angeles Times Book Review*

"The electrons fairly leap as Jonnes personifies that high-voltage history with a three-wired account." —*Johns Hopkins Magazine*

"[*Empires of Light*] moves seamlessly back and forth in time.... Jonnes is a fine biographer and an excellent scientific and industrial historian. She's done a superb job of telling an important story." —*Rocky Mountain News*

"Compelling ... Like the late Stephen Ambrose, historian Jill Jonnes paints her story on a broad canvas and populates it with titans." —*BookPage*

"Fascinating." —*The Buffalo News*

"With *Empires of Light*, Jill Jonnes joins the genre of academicians who truly document for the nation's collective memory the significant struggles that led to commonplace conveniences of today." —*Chicago Tribune*

"[*Empires of Light*] is fascinating reading for a general audience as well as those with a more specialized interest." —*Booklist*

"[Jill Jonnes] brings [Edison, Tesla, and Westinghouse] to life through cumulative biographical detail." —*Boston Sunday Globe*

"A crackerjack account of the race for electrification . . . [*Empires of Light*] is a story of the collision of business and technology, and Jonnes tells it well." —*San Francisco Chronicle* (Best Books of 2003)

"Jonnes's book makes us think about the dramatic changes electricity brought." —*Milwaukee Journal Sentinel*

"Jill Jonnes has delivered an absorbing tale." —*The San Diego Union-Tribune*

"Jonnes lucidly lays out the technical issues, paying plenty of attention to the personalities involved to liven things up for the general reader." —Newsday.com

"Thoughtful and well paced." —*Kirkus Reviews*

"Jonnes serves up plenty of color in an engaging and relaxed style." —*Publishers Weekly*

"Historian Jill Jonnes re-creates this venemous rivalry in a delightful book that may remind readers of E. L. Doctorow's *Ragtime*. . . . But *Empires of Light* is no fiction; it's a meticulously researched narrative in which famous people go baying after an elusive goal: to power cities by harnessing a hidden force wrested from the atmosphere." —*Discover*

"Jill Jonnes's *Empires of Light* is the most exciting science/business adventure to come out in the past decade. Once she gets past the initial discoveries of the properties of electricity, her brilliant storytelling pulls the reader into a gripping, real-life turn-of-the-century tale full of twists, turns, ironies, dirty tricks, breakthroughs, challenges, accomplishments, tragedies and triumphs." —*Houston Chronicle*

"A thoroughly engaging and highly informative account of three inventors who pioneered the production and distribution of electricity. Without these three engineers, the world would simply not be what we know today."

—HENRY PETROSKI,
author of *The Evolution of Useful Things*

"Jill Jonnes's *Empires of Light* is the captivating—no, let's say electrifying—saga of the War of the Electric Currents fought at the close of the nineteenth century with typical Gilded Age excess by Thomas Edison, Nikola Tesla, and George Westinghouse. From the electrification of J. P. Morgan's New York mansion to Westinghouse's subjugation of Niagara Falls, Jonnes explains in human terms how alternating current achieved dominance over direct current, a victory of incalculable importance in the history of the world—and she tells the story with great, at times even macabre, verve, as in her account of the invention of the electric chair and its horrifying first use. Along the way, she solves numerous little mysteries of electric power, among them why Broadway became nicknamed 'The Great White Way.'"

—ERIK LARSON, author of the
bestselling *The Devil in the White City*

"*Empires of Light* is a fascinating and vivid portrait of a tumultuous era. In a fast-paced narrative, Jill Jonnes re-creates the personalities, technologies, and corporate intrigues that changed America by—literally—electrifying the nation."

—LAUREN BELFER, author of *City of Light*

PHOTO: © PEGGY FOX

JILL JONNES is an author and historian with a Ph.D. from Johns Hopkins University. She has received awards from the Ford Foundation and the National Endowment for the Humanities and lives with her family in Baltimore, Maryland.

EMPIRES *of* LIGHT

EMPIRES *of* LIGHT

Edison, Tesla, Westinghouse, and the Race to Electrify the World

Jill Jonnes

Random House Trade Paperbacks
New York

2004 Random House Trade Paperback Edition

 Published in the United States by Random House Trade Paperbacks, an imprint of The Random House Publishing Group, a division of Random House, Inc., New York, and simultaneously in Canada by Random House of Canada Limited, Toronto.

This work was originally published in hardcover and in slightly different form by Random House, an imprint of The Random House Publishing Group, a division of Random House, Inc., in 2003.

All diagrams of scientific apparatus courtesy of John Ross

Library of Congress Cataloging-in-Publication Data

Jonnes, Jill

Empires of light: Edison, Tesla, Westinghouse, and the race to electrify the world/Jill Jonnes.

p. cm.

Includes bibliographical references and index.

ISBN 0-375-75884-4

1. Electric engineering—History. 2. Electrification—History. 3. Electric power—History. 4. Edison, Thomas A. (Thomas Alva), 1847–1931. 5. Tesla, Nikola, 1856–1943. 6. Westinghouse, George, 1846–1914. 7. Competition—United States. I. Title.

TK18 .J66 2003 621.3'09—dc21 2002031866

Random House website address: www.atrandom.com

Printed in the United States of America

9 8 7

Book design by Mercedes Everett

For my husband, Christopher Ross

Great are the powers of electricity. . . . It makes millionaires. It paints devils' tails in the air and floats placidly in the waters of the earth. It hides in the air. It creeps into every living thing. . . . Last night it nestled in the sherry. It lurked in the pale Rhine wine. It hid in the claret and sparkled in the champagne. It trembled in the sorbet electrique. . . . Small wonder that the taste was thrilled and the man who sipped was electrified. . . . Energy begets energy.

—***Buffalo Morning Express,***
January 13, 1897,
describing the banquet
celebrating the city's first
electricity from the
Niagara Falls Power Company

Contents

List of Scientific Diagrams

Introduction

Great, indeed, is the power of electricity. And in the final decades of the nineteenth century, three titans of America's Gilded Age were among the Promethean few who dreamed of the possibilities hidden in this ethereal force of nature—its awesome power visible only in the wild rumble and slash of electrical storms. Each titan was determined to master the "mysterious fluid." Each vied to construct an empire of light and energy on a new and monumental scale; each envisioned radiant enterprises that would straddle the globe, illuminating the inky night and easing forever the burden of brute labor. Thomas Alva Edison was the best known of these dreamers in 1879. The nation's greatest inventor, Edison was creator of the incandescent light bulb and mastermind of the world's first incandescent light network. Then there was Nikola Tesla, the elegant, highly eccentric electrical wizard who revolutionized the generation and delivery of electricity. Tesla was the Serbian immigrant dreamer who foresaw using the vibrating waves of the earth itself to generate unlimited power and communications. The final member of this trio was George Westinghouse, the charismatic Pittsburgh inventor and tough corporate entrepreneur. He built up company after company, an industrial idealist who imagined a world powered by cheap and plentiful electricity. All his working life, he strove heart and soul to create that electrified world.

This is also the story of the nascent years of the electric power industry and the rise of a new technology that completely transformed society, a tale told largely through these three visionary figures, their triumphs, their blunders, their caustic feuds. As each struggled to make real his electrical dreams and dominate "the subtle and vivifying current," the stage was set for one of the most unusual and vicious battles in American corporate history, the War of the Electric Currents. This war pitted Thomas Edison and his tried-and-true technology of DC (direct current) against Westinghouse and Tesla's new and experimental AC (alternating current). It is the archetypal corporate struggle, a modern industrial epic where American business titans battled to dominate and control a world-changing technology, to create whole new Empires of Light. In a time of nineteenth-century Darwinian harshness, these new technologies drove the relentless growth of large and complex corporations, the economic basis for a century of astonishing societal and material change. The rise of unchecked capitalism and large corporations, in turn, forced the nation to confront its entire form of governance. *Empires of Light* is a story that resonates strongly in our times.

EMPIRES *of* LIGHT

CHAPTER 1

"Morgan's House Was Lighted Up Last Night"

In the late spring of 1882, Thomas Alva Edison, world famous as the folksy genius who had invented the improved telegraph and telephone, the amazing talking phonograph, and the incandescent light bulb, would shamble in occasionally to the hushed, formal suites of Drexel, Morgan & Company at 23 Wall Street, an imposing white marble Renaissance palace of mammon. There in a glass-walled back office, J. Pierpont Morgan presided at an oversize rolltop desk. The autocratic senior partner wore a banker's black suit, starched snowy shirt, wing collar, and fine gray silk ascot. His expensive, ever-present Havana cigar made the air smoky, redolent of privilege and power. Morgan's investment firm was partially bankrolling Edison's fevered building of America's first incandescent electric lighting system in the crowded commercial blocks of lower Manhattan. When Edison visited Drexel, Morgan, the clean-shaven, still boyish inventor loved to disparage the office's gaslight globes as burning a "vile poison." But soon the gaslight would be gone, preempted by Edison's beloved clean electric light.

Edison, thirty-five, was already a celebrated figure in the downtown streets, recognizable in his signature slouch-brim hat or battered stovepipe, shabby shirt, bright neckerchief, and frayed black

---

Library of the J. Pierpont Morgan house

Prince Albert coat. He and his crews were logging dusty eighteen-hour shifts as they pushed to finish the far-behind-schedule Pearl Street Station generating plant and install (only at night) fourteen miles of just-below-the-street electrical conduits. All morning and afternoon pedestrians ebbed and flowed through the financial neighborhood, dark-suited men sporting shiny top hats or black bowlers, clutching their canes. "Bank messengers, with bags filled with coin, greenbacks, bills of exchange, bonds and stocks, hurry along," wrote one contemporary of hustling-bustling Wall Street, "keeping a firm grip upon their bags and eying each person they pass warily, office boys, telegraph boys with yellow envelopes containing messages from all quarters of the globe, dart here and there through the throng."[1] These acolytes of the high-toned, handsome financial district shared the jammed nearby streets with horse-drawn trolleys, heavy delivery wagons, dog-drawn rag carts, noisy oyster sellers, and small boys hawking any one of the city's dozens of newspapers. Everywhere, with the weather warming up, the city's streets reeked of horse piss and dung left daily by the 150,000 horses pulling the city's trams, trucks, Broadway stages, and fancy rigs. At night, when Edison most liked to work, he could be found with his Irish crews laying trenches somewhere near Pearl Street, already dirty with grease and tar, or tinkering with the six jumbo dynamos installed up on the reinforced second floor.

That late spring and summer, Edison had occasion to confer with J. Pierpont on another small but important job. In his office, Morgan cultivated a renowned ferocity: the gruff, impatient bark, the famed glare that challenged visitors of any rank to intrude. Other wealthy men in this most hirsute of eras flaunted complex and flamboyant beards and mustachios, but the forty-five-year-old Morgan sported only a plain, trimmed mustache. J. Pierpont Morgan had been raised an old money gentleman, conservative and stern in manner and habits. But the America of the 1880s was changing rapidly, daring men and women to dream bold dreams, to grasp for great ventures and great wealth. Just a few blocks south, the Roeblings' magnificent East River Bridge was nearing completion after thirteen arduous years, a soaring engineering marvel of suspension, floating across the shimmering New York waters. Nearby, the elevated railroads with their small belching steam engines chugged stolidly along, high above the chaos and stench of Manhattan's tangled traffic, astounding visitors with their efficient moving of tens of thousands of workers as they snaked north between tenements and offices and then out to the far bucolic reaches of the city. The miracle of the great Atlantic cable flashed telegrams across the coldest depths of the ocean. Where once letters from Pierpont's father in the London office took weeks to arrive, now telegrams pulsed through in mere minutes. The railroads had become mighty, creating new cities where there had been only marshland or prairie. In just the past year, they had laid an astounding ten thousand miles of track. The 1880 census showed fifty million Americans. Morgan, unlike many of his old money peers, relished this new temper of the times, admired men like Edison who were bold, ambitious, hardworking, confident.

Late that spring, Morgan, who had just returned from a long European tour, had briefly put aside his considerable business concerns and announced to Edison an audacious decision. He was going to personally showcase the advantages of Edison's pioneering incandescent light in his elegant Madison Avenue brownstone, just then in the throes of top-to-bottom renovation. Morgan's Italianate mansion would become, thereby, the first private residence in New York to be illuminated solely by electricity. This was, of course, no simple matter. Nonetheless, the imperious Morgan wanted the electricity installed and working by the time he, his wife, Fanny, and their three teenage children moved in that fall from their country estate, Cragston, up the Hudson River. Edison was delighted to oblige, for it would be a great coup to have Morgan's personal imprimatur on what many dismissed as a dangerous and exotic novelty. Whatever people thought of J. P. Morgan, no one thought him a fool. Money men had learned that he was decisive, intelligent, and swift of action, and above all, he kept his word, no small matter when spectral figures liked Jay Gould preyed upon the stock market.

And so, as the shad were about to make their annual run up the Hudson River, a crew of Edison workers clopped up in a horse-drawn wagon to Morgan's nearly renovated mansion at 219 Madison on the northeast corner of 36th Street. They laboriously excavated a large earthen cellar beneath the wooden stable, their shovels rhythmically slinging dirt and rocks into a growing pile. Within the musty space of the dirt cellar, they installed a squat steam engine and boiler to power two electric generators, all of which displaced Morgan's carriage horses to a nearby stable. The men also dug a ditch connecting the new cellar to the house, lined it with bricks, laid in the electrical wires, and bricked it over. Inside the mansion, decorator Christian Herter supervised the snaking of insulated electrical wires up through the elaborately wood-paneled and plastered walls where ordinarily the gas lines would have gone. These wires were then threaded through to every space in the mansion, and new electrical fixtures were installed. In some rooms electrical wires hung straight down every few feet from small holes in the tall ceilings, sprouting at their tips several small light bulbs.

On Thursday, June 8, 1882, Edison Electric Company president Major Sherbourne Eaton wrote Edison, "Morgan's house was lighted up last night. I was not there but I am told that the light was satisfactory and that Morgan was delighted. The armature of the 250 light [bulb] machine sparked badly. It will have to be changed at once. Vail took charge of that. Herter was present and declared himself entirely satisfied. Morgan is pleased with everything but Herter's fixtures."[2] By fall, as the New York social season opened, the Wall Street financier and his family were installed in their new home with its 385 electric lights, casting a soft, even, incandescent glow everywhere, from the servants' halls and butler's pantry to the bedrooms and the "Japanese manner" reception room and sitting room. The Romanesque dining room with its high oak paneling was particularly striking, for there electric lights cast a lovely jeweled radiance through the twelve-foot-square stained-glass skylight.

The deluxe *Artistic Houses* rhapsodized about every rich and costly detail of Morgan's newly renovated brownstone residence, gushing especially about the vast and splendiferous terra-cotta drawing room, where "a breath from the Graeco-Roman epoch of Italia seems to have left its faint impress on the walls, or rather its faint fragrance in the atmosphere . . . amid the aroma of perfect taste." This must have pleased Morgan, who disdained the obvious vulgarity of many of the new Gilded Age millionaires. His house was meant to convey an aura of money and power, subtly burnished by his European education, culture, and worldly intelligence. What was genuinely new and unique was Edison's electric light. "Each room is supplied with it, and, in order to illuminate a room, you have simply to turn a knob as you enter. By turning a knob near the head of his bed, Mr. Morgan is able to light instantaneously the hall and every room on the first floor, basement, and cellar—a valuable precaution in case of the arrival of burglars."[3] This assumed burglars did not prowl and enter in the middle of the night. Because, as Morgan's son-in-law Herbert Satterlee explained in a memoir of Morgan, "The generator had to be run by an expert engineer who came on duty at three P.M. and got up steam, so that at any time after four o'clock on a winter's afternoon the lights could be turned on. This man went off duty at 11 P.M. It was natural that the family should often forget to watch the clock, and while visitors were still in the house, or possibly a game of cards was going on, the lights would die down and go out."[4] Then there was a careful groping about in the sudden murk to light beeswax candles and kerosene lamps.

Yet that was the least of Morgan's problems as a proud pioneer consumer of electrical power. Each silvery winter afternoon, the noise of the city day ebbed away in the genteel and moneyed streets on Murray Hill. Then the delicious still of the indigo evening mingled only with the occasional soothing clip-clop of passing horses and broughams. As the handsome houses lit up their gaslights, all was quiet. But now, when the sun dipped below the horizon, Mr. Morgan's steam engine and electrical generators roared to life, shattering the descending blessed calm. These powerful machines clanked and throbbed so intensely, Mrs. James Brown next door complained that her whole house was vibrating. And that was not all. The infernal steam engine contraption, operating as it did on coal, also belched noxious fumes and smoke. Mrs. Brown reported that this was permeating her pantry, leaving her silver tarnished. Mr. Morgan reassured his aggrieved neighbor's husband that an "expert" from

Edison's company "will call and see from personal observation and consultation what the features are which cause you annoyance.... I need scarcely add that I shall spare neither exertion nor expense" to tame the overwrought machines.[5]

Just after Christmas, when three weeks had passed and nary an Edison man had materialized to right the situation, Morgan wrote indignantly to Sherbourne Eaton: "I must frankly say that I consider the whole thing an outrage to me, as well as the neighbors—& am unwilling to stand it any longer. Please let the matter have immediate attention."[6] Finally an Edison crew appeared and solved the problem by underpinning the machines with India rubber pads, lining the stable with felt, and further cushioning the whole installation with sandbags. Then yet another ditch was dug across the yard, this to funnel the coal smoke from the steam engine into the mansion's chimney. Now, a new kind of noise impinged. Reported son-in-law Satterlee, "In the winter when the snow melted above the brick conduit, all the stray cats in the neighborhood gathered on this warm strip in great numbers, and their yowling gave grounds for more complaints."[7] And there was, of course, the intermittent annoyance of wires short-circuiting and the generator occasionally malfunctioning.

All this was understandably trying to the Morgan family. Yet Morgan was surprisingly patient. As an investment banker who had backed many railroads, which were continually absorbing new technologies, he seemed quite accepting that problems large and small were inevitable. Had it not taken three arduous tries before the Atlantic cable was properly laid and began to work reliably? Finally, however, in the fall of 1883, J. Pierpont requested Edward H. Johnson, one of Edison's top executives, to please come have a look at the mansion's less-than-satisfactory year-old electrical arrangements. Johnson was not happy about going to Morgan's, but in a fledgling industry desperate for capital and credibility, he had little choice. Morgan's firm was a significant power in Wall Street. In a brief reminiscence written for Satterlee's book, Johnson said that "after thoroughly canvassing the lighting of the house," he found the system already outmoded. Electric light technology was advancing that quickly. "Mr.

Morgan inquired of me what I thought of it. I asked if he wished an honest and candid reply. He said he did. I said, 'If it was my own I would throw the whole D—— thing into the street.' [Replied Morgan,] 'That is precisely what Mrs. Morgan says.' " The next day, when Morgan was at his office reviewing balance sheets and surveying all that happened through his cigar haze, he summoned Johnson and asked him to go up to the mansion personally and redo the electric. A reluctant Johnson agreed.

As part of the new, upgraded electrical lighting of the Morgan residence, Johnson decided to improvise "an arrangement for giving light on Mr. Morgan's library table by means of concealed wires in the floor and contact spuds fixed in the legs of the table to penetrate the heavy and costly rug which covered the floor." The next morning quite early, Johnson received another summons to the Morgan manse. As he headed swiftly toward Madison Avenue and 36th, he had a queasy feeling something had gone awry. Upon entering Morgan's magnificent mosaic vestibule and removing his derby, he took one whiff of the air and his worst fears were confirmed. "The house was pervaded by a strong smell of wet, burned wood and burned carpet." The servant who had answered the door escorted Johnson to the library. "The library floor was torn up in several places; and in the centre of the room was the partly burned desk and burned rug and other charred objects piled in a heap. . . . One of the spuds [beneath the library table] had become bent or broken and an imperfect contact was made and a fire ensued completely wrecking the beautiful room. The family were at the opera at the time." Johnson surveyed the soggy, blackened detritus, his spirits sinking rapidly. For while J. P. Morgan did not yet fully dominate American finance as he eventually would, he was still a famously impatient and ill-tempered man and growing ever more influential among the all-important New York money men. Years later, that moment was still vivid to Johnson. "It was a dismal scene. . . . Suddenly I heard footsteps and Mr. Morgan appeared in the doorway with a newspaper in his hand and looked at me over the tops of his glasses.

" 'Well?' he said.

"I had formulated an explanation, and was prepared to make an elaborate excuse. Just as I opened my mouth to speak, Mrs. Morgan appeared behind Mr. Morgan, and as I caught her eye she put her finger on her lips and then vanished down the hall. I said nothing but looked at the heap of debris.

"After a minute's silence Mr. Morgan said, 'Well, what are you going to do about it?'

"I answered, 'Mr. Morgan, the trouble is not inherent in the thing itself. It is my own fault, and I will put it in good working order so it will be perfectly safe.'

"He said, 'How long will it take to fix it?'

"I answered, 'I will do it right away.'

" 'Alright,' he replied, 'see that you do.' "[8]

Morgan's son-in-law wrote that the banker was subsequently so delighted with his electricity that he "gave a reception, and about four hundred guests came to the house and marveled at the convenience and simplicity of the system." Two of the guests, California gold rush millionaire Darius Ogden Mills and his son-in-law, *New-York Daily Tribune* publisher Whitelaw Reid, promptly contacted the Edison Electric Illuminating Company to have their own houses electrified. Mr. Mills, his shrewd face wreathed with his signature white muttonchop sideburns, also appeared the next morning at Drexel, Morgan and ordered his broker there to buy him a thousand shares of Edison Electric stock. Morgan, who always kept a keen eye on the office's comings and goings, intercepted Mills as he was leaving. He inquired why Mills, who was a much admired investor, was there. When he heard, Morgan, whose firm was Edison's lead banker, began then and there to match every order Mills made for Edison stock with an equal order of his own personal stock. Wrote Satterlee of Pierpont's pivotal role, "His faith in the new industry, his advice, and his constant financial support were the factors that led to its spectacular development; otherwise it might have taken many more years for it to reach its tremendous proportions."[9]

So pleased was Morgan with his electric alternative to what Edison gleefully damned as "the vile poison" gaslight that he had Johnson install electric lights in the rectory of his church, St. George's, as well as in the church gym, to make it easier to use in the winter. He also had a family friend's school for young children wired. Explained Satterlee, "Pierpont sent Johnson around with his mechanics and electricians, in the same way that he would send a basket of his best peaches or grapes from 'Cragston' to those who did not have orchards or grapevines of their own. Nothing in life pleased him more than giving luxuries to friends who could not have them." Morgan had his own mansion rewired several times. "He never seemed to care how much the walls and floors of the house were torn up," wrote Satterlee, "so long as the latest and most up-to-date devices were put in before other people had them." Morgan once joked, "I hope the Edison Company appreciate the value of my house as an experimental station."[10]

Satterlee's affectionate portrait fondly conveys Morgan the enthusiast, who put his money and considerable influence behind the new electric light. But there was a far darker side to Morgan. As a conservative financier, he was deeply uncomfortable with the roller-coaster chaos and competition of aborning industrial capitalism. As the years passed, Morgan concluded that he much preferred the predictability of monopolies he controlled. And, in truth, Morgan's conservative soul flinched again and again in the early years at the huge and risky amounts of capital the radical new technology of electricity would turn out to require. In truth, it was not Morgan but Thomas Edison who took the biggest risks when it came to building the first central station in Manhattan. All this made Morgan even more determined that the early natural monopoly of Edison Electric Light and its handful of manufacturing companies should be sustained. As Morgan's power grew, so did his passion for control, and it would govern much of his ruthless strategy in the coming bitter struggle over electricity.

For the very reason that electric lighting in the early 1880s was still such an expensive and exotic novelty, it immediately became a status symbol for the very rich and the adventurous. And there was no one richer in 1881 than William H. Vanderbilt, president and majority owner of the powerful New York Central Railroad. Vanderbilt would eventually be, along with certain Drexel and Western

Union partners, one of Edison's significant (but later) backers in electric lighting. A big, stolid man with luxuriant curling mutton-chops framing a bland face, Vanderbilt had earned the sobriquet "Colossus of Roads." Having inherited $100 million from his notorious skinflint father, Commodore Cornelius Vanderbilt, the mild-mannered William had handily doubled this sum by his own sagacious flair for railroads and stocks. Vanderbilt had been out to Edison's Menlo Park laboratory in 1880 and admired the soft glow of the light bulbs in the cold New Jersey evenings. More impressive yet was Edison's splendid new office at 65 Fifth Avenue, electrically ablaze with numerous ceiling electroliers and lamps. One day in the spring of 1882, Vanderbilt himself came to "65" and was quite taken with its amazing effulgence of incandescence. Always the eminently practical businessman, Vanderbilt desired to experience himself this new technology that he was being asked to invest in. With his deluxe mansion nearing completion, Vanderbilt had offered his house (before Morgan did) as the first in New York to be illuminated with electricity. The emblems of status may shift, but human nature generally delights in being first.

Even before its completion, William H. Vanderbilt's new "box house" was one of the most talked-about structures in the United States, a mansion as monumental as the owner's fortune. And the fortune was, in turn, a testament to the new and mighty railroad corporations, private industrial entities whose size, complexity, power, and potential for profit dwarfed all that had come before. Vanderbilt, quite used to being boss of thousands of rail workers, thought nothing of engaging a seeming army of six hundred workmen and sixty foreign artists and craftsmen, who swarmed about Fifth Avenue and 51st Street, building and ornamenting with antlike alacrity a palatial triple mansion, a luxurious edifice of excess (such as Morgan found quite vulgar). For two years, New Yorkers had watched agog as the mansion was completed in record time. (Word was it had cost $6 or $7 million, at a time when four-fifths of the nation earned $500 or less a year.) Entered through towering Italian sculptured bronze doors, Vanderbilt's dream house featured everything from Venetian-style rooms with jewel-encrusted wallpaper to a subdued bamboo-lined Japanese parlor. William, with his great belly and luxuriant muttonchops, would live with his beloved wife in fifty-eight-room opulence in one-half, while two daughters and their families each occupied the two adjoining mansions. So the self-assured Edison and his people had eagerly bustled in in the spring of 1882, pleased as punch to be showcased amid such grandeur. They installed a stand-alone generator and wired the huge mansion, as well as its two-story skylighted art gallery hung with large, expensive French pastoral scenes.

Why, then, a reporter from Saint Louis asked the ever loquacious Thomas A. Edison not long after, as the inventor confidently regaled him with the imminent downfall of the gas industry, was the Vanderbilt mansion now lit with gaslight? "He never tried our light," asserted Edison. "The first night when the engine ran it made a noise, as all new engines will, and Mrs. Vanderbilt complained of it, and that settled it."[11] In fact, Edison later elaborated, "Mrs. Vanderbilt became hysterical. . . . We told her we had a plant in the cellar, and when she learned we had a boiler there she said she would not occupy the house. She would not live over a boiler. We had to take the whole installation out."[12] But was it not true, pressed the journalist, that the electric wires had also started a fire? "There is nothing to it except just this: In running temporary wires in different directions just to show Mr. Vanderbilt the effects, one of them came into contact with a burglar alarm wire, and this one, being overheated, charred some of the gold-thread wires of the cloth wallpaper which it ran into—that is all."[13] While Edison belittled such high-profile incidents and even implied, it seemed, that the nation's richest citizen might be a bit henpecked and lily-livered, his rivals in the gas lighting industry made sure the press knew the sorry details of this failure of the newfangled electric light.

This "Vanderbilt incident" was why chroniclers of the electric light were careful to describe Mr. Morgan's house as the first in New York to be *successfully* wired for electricity. Both these stories—the burnt Morgan library, the small Vanderbilt fire—were seen as telling commentary on what made Morgan—as opposed to Vanderbilt—such an important figure in the early years of the electric light. Morgan was lauded as courageous and persevering in the face of serious setbacks, a true pioneer who put the full force of his financial and personal influence behind the new technology (though nowhere near enough, if you asked Edison). In contrast, Vanderbilt had retreated from personal endorsement at the first sign of trouble, but later, when success seemed more certain, he came in as a major but passive investor.

Both stories take us back to that shimmering moment when electricity—now an utter commonplace—was introduced into the gaslit household. Wealthy, cultivated women in floor-length, rustling dresses delighted in showing their friends how if you just turned a knob on the wall, the room's clear incandescent bulbs began almost magically to glow, casting an even, clear light. Unlike candles, the electric light did not burn down or become smoky. Unlike gaslight, there was no slight odor, no eating up of a room's oxygen, no wick to trim or smoked-up glass globes to be cleaned. Many months went by before the light bulbs burned out. But the introduction of electricity in the home was also fraught with the peril of electric shock or fire, which gave a certain delicious frisson to the whole exotic, expensive enterprise. In the early 1880s, electricity was still a wondrous and "mysterious fluid" resonant with glamour, status, and danger.

But for the visionary capitalist, electricity possessed other, more practical allures. Already this astonishing invisible "agency" had birthed two radically new technologies—the telegraph and the telephone—that had forever compressed and altered the age-old realities of time and distance. The perspicacious were already reaping fortunes. The most farsighted were tantalized by even greater electrical prizes. Who would further harness electricity to light the nation's streets, its dim factories, and all those millions of households, dramatically transforming man's age-old sense of day and night? He who could bestow more hours to the day—for work, for play—would be wonderfully rewarded. Of even greater moment in these commercial times, who would harness electricity to operate work-saving machines, mechanisms artfully reinvented to liberate humankind from the hard toil of farm and factory? He who could unleash the full, only-dreamed-of potential of electricity *and* control this awesome invisible power would become wealthy and powerful indeed. Was it any wonder the War of the Electric Currents would be so fiercely waged?

CHAPTER 2

"Endeavor to Make It Useful"

The elusive, invisible mystery of electricity became the subject of specific, recorded inquiry in the dawning days of the Golden Age of ancient Greece. In 600 B.C. Thales, erudite philosopher and astronomer in the thriving Ionian port of Miletus, observed the special qualities of the rare yellow orange amber, jewel-like in its hardness and transparency. If rubbed briskly with a cloth, Thales showed, amber seemed to come alive, causing light objects—like feathers, straw, or leaves—to fly toward it, cling, and then gently detach and float away. Amber was similar to a magnet in its qualities, yet it was not a lodestone. As a youth, Thales of Miletus had studied in the sacred Egyptian cities of Memphis and Thebes. Perhaps it was there, under the burning sun, that this earliest of Greek philosophers first learned from the priests about the prized amber, with its seeming possession of a soul.

In the myth of Phaeton, handsome mortal son of Apollo the sun god, the Greeks explained amber's magic. Phaeton aspired to drive his father's lustrous golden chariot on his own through the shimmering sky, and his father, to prove his love, rashly agreed. Soon after Phaeton took the reins, Apollo's formidable steeds sensed his youthful alarm and veered rebelliously off course, racing first far

Michael Faraday

into the darkest heavens and then too near the earth, scorching the land, setting whole cities and nations afire, turning fruitful Libya into desiccated desert, roiling the oceans, and splitting open continents. Jupiter, enraged by such hubris and wanton destruction, hurled a fatal thunderbolt at Phaeton, who fell lifeless from the sky, his burning hair blazing like a shooting star across the darkened sky. His sisters, the Heliades, gathered about their dead brother, mourning so bitterly that finally Jupiter, taking pity, turned these lissome girls into ever sighing poplars and their tears into translucent amber that tumbled into a passing stream. This bittersweet myth explained amber as a mysterious godly gift. We now know that gleaming amber is but prosaic fossilized tree resin. But amber's rare qualities—the creation of static electricity—were sufficient to pique human curiosity and yield up the tiniest clue about the mysterious power of electricity.

Almost two millennia would pass before Thales's original observations about amber were enlarged upon. In 1600, the much esteemed London physician and philosopher William Gilbert was appointed chief doctor to the strong-willed and aged Virgin Queen, Elizabeth I. Gilbert not only replicated Thales's amber experiments, he went far beyond them. An admired officer of the conservative Royal College of Physicians, Gilbert was also actively engaged in scientific investigations. However, he held off publishing his challenging treatise, *De Magnete*, until the pinnacle of his career, when he was president of the Royal College and firmly ensconced as the popular elderly queen's personal doctor. The full title (translated into English from the original Latin) was *On the Magnet, Magnetick Bodies Also, and on the Great Magnet the Earth; a New Philosophy Demonstrated by Many Arguments and Experiments.* Gilbert praised his own "new style of philosophizing," his bold and "unheard-of doctrines," and warned that his great life's work was not for "smatterers, learned idiots, grammatists, sophists, wranglers and perverse little folk." The earth's interior, posited Gilbert, was "a pure, continuous magnetic core, which orients our globe in the heavens just as it swings a compass needle to the north."[1] This explained many actions of magnets, including their noticeable "lines of force."

These were not idle intellectual inquiries Gilbert was pursuing, but questions of tremendous commercial moment. England was a dominant sea power in avid pursuit of the wealth of nations, and Queen Elizabeth was seeking to improve the arts of navigation. Therefore, anything that shed light on the nature of magnetism and compass reading was important. A famous painting shows William Gilbert, arrayed in white ruff and dark velvet robes, at his handsome London residence, Wingfield, demonstrating electrics and magnetics for his sovereign monarch and her court. The seated Queen Elizabeth, who favored exquisite jewel-encrusted, fur-trimmed gowns, watches intently, as do two of Her Royal Majesty's most famous explorers and navigators, Sir Walter Raleigh and Sir Francis Drake. Gilbert is standing before them, showing the strange attractions of amber, magnets, and various substances. The very word *electric* was coined by Gilbert, who played on the Greek word for amber, *elektron,* to come up with a term to describe amber's attracting qualities for certain materials.

Like a good philosopher and experimenter committed to truly observing and explaining factually what happened in nature, Gilbert did not rest with amber, but tested all kinds of materials, finding that glass, rock crystal, sulfur, sealing wax, and some minerals, when rubbed, also became "electric." This was Gilbert's great electrical discovery and contribution: that numerous hard materials—not just amber—could be electrified when rubbed. (What Gilbert could not know, of course, was that his rubbing created *charged* electrons—positive or negative—that attracted like or repelled opposite. When these charges flow along a conductive material, they become an electric current.) He used a pivoted (nonmagnetic) light gilt needle, his "versorium," or electroscope, to study which materials had electrostatic attraction. (The needle swung toward or away, depending on whether the charge was positive or negative.) Moreover, his further inquiries into these "electrics" delineated how the weather, water, olive oil, and so on affected their "electricity." The good doctor deduced that what activated the numerous motions of attracting and repulsing was an invisible watery substance he termed "electrical effluvia." In 1603 Gilbert's great patroness, Queen Elizabeth, died, and the author of *De Magnete* soon followed. His legacy was the discovery that electricity could be generated and his admonition that philosophers must "seek knowledge not from books only but from things themselves."

For more than half a century, human knowledge and understanding of electricity progressed little further. Philosophers throughout Europe remained fascinated by electricity, and *De Magnete* was studied carefully and its experiments copied. But the next major electrical advance came almost unwittingly (as often happens in science), from the Lutheran mayor of Magdeburg, a small free-trading city in the Holy Roman Empire (today Germany) burned to the ground by marauding Swedes in 1631. The mayor, Otto von Guericke, was a well-educated scion of a leading family, and he devoted his middle decades to rebuilding and reviving his crushed city and retrieving through treaties its free status. The overburdened von Guericke found soothing solace, "a gate of tranquility," in astronomy. He decided to try to replicate the earth's sphere and study a complex system of "virtues" he believed swirled in and about it. He created a solid sulfur ball the size of an infant's head, which he placed in a sturdy wooden frame and rotated by an attached handle. When this twirling yellow globe (later ones were far larger and incorporated more minerals) was rubbed, it became electrostatic and attracted and then repelled many light objects, typically a feather.

The convivial burgomaster enjoyed amusing his guests by walking about with a small sphere steadily propelling before him a buoyant feather, ultimately steering the feather to cling to some visitor's nose. When the sphere really got whirling and was rubbed, it began to glow and throw off sparks. With his globe, von Guericke inadvertently showed that one could produce notable amounts of electricity, yet he did not advance actual knowledge. In Germany in the late seventeenth century, von Guericke happily made and dispensed his globe to others, but no British (or other) philosopher seems to have interested himself in its potential for electrical experiments. That did not transpire until 1709, when Englishman Francis Hawksbee, curator of instruments at the Royal Society, created a similar object, also an electrostatic machine. His was a mounted hollow glass sphere that could be cranked to high speeds and rubbed to create electricity and eventually an eerie light inside, a feat shown to great acclaim. Hawksbee thought to add a "rubber," a short adjustable column with a stuffed leather piece that pressed steadily on the glass (in place of a hand) to provide friction. Electrostatic machines like this could generate sizable sparks and became the standard source of electricity for experiments and entertainment during the early eighteenth century.

The next true electrical advance came from Stephen Gray, a modest native of Canterbury on the Stour River, then as now home of the great Anglican cathedral. Gray was from a family of dyers, but he acquired enough education to read Latin and become a friend to the astronomer royal in Greenwich. While Gray earned his daily bread as a dyer, his inner being was dedicated to science and experiment. He sufficiently impressed the professors at Cambridge with his stargazing reports that at age forty-one they hired him to work at Trinity College's new observatory. At the university Gray discovered the mysterious delights of electricity, but also "mercenary" flaws in his employers. He retreated to Canterbury for a few years and then sought residency at the Charterhouse in London, a day school for poor boys and a selective residence for impecunious gentlemen. Knowing well England's continuing obsession with practical advances in sea-going, Gray assured the trustees that if admitted, he intended to devote himself to "inquiries Relating to Astronomy and navigation and might haply find out something that might be of use."[2] Whatever else Gray did in his first decade as a genteel Charterhouse pensioner, by 1729 he began to work on seeing how far he could send the "electrical vertue." Using an electrified glass tube, he tested all kinds of substances, finding metal wire to be excellent for conveying electricity, as was plain packthread. Having succeeded in sending electricity fifty-two feet, in June of 1729 Gray repaired to the grander spaces of a wealthy friend's country estate. There they soon transmitted electricity 765 feet and methodically worked their way through all manner of substances, hollow or solid, determining which conducted electricity well or not at all. Gray also noticed that if you suspended an iron rod pointed at both ends from silk threads and touched it with an electrified glass tube, it briefly gave off cones of light, a wondrous and memorable result. Gray further remarked that great sparks of electricity made noises not unlike that produced by thunder.

The most memorable, original, and flashy of Gray's experiments, and one that was delightedly replicated for years, was that of the dangling Electrified Boy. On April 8, 1730, Gray constructed a stout wooden frame from which he suspended via strong silk strings—as if he were a very large bird—a forty-seven-pound boy, probably one of the Charterhouse charity students. All parts of the boy were thickly covered in nonconducting clothes, leaving only his head and hands and a few toes naked. In one outstretched hand, the boy held a wand with a dangling ivory ball. When Gray touched the boy on the backs of his naked toes with an electrified glass tube or vial, the electricity traveled to the boy's head (where his hair stood on end) and his outstretched hands. Then three feathery-light piles of brass leaves under the boy rose up in three impressive little clouds, before sinking back down. One pile rose only to the ivory ball, showing that the electricity had continued on beyond the hand, one pile rose as high as the face, and the third rose to the bare hand, which hung lower than the head. A thrilling variation had the dangling boy stretch out his finger and touch another person. This gesture transmitted a noticeable spark and a shock, *if* the standing person was on a conducting material. Gray found he could also elicit sparks from the end of the boy's nose. The presence and nature of the "electrical vertue" was thus dramatically made visible. Pensioner Gray had invaluably demonstrated to the world much about the conductive qualities of electricity. Human nature being what it is, frolicsome philosophers could not resist many amusing versions of the Electrified Boy demonstration. Men were soon bestowing electrified kisses on (literally shocked) women. One waggish French host wired the chairs at his dinner table so that his astonished guests suddenly found sparks flashing from their forks!

For the next major electrical leap forward, we move across the English Channel to the Continent. All over England and Europe in this self-proclaimed Age of Enlightenment, the sages of the day were studiously observing the natural world around them, offering up theory upon theory to explain things previously taken for granted, ignored, or explained by myths or magic. Electricity was one of the most mesmerizing and entertaining of nature's multitude of puzzling phenomena. Up until this time, philosophers had learned from the royal physician William Gilbert that there *was* electricity, what he called "electrical effluvia." From von Guericke and Hawksbee, people had learned to make and operate electrostatic machines that more efficiently produced electricity. From pensioner Stephen Gray came a serious study of which materials were conductors and then how far electricity could be conducted. In this era, even the most dedicated and serious scientists felt it was their role to provide awe-inspiring electrical demonstrations that also entertained.

At this stage, the pursuit of electrical knowledge was restricted by the limited amounts of electricity experimenters could actually generate. The first solution to *storing* electricity was stumbled upon in Leyden, Holland. Professor Pieter van Musschenbroek's lawyer friend, Andreas Cuneus, had come by the laboratory to divert himself with various electrical experiments. One involved filling a jar with water and electrifying it by touching a wire sitting in the water with an electrified glass vial. Cuneus then touched the protruding wire and received a bad jolt from the stored electricity. He mentioned this to Musschenbroek, who in mid-January 1746 electrified the jar of water (via the wire) with one of his great whirling globes, not a small electrified glass vial. He wrote several days later to an academic in Paris to explain "a new but terrible experiment, which I advise you never to try yourself, nor would I, who have experienced it and survived by the Grace of God, do it again for all the kingdom of France." Musschenbroek warned that when he touched the metal protruding from the jar, he was "struck with such force that my whole body quivered like someone hit by lightning . . . the arm and entire body are affected so terribly I can't describe it. I thought I was done for."[3] The sensational news of the Leyden jar spread like electrical wildfire, causing every other serious electrical philosopher to construct one and experience its fearsome jolts. Soon it was found that lining the jar halfway up with metal foil inside and out helped retain the electricity, while the power of several jars combined created a shock strong enough to kill a sparrow. For the first time, scientists could actually store these wondrous electrical forces.

The standard Leyden jar consisted of a squat container of thin

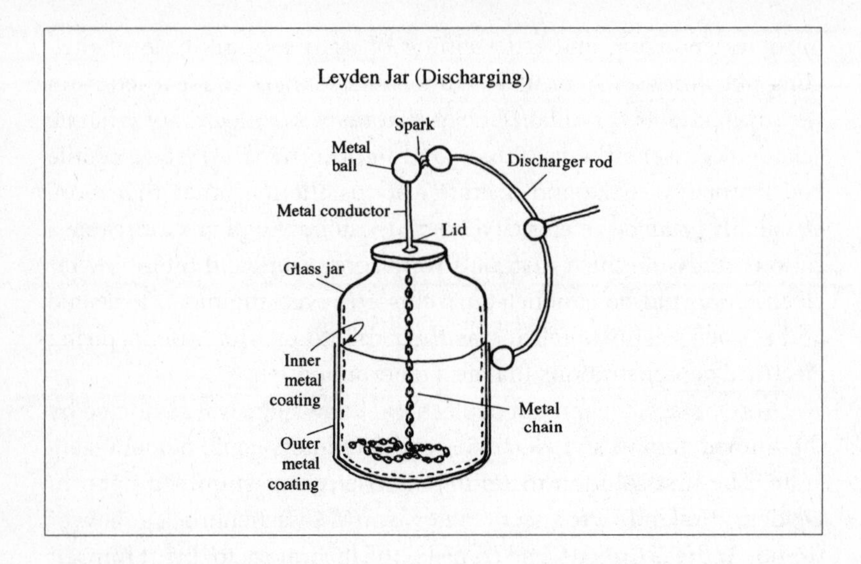

glass encased inside and out on its lower two-thirds with a metal foil coating. A brass wire topped by a small brass ball protruded from the snug (nonconducting) cork lid and then also hung down into the water-filled jar. Electricity generated by electrostatic machines was directed into the glass jar via the ball and wire, traveling down the metal wire or, later, down a long metal chain, and then charged the metal foil and water. When a philosopher needed electricity, he touched the ball and out it flowed in a great jolt. Leyden jars could retain their electrical charge for as long as several days, depending on insulation and storage conditions.

While the philosophers in the royal societies and universities observed, measured, and cogitated on their electrical findings in respectable learned journals, they also continued to use electricity (such as it was understood) for pseudomagical entertainments and diversions. With the invention of the Leyden jar, many a traveling huckster took to charging good money to the bold and brave for a chance to experience the man-made lightning that sparked forth from the jar. At the Versailles court of Louis XV, the relentless search for novelty and amusement led the Abbé Nollet, a serious student of the electrical arts, to begin arranging spectacular electrical displays. Most memorable was the assembling of 180 gendarmes in a big circle in the Grande Gallerie, each holding the next man's hand. Nollet, attired in powdered, coiffed peruke and a long, pinch-waisted coat with fashionable big cuffs, wished to see how far the electrical shock would travel. The electrostatic machine was revved up, charging the Leyden jar. The gendarme at the head was holding the Leyden jar. When he touched the brass ball, 179 others leaped into the air with a strange precision. "It is singular to see the multitude of different gestures," wrote the good abbé, "and to hear the instantaneous exclamation of those surprised by the shock."[4] The king and his jaded courtiers found the leaping gendarmes both marvelous and hilarious. It was also a palpable demonstration that electricity traveled (though they could not know it) at the speed of light. Having diverted the court, Nollet satisfied his scientific soul that this was not a fluke by repeating the test first on two hundred robed Carthusian monks at their monastery in Paris and then a great circle of six hundred people at the Collège de Navarre.

The mysteries and marvels of electricity absorbed many of the brightest intellects of early-eighteenth-century Europe, but the next electrical advances emanated from a very unlikely venue, the British colonies in North America. In the attractive Quaker port city of Philadelphia, laid out in a spacious grid, Benjamin Franklin, the respected editor of the popular *Poor Richard's Almanack,* had attained sufficient prosperity to devote himself largely to philosophical matters. Little involved in his printing business and not yet caught up in politics, in 1744 Franklin saw in Boston a Scottish lecturer demonstrate the popular dangling Electrified Boy. The next fall, a Quaker merchant friend in London, Peter Collinson, sent along various electrical apparatus and writings. Franklin was soon hooked. Months after European philosophers had become enthralled with the new possibilities of the Leyden jar and were hotly debating how and why it could store electrical charges, Franklin and his experimenting colleagues had one in hand. The genial Franklin, forty, his brown hair shoulder length, attired in plain breeches and long jacket, held electrical court at his small Market Street house that also served as his printing shop, library, and now laboratory.

By the spring of 1747, Franklin was writing Collinson, "I never before was engaged in any study that so totally engrossed my attention and my time as this has done lately. What with making experiments when I can be alone, and repeating them to my friends and acquaintances, who, from the novelty of the thing, come continually in crowds to see them, I have, during some months past, had leisure for little else." Curious neighbors and strangers jammed in to watch Franklin generate electricity with his fast-turning glass cylinder, capture those flaring sparks in his Leyden jar, and then demonstrate the many puzzling ways electricity seemed to work. At one point Franklin wrote his friend in London, "If there is no other use discovered of electricity, this however is considerable, that it may make a vain man humble."[5] It was a most confounding effluvium.

In fact, Franklin made considerable strides as he applied his tremendous curiosity and intellect to understanding electricity, conducting hundreds of experiments aimed at teasing out specific electrical qualities. He also devised numerous "recreations," including the "counterfeit spider," a black arachnid that came "alive" as it danced about in an electrical field; a book that appeared aflame; and a most amusing electrical fish, a thin gold leaf in a tapered shape. "If you take it by the tail, and hold it at a foot or greater horizontal distance from the prime conductor, it will, when let go, fly to it with a brisk but wavering motion, like that of an eel through water."[6] Perhaps most telling of future events was a "magical" picture showing the British king wearing his crown. Those who tried to remove the crown received a notable shock. Such enterprises were both playful and instructive, and Franklin was the first to assert that there were not different kinds of electricity—as many believed—but one basic fluid with positive and negative qualities. Yet Franklin, being ever the practical soul, was "chagrined that we have been hitherto able to produce nothing in this way of use to mankind."[7]

Franklin fluctuated between a humble awe at the small revelations of nature's secrets and a giddy glee at his newfound skills. He and his merry band of electrical experimenters had sufficiently mastered the production and storage of electricity to plan a most unusual and festive dinner down by the Schuylkill River prior to dispersing for the muggy summer months. In April 1749, Franklin jovially described their plan: "A turkey is to be killed for our dinner by the electrical shock, and roasted by the electrical jack, before a fire kindled by the electrified bottle; when the health of all the famous electricians in England, Holland, France, and Germany are to be drank [*sic*] in electrified bumpers, under the discharge of guns from the electrical battery."[8]

Like other philosophers before him, Benjamin Franklin suspected that lightning was simply a massive jolt of electricity. And by early 1750, he believed he had come up with a practical and lifesaving object that could also test out his lightning-is-electricity theory. Franklin suggested erecting tall, pointed metal poles on tall buildings to conduct the lightning down to the ground, where it would dissipate into the dirt, thus protecting the structures from fire. For the actual experiment, the lightning off these poles could be collected into Leyden jars. He explained, "The effect of [points is] truly wonderful; and, from what I have observed on experiments, I am of opinion that houses, ships, and even towers and churches may be effectually secured from the strikes of lightning by their means."[9] However, people were understandably reluctant to tempt lightning, a known creator of conflagrations, onto their property. This same year around Christmas, Franklin himself, failing to exercise caution, was knocked senseless just by man-made electricity intended to kill a turkey. Franklin described taking "a universal blow throughout my whole body from head to foot, which seemed within as well as without. . . . I had a numbness in my arms and the back of my neck which continued til the next morning but wore off."[10]

In the next two years, Benjamin Franklin found his talents and time given over more and more to pressing public affairs. Yet he was still thinking deeply about the mysteries of electricity and in September of 1752 finally had occasion to carry out a long-planned experiment. On an oppressive and muggy afternoon, as a thunderstorm began building, its angry dark clouds boiling up ominously on the horizon, Franklin and his grown son hurried through the heat out into the fields still flecked with goldenrod and wildflowers. Franklin carried a simple kite he had constructed using silk handkerchiefs and cross-sticks. At the very top of the kite, he had affixed a foot-long piece of metal wire to serve as a conductor. At the bottom, he had attached a hempen string and from that a ribbon of silk (a known nonconductor). Where the hemp and silk met, Franklin had hung a metal key. As the storm winds gusted violently through, Franklin maneuvered his kite into the air, whereupon he and his son sought shelter from the downpour in a small wooden structure. The kite raced high up into the gray clouds, while the nearby trees rocked in the storm gusts. In the distance, lightning vibrated brilliantly across the darkening sky. Franklin and his son watched the hemp, but there was no sign of electrical life. Wet and losing hope, they suddenly saw what they were looking for—the loose threads of the hempen string rising as if electrified. Franklin, knocked out by electricity one time, gingerly touched his knuckle to the metal key and saw and felt a distinct electric spark. Once the string was wet, the electricity from the passing lightning storm flowed steadily down it. Franklin was just lucky no lightning struck his kite directly. And so it was that Franklin, by this brilliant experiment, entered the pantheon of electrical giants, philosophers who signally broadened and advanced the human understanding of this powerful, puzzling, invisible energy. No more did electricians have to speculate. They now knew, thanks to Benjamin Franklin, that lightning was electricity similar to—or maybe the same as—that which men were making with their electrostatic machines. Franklin was hailed as an eighteenth-century Prometheus, one who had literally stolen fire from the heavens and lived to tell the tale.

What Franklin did not know as his kite bucked through the electrified heavens that warm late-summer afternoon was that his book *Experiments and Observations on Electricity* had been translated into French and portions read aloud earlier that spring at a Parisian scientific meeting. The many fractious French scientists, hearing these fascinating ideas for experiments with lightning rods and kites, promptly set off to try them. Three installed a forty-foot-tall lightning rod in a beauteous Gallic garden out in the countryside of Marly-le-roi. The pole, enshrined in a simple wooden structure, stood atop a three-legged stool set in three wine bottles. On May 10, 1752, a thunderclap was heard and an assistant rushed up with a Leyden jar just as a lightning storm began. He successfully drew off large sparks from the pole. As the heavens opened and hail pummeled down, a curious crowd ignored their ever wetter clothes to watch as the lightning was drawn off again and again into the jar. And so, much to Benjamin Franklin's amazement, not long after he had flown his kite, he discovered that French electricians had beaten him to it! They had performed his proposed experiments from his book first.

But their dramatic success and wild enthusiasm propelled Franklin to glittering fame and celebrity throughout the United States and Europe. The king of France, no less, applauded this astonished (but delighted) colonial tradesman for his scientific brilliance. Harvard and Yale bestowed honorary degrees, the prestigious Royal Society in London conferred the Sir Godfrey Copley gold medal and inducted the Philadelphian and fellow British subject into its exalted ranks. As one Italian scientist said, "Who would have ever imagined that Electricity would have learned cultivators in North America?"[11] Word spread of these Promethean wonders, and Georg Richman, a Swedish scientist living in St. Petersburg in Russia, sought to replicate them. He held aloft a wire-tipped pole during a lightning storm and had the misfortune to attract a full hit of lightning. He was electrocuted on the spot, becoming in 1753 the first person to die during an electrical experiment, a martyr to the young science.

Predictably, just as those seriously investigating electricity used it also to divert and amuse, so was it quickly hailed as a medical boon. As one Italian scientist said in 1746, "No thought could occur more readily than this, the moment people saw such light flashing from the body, limbs, and skin, and felt the stings, painful blows and sharp stimuli that penetrated almost into the bones when the lights appeared."[12] Real doctors (such as they were in those days) and itinerant quacks promoted electricity as the cure for constipation, nervous disorders, sexual maladies and infertility, sciatica, rheumatism, lacrimation, and herpes, to name a few ailments. In Italy, three medical schools opened devoted just to healing through electricity. Italy (then part of the Austrian empire) would also be the scene of one of the greatest modern advances in electricity, the consequence of a bitter scientific dispute between the quiet, retiring Luigi Galvani, esteemed physician and anatomist at the University of Bologna, and Alessandro Volta, respected professor of physics at the ancient University of Pavia. Like almost every other scientist of their age, each man was deeply fascinated by electricity, Galvani wondering about its possible role in the workings of the body's nerves and muscles, Volta looking at the basic nature of electricity and its chemical interactions.

This famous scientific feud had its origins in a series of experiments performed by Luigi Galvani, forty-four years old, handsome, clean-shaven, and coiffed in the era's gentlemanly fashion—white peruke, fine lace jabot at his neck, and long dark coat and breeches. On January 26, 1781, Galvani was dissecting a pair of large frog's legs with some assistants. Nearby, an electrostatic machine was being turned. Galvani's assistant noticed with astonishment that when he touched the frog's leg at a particular spot with a scalpel, the leg—unattached to any frog—jerked. Galvani wrote later, "I immediately repeated the experiment. I touched the other end of the crural nerve with the point of my scalpel, while my assistant drew sparks from the electrical machine. At each moment when sparks occurred the muscle was seized with convulsions."[13] Sensitized to this phenomenon of jerking frog's legs, Galvani now noticed that prepared frogs suspended by brass hooks and hung on an iron trellis in his backyard also jerked when the brass hook was pushed against the trellis. Galvani became convinced that what he was seeing was "animal electricity" completing a circuit with these metallic instruments. "The idea grew that in the animal itself there was an indwelling electricity."[14] For the next decade, Galvani quietly worked on a variety of experiments with his frog's legs and published a series of papers in Latin (the scientific language of the day), culminating in 1791 with "On the Effect of Electricity on the Motion of the Muscles." Just as when Benjamin Franklin's work became known, hundreds of dedicated electrical philosophers rushed to replicate Galvani's "animal electricity" experiments.

Among them was physicist Alessandro Volta of Pavia, forty-six, a stern-visaged, clean-shaven man with a thin fringe of dark hair. Professor Volta's electrical accomplishments were already so stellar that he had been inducted into London's prestigious Royal Society. Son of an impecunious nobleman, Volta had grown up on Lake Como and shown a precocious ability in science and mathematics. His major discovery up to this time was the highly sensitive "condensing electroscope" that measured electric charge. While Volta initially accepted Galvani's striking discovery, applauding "the fine and grand discovery of an animal electricity, properly so called," as he began to replicate the experiments on the frog's legs he became more and more skeptical.[15] Ultimately, he became convinced that the electricity was coming not from the frogs, but from the metal, and in 1794 he challenged Galvani to refute his, Volta's, findings that the electricity was actually "metallic." Galvani's many supporters, led by his nephew Aldini, came back with a frog experiment that showed a distinct jerk even when no metal was involved. The always reserved Galvani himself was still mourning the loss in 1790 of his beloved wife, a learned woman who had worked on these experiments with him. The feud grew sufficiently bitter that Volta would write to one friend, "I know those gentlemen want me dead, but I'll be damned if I'll oblige them."[16] So the animal versus metallic electricity dispute dragged on, with each scientist and his followers throughout the Continent performing many subsequent experiments that often did little to illuminate the situation but seemed important or decisive at the time.

Then in 1796 Napoleon Bonaparte swept over the Alps with the French army, and these Austrian citizens found themselves subjects of a new revolutionary Cisalpine Republic. Luigi Galvani loyally refused over the next two years to take an oath of allegiance to the new government, and in late April 1798 he was expelled from the University of Bologna, where he had happily practiced anatomy for thirty-five years. In early December, Luigi Galvani, sixty-one, died, politically ostracized and penniless. Professor Volta, after initial protests, reached some accommodation with the Napoleonic regime and continued on at the University of Pavia.

As leader of the "metallic" electricity forces, Alessandro Volta had been methodically testing dissimilar metals and measuring their electric charge with his sensitive electroscope, determining if each was positive or negative. He also observed that the electrical charge was noticeably stronger when his finger touched the metals, which he deduced was the effect of saline moisture. On March 20, 1800, Volta wrote a letter in French to Sir Joseph Banks, president of London's Royal Society, saying, "After a long silence, for which I shall offer no apology, I have the pleasure of communicating to you . . . some striking results I have obtained in pursuing my experiments on electricity excited by the mere mutual contact of different kinds of metal, and even by that of other conductors, also different from each other, either liquid or containing some liquid, to which they are properly indebted for their conducting power. . . . This apparatus to which I allude, and which will, no doubt, astonish you, is only the assemblage of a number of good conductors."[17]

Drawing on his carefully observed knowledge about which materials did and did not conduct electricity, Volta assembled atop one another inch-wide pairs of copper disks and zinc disks separated from other such disks by a cloth disk or pasteboard soaked with salt water. When these disks were all touching one another, the copper lost electrons to the saline-soaked cloth, while the zinc *gained* electrons from the same wet cloth. As the zinc dissolved, hydrogen gas was produced at the surface of the copper. The resulting electrical charges flowed out in a steady direct current along the wires. This

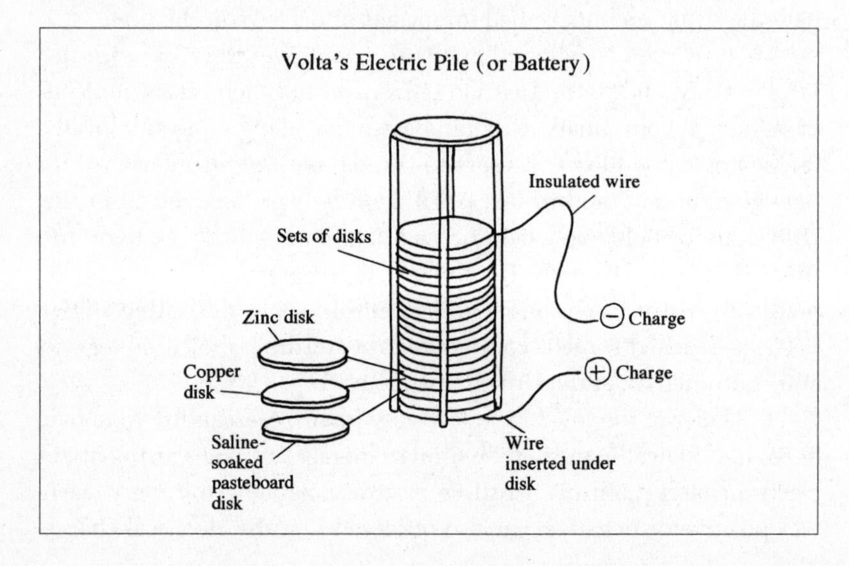

electrochemical reaction produced the first steady generation of man-made current, flowing from the battery on two insulated wires, one on top and one on the bottom.

Here was the first primitive battery able to deliver a continual, steady electrical charge, in contrast with the electrostatic machines or the Leyden jar, which delivered their electricity in high-voltage static bursts or jolts. His pile, said Volta, operated "without ceasing, and its charge re-establishes itself after each explosion. It operates, in a word, by an indestructible charge, by a perpetual action or impulse in the electric fluid." Actually, the Volta battery pile would stop generating electricity once the saline liquid dried up or the metal was all dissolved. For while Volta believed it was solely the contact of dissimilar metals that powered his battery, it was in fact the electrochemical interactions between these metals and the liquids. The longer the battery ran, the more these materials diminished. Volta also designed an alternate version of his battery, which he called his "crown of cups," that featured brine-filled cups connected by alternate strips of zinc and silver and metal wire.

With France and England at war, Volta's communication took some time to reach London and was not read before the Royal Society until June 26, 1800. Volta's brilliant and pathbreaking work was hailed throughout Europe and America. A contemporary declared the battery "the most wonderful apparatus that has ever come from the hand of man, not excluding even the telescope or the steam engine." Other scientists quickly replicated Volta's "electrical pile," building larger and larger versions. Ironically, the steady current of electricity generated by Volta's battery—steady direct current—came to be known as Galvanic, thus verbally immortalizing his electrical foe. Volta, in turn, was linguistically enshrined by the term *volt,* which measures the electrical force of a current. In the immediate aftermath of his epochal electrical triumph, Volta was much honored and richly rewarded, becoming a count in the Napoleonic government. Newly married, famous, and wealthy, he then played little role in further advancing his battery or electrical philosophy.

Other scientists scrambled to construct ever bigger, more powerful Voltaic batteries to generate more powerful, longer-lasting Galvanic currents. Over in London at the new Royal Institution on Albemarle Street, already "the world's greatest showplace for the popularization of science," the dashing young chemist Humphry Davy arranged to have constructed progressively larger batteries in the institution's basement laboratory. In the first decade of the nineteenth century, Davy applied these big batteries to his pathbreaking work establishing that electrochemical reactions were the basis of electricity. The Royal Institution was opened in 1799 to advance and apply science "to the common purposes of life," and its initial patrons were the distinguished English grandee Sir Joseph Banks, president of the Royal Society, and the controversial inventor-statesman Count Rumsford, an American-born "opportunist, womaniser, philanthropist, egotistical bore, soldier of fortune, military and technical adviser."[18] It was Count Rumsford who hired Davy, a handsome and charismatic young chemist who was beginning his swift rise to scientific and intellectual glory. Davy's public lectures at the Royal Institution combined such erudite pyrotechnics and dazzling showmanship, hundreds of well-dressed personages soon flocked into the amphitheater to watch and listen as this effervescent Cornishman with his tumbling locks brought to life the abstruse mysteries of electricity or chemistry. The poet Samuel Taylor Coleridge, friend and admirer, attended these crowded scientific affairs and claimed in awe that if Davy had not been "the first chemist of the age, he would have been the first poet."

From the start, Davy believed that the electricity produced by Alessandro Volta's battery came from electrochemical interactions. Moreover, once Davy had constructed a sufficiently powerful battery in October 1807, he demonstrated that the converse was true—chemical compounds could be decomposed into their basic elements by electricity. Davy decomposed alkalies into potash and soda ash with electricity, then further extracted entirely new elements, potassium and sodium. Davy, this son of a humble wood-carver, had by now burnished his innate nobility by marrying a wealthy heiress. Fortune was soon followed by public honors, including a knighthood. He became Sir Humphry Davy. Sir Humphry's passion, panache, and accomplishment allowed him to travel in the highest social circles, whether among dukes and lords at their great country estates or among the era's most brilliant artists and thinkers, immortals like the poet William Wordsworth and the incomparable painter of landscapes and light Joseph Turner. In 1808, Davy, by now director of the Royal Institution, had constructed a gigantic battery of two thousand pairs of plates in the basement laboratory. Applying the intense electrical energy produced by this massive battery to alkaline earth, Davy extracted more new elements—magnesium, calcium, barium, and strontium.

But most visually spectacular of Sir Humphry's electrical researches was the arc light. In 1809, he gave one of his most literally dazzling lectures. He stood before his usual large enthralled audience and held up two thin charcoal sticks, which served as conductors of electricity. One stick was then connected to a powerful Voltaic pile. When the electricity began to flow through the first stick, Sir Humphry touched it to the top of the second stick. A brilliant spark appeared where the two met. What amazed those present was that as Sir Humphry pulled the two sticks slightly apart, the spark grew larger and the electricity traveled in a dazzling arc of light between the two slender carbon rods. This almost painfully bright blue white light would burn until either the carbons or the electricity ran out. Sir Humphry's thrilling demonstration set off determined efforts to produce a commercial arc light, but properly calibrating the carbons created many troubles and no battery could economically run bright lights for the many hours needed to be practical.

And there the art of electricity would linger, as frustrated scientists and philosophers tried year after year to unlock and decipher electricity's multitude of remaining secrets, especially what many suspected was its relationship to magnetism. The early days of dedicated individual philosophers working on electricity in their homes—men like Benjamin Franklin, Otto von Guericke, and Stephen Gray—were slowly giving way now to the modern era of university-trained scientists operating out of university laboratories or special institutes. Mastering the growing mass of specialized knowledge was time-consuming, while few individuals could afford to construct the ever more complex equipment needed to study electricity, especially the gigantic Voltaic piles. Moreover, national leaders were well convinced of the vital importance of science and technology to national wealth and well-being. So it was not surprising that the next great electrical breakthrough came out of a European university.

In the spring of 1820, Hans Christian Oersted, forty-three, physics professor at the University of Copenhagen, was giving a private lecture on electricity to a group of advanced students. The professor, a tall, sturdy man with dark curls and thick sideburns, was dressed in a long black coat, vest, and high-collared white shirt and cravat. Before him on his wooden laboratory table, Professor Oersted had set up a small Voltaic battery. In his hand, he held a charged electrical wire, intending to make a point about heating platinum wire using electrical current. But as he prepared to apply the wire, the professor noticed the wildly swinging magnetic needle on his large desk compass. When he moved the wire above and around the compass, the needle responded strongly—as if to a magnet—and in very specific ways to the electric current.

Oersted, like Sir Humphry Davy, was a self-made scientist, a country boy of little means whose love for learning and intellectual prowess had earned him scholarships and finally a university appointment. Fluent in several languages, Oersted was trained in pharmacy (thereby having access to many chemicals), had traveled widely to meet other scientists, and was also among the many striving to establish the relationship between magnetism and electricity. In an 1813 book, he had written, "An attempt should be made to see if electricity in its latent stage, has any action on a magnet as such."[19] Now here, seven years later, by his astute observation of the swinging compass needle, Professor Oersted had stumbled across the long-sought clue. In subsequent experiments, Oersted found that the compass needle swung about and took up a position at right angles to the charged wire. If he reversed the flow of electricity, the needle deflected in the opposite direction. In short, electric current produced its own magnetic field and, by means of that field, a force.

On July 21, 1820, Han Christian Oersted announced his discovery of electromagnetism in a four-page paper written in Latin, whose (translated) title was "Experiments on the Effect of an Electric Current on the Magnetic Needle." He sent copies of his monograph to every important university, learned society, and electrical scholar in Europe. Oersted wanted full credit for making this historic discovery, which unraveled further the mysteries of this invisible but potent force, electricity. As word of his great breakthrough slowly spread in this era before telegraph, telephone, or train—Sir Humphry learned of it only in October—Han Christian Oersted was hailed and feted as an electrical genius, as had been Benjamin Franklin and Alessandro Volta before him. Oersted was awarded the Royal Society's Copley gold medal and inducted into many scientific societies.

When Parisian professor of mathematics André Marie Ampère read of Oersted's experiments, he was highly skeptical. Ampère, a deeply religious man whose personal life was bedeviled by disastrous marriages and delinquent children, sought solace in mathematics, chemistry, and science. Ampère readily replicated Oersted's work, then proved that the strength of the magnetic field intensified with the rise of the power of the electric current. Ampère also showed that currents flowing parallel in the same direction attract each other and currents flowing in opposite directions repel. Over the next decade, this new understanding—that a current-carrying wire created a magnetic field around it—led to the creation of ever more powerful electromagnets. In Albany, New York, American engineer and inventor Joseph Henry was one of the pioneers. He used his wife's old silk dresses to insulate wires, enabling him to wrap more than one coil around a piece of iron. The more layers of coils, the more powerful the electromagnet. This work would earn him a professorship at Princeton University. Henry, who later became the first head of the new Smithsonian Institution in Washington, D.C., devised gigantic electromagnets by wrapping hundreds of feet of insulated electrical wire around huge horseshoe-shaped pieces of iron. Demonstrations with these odd-looking instruments elicited thrilled gasps from audiences, for these electromagnets could lift loads as heavy as a ton. Yet despite all these advances in the understanding of electricity, it had few useful applications.

Oersted's discovery of electromagnetism and Ampère's expansion of that work generated a deluge of fresh electrical research and an ensuing avalanche of new scientific articles. At the Royal Institution near Piccadilly Circus, the assistant Sir Humphry Davy had hired back in 1813, Michael Faraday, had become England's foremost analytical chemist, a man renowned for his brilliance, originality, hard work, gentle nature, and sheer scientific productivity. (Many would later say that Sir Humphry's greatest scientific contribution was hiring the unknown Faraday.) Though Michael Faraday had evinced little interest in electricity, in the spring of 1821 the editor of the *Annals of Philosophy* suggested that Faraday apply his much admired intelligence to authoring a survey article on the new science of electromagnetism, thus luring the thirty-year-old philosopher into this fascinating and puzzling field.

Faraday looked more like a poet of the Romantic school than a scientist, for he had a wonderfully handsome, gentle face with a high brow, dark, intelligent eyes, and a headful of thick, wavy curls parted in the middle. Faraday had *not,* like Oersted, come to science through a university. Indeed, his formal education ended at age twelve when he was apprenticed for seven years to a bookbinder. Faraday had completed his service and was just starting that career when a friendly customer bestowed upon him coveted tickets to Sir Humphry's highly popular lecture series entitled "The Elements of Chemical Philosophy." Faraday was so smitten by what he heard and saw at the Royal Institution that he amplified the exquisite notes he had taken during the quartet of talks, made numerous illustrations, compiled an index, and bound it all together into a lovely little book. This he sent along to his new idol, Sir Humphry Davy. Later Faraday would write, "My desire to escape from trade, which I thought vicious and selfish, and to enter into the services of Science . . . induced me at last to take the bold and simple step of writing to Sir H. Davy."[20] Sir Humphry, having risen to magnificent heights from his own humble beginnings, had been sufficiently impressed by the ambition, intelligence, and ardor of this twenty-two-year-old blacksmith's son (and his jewel of a book) to hire Michael Faraday as his assistant. The job paid £100 a year, along with two upstairs rooms at the institution and a supply of coal and candles. So while Faraday was not a university man, he was ensconced in a prestigious and well-funded institute. Such was Faraday's natural scientific brilliance that his discovery of benzene and related chemicals pioneered the way for the new and important aniline dye industry, even as he also advanced the liquefaction of gases. By 1824, he was a Fellow of the Royal Society. The next year, at age thirty-three, Michael Faraday was appointed director of the Royal Institution's laboratory.

Once he delved into the great mass of books and articles on electricity, trying to sort out what was important for his survey article, Faraday was thoroughly ensnared by the conundrum of this invisible force. By 1822, Faraday had written in his laboratory notebook, "Convert magnetism into electricity."[21] He tried on four occasions during the 1820s, even as he was making stunning breakthroughs in chemistry, to figure out some means for converting magnetism to electricity, but he was stymied. Then, on August 29, 1831, Faraday, dressed as ever in his plain black cutaway coat, high-waisted pants, high collar, and plain cravat, noticed a weird effect that provided the vital electrical clue.

Faraday had an iron ring. On one side he had wound a coil of insulated wire that was attached to a battery. On the other side of the ring he wound a second coil of insulated wire, and that was attached to a galvanometer, which measures small electric currents. When Faraday activated the battery to electrify the first coil, he looked hopefully for signs of life in the second coil but saw none. What did

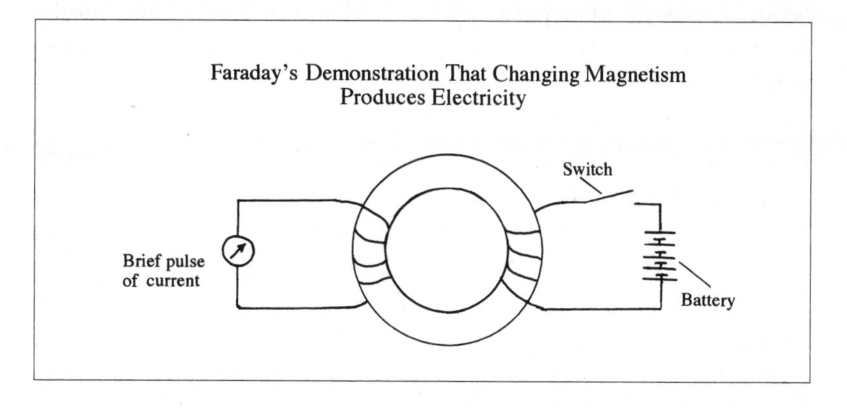

catch his eye was that the galvanometer registered a feeble and momentary current when the battery was attached or detached. Here was the first clue: A *change* in the charged coil's magnetic field—starting and stopping it by activating the battery—briefly created current in the second wire coil. Faraday's friend and colleague John Tyndall would later ponder Faraday's genius and explain it thus: "He united vast strength with perfect flexibility. His momentum was that of a river, which combines weight and directness with the ability to yield to the flexures of its bed. The intentness of his vision in any direction did not apparently diminish his power of perception in other directions; and when he attacked a subject, expecting results, he had the faculty of keeping his mind alert, so that results different from those which he expected should not escape him through preoccupation."[22] Faraday had registered those tiny movements on the galvanometer.

Over the next couple of months, as autumn brought the usual rain and gloom to London, Faraday found a day here and there to pursue these intriguing blips by the galvanometer. In his first experiment, he wound a coil of wire around a straight iron core. He took two long bar magnets and held them in a V and used the wrapped iron core as the third side of the triangle. When the V of the magnets was separated, breaking the magnetic circuit, current was induced in the coil. Noted Faraday in his laboratory journal, "Hence, distinct conversion of Magnetism into Electricity." To further test this, he took a pencil-shaped magnet and simply moved it in and out of coiled wire. The moving magnetic field created a brief current in the wire. But Faraday was interested in producing a continuous current, not just brief bursts.

So he set up a simple twelve-inch copper disk on an axle that revolved between the opposite poles of a permanent magnet. On one side of the copper disk, a wire ran from the axle to a galvanometer. Then another wire led from the galvanometer to a metallic conductor held against the rim of the copper disk. When the copper disk revolved, disturbing and thus changing the magnetic field of the magnet, the galvanometer registered a continuous electric current. Wrote Faraday in his minutely detailed laboratory notebooks, "Here

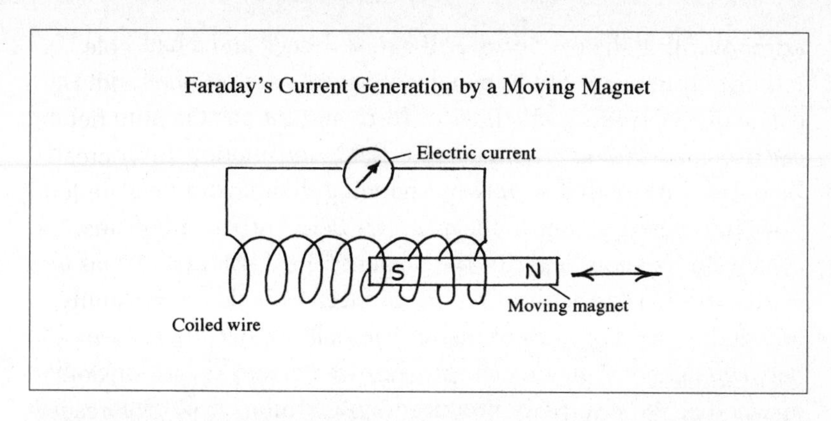

therefore was demonstrated the production of a permanent current of electricity by ordinary magnets." Simply stated, Faraday's law of electromagnetic induction says that an "electric current is set up in a closed circuit by a changing magnetic field."[23]

Faraday modestly described his epochal discovery of the world's first electric dynamo as "A New Electrical Machine" in a paper before the Royal Society in London on November 24, 1831. Oersted had shown how to create magnets with electricity, while Faraday had revealed the other, even more mysterious and momentous half of electromagnetism, how to generate electricity with magnets. As marvelous as was Michael Faraday's further deciphering of electricity, no one at the time envisioned its extraordinary ultimate consequences, for who could imagine that this was the foundation of the modern electrical industry? Unlike Volta, who largely rested on his laurels

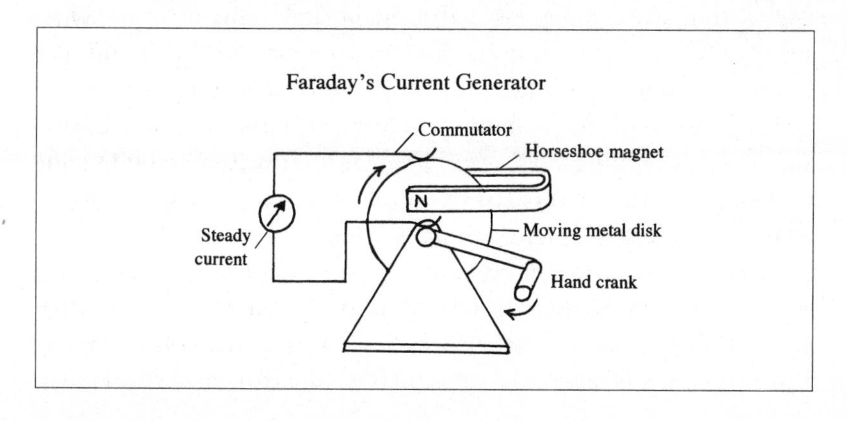

after inventing the battery, over the next decade and a half Faraday's "almost intuitive insight into many of nature's secrets" led him to elucidate on electromagnetic lines of force and the relationship among electric current flow, the magnetic field, and motion through that field. He showed that simply by changing the magnitude of current flow, you caused change in the magnetic field, and that in turn caused current in any conductor in the magnetic field. Faraday would also clarify the electrochemical nature of electricity, determine the specific inductive capacities of many materials, establish a relationship between light and magnetism, and resolve the long debated question of whether the electricity produced by lightning, electrostatics, batteries, and his generator were all the same invisible entity. They were.[24] As one biographer writes, "His versatility, originality, intellectual energy and sheer stamina leave us in awe."[25]

Faraday's own life, work, and stature became an inspiration and model for successive generations of scientists. Believing that Sir Humphry's wealth and titled eminence distracted from his wholehearted pursuit of science, Michael Faraday politely turned down time-consuming titles, opportunities to earn a fortune, and all the socializing attendant on honors and wealth. A devout member of the small Sandemanian Christian sect, he lived modestly, quietly, and happily with his beloved wife upstairs at the Royal Institution. But down in the basement laboratory Michael Faraday was a veritable lion, a passionate and brilliant scientist of rare energy able to select and focus on the most meaningful, discerning problems. His scientific output was prodigious and fundamental, influencing peers in many fields. His laboratory notebooks set a standard of beautifully observed detail, organization, and honest record keeping. The charm of his prolific writings—and his readiness to admit his many laboratory failures on the road to experimental success—earned him wide and enduring readership. His three-volume *Experimental Researches in Electricity and Magnetism* remains a classic.

In the 1830s, with Sir Humphry Davy dead in 1829 at the young age of fifty-two, Faraday truly took over the running of the Royal Institution. One of his first acts was to inaugurate the Friday evening discourses, as well as special Christmas lectures for children. Faraday, whose whole life course was radically and joyfully altered by his attendance at Sir Humphry's famously enthralling lectures, viewed these public events as highly important. Who could say which child might embrace a life of science after a Christmas lecture or which influential and enthusiastic member of the Friday night audience might decide to shower grateful guineas on the Royal Institution? In the age when laboratory science was truly coming to the fore, Michael Faraday was its greatest sage and prophet. He was fittingly also the institution's most scintillating and mesmerizing speaker, his handsome face full of passion, hair flying poetically as he moved fluidly about to show his experiments before the packed amphitheater. The Friday evening lectures began promptly at 9:00 P.M. before an expectant, educated audience dressed formally as for the opera. Recalled one fan, "His audience took fire with him, and every face was flushed."[26] Faraday's friend Tyndall wrote, "He exercised a magic on his hearers which often sent them away persuaded that they knew all about a subject of which they knew but little."[27] When the lecture ended promptly at 10:00 P.M., the animated audience drifted to the institution's magnificent two-tiered library, there to imbibe refreshments, view an exhibition based on the evening's topic, and marvel at science. Faraday's 1849 Christmas lecture for children, "The Chemical History of a Candle," is still read.

Under Michael Faraday's ardent leadership, the Royal Institution became one of England's most important social and intellectual centers when that nation was powerfully ascendant, attracting many eminent Victorians and luminaries, including Charles Dickens, Charles Darwin, and T. H. Huxley. Wrote one Faraday biographer, "Such was the prodigality of his output and the diversity of his skills that modern chemists, no less than physicists, engineers, and material scientists, regard him as the founder of their subjects: some sciences and technologies owe their very existence to his work. . . . He bequeathed to posterity a greater body of pure scientific achievement than any other physical scientist, and the practical consequences of his discoveries have profoundly influenced the nature of civilised life."[28] Faraday was uninterested in spending his own time making anything specifically practical or useful. "A philosopher," Faraday explained, "should be a man willing to listen to every suggestion but determined to judge for himself. He should not be biased by appearances, have no favourite hypothesis, be of no school and in doctrine have no master.... Truth should be his primary object. If these qualities be added to industry, he may indeed hope to walk within the veil of the temple of nature."[29] And so, dedicated to the higher calling of Truth, Michael Faraday had little patience for utility. After he had demonstrated a new chemical process or opened a new electromagnetic realm and the inevitable question followed, "What is its use?" Faraday liked to quote Benjamin Franklin, who had famously replied: " 'What is the use of an infant?' The answer of the experimentalist is, 'Endeavor to make it useful.' "

In the ensuing years, scientists and inventors in England, Belgium, France, Germany, Italy, the United States, and every Western nation all mightily endeavored to make electricity useful, exerting their mental faculties to the utmost in the wake of Faraday's magisterial work. Electroplating was at this time electricity's one practical industrial and commercial application. However, batteries were steadily improving, so that by the 1840s electricity wrought its first miraculous revolution, becoming the basis for a workable telegraph. These middle decades of the nineteenth century saw remarkable, rapid technological advance, with steam engines, railroads, and the telegraph demolishing all traditional notions of power, space, and time. In these same years, coal-gas lighting became a cheap and convenient illuminant indoors in urban homes, offices, and some factories, displacing whale oil and candles. Distributed from a central gashouse to buildings and sidewalks via underground pipes—much the way water was—coal-gas lighting spread quickly in big cities. On cloudy or moonless nights, gas lamps now supplanted whale oil lamps on major metropolitan streets.

The coming of gaslight, like the coming of the railroads and the telegraph, noticeably altered the aeons-old rhythms of time and place. "A new age had begun for sociality and corporate pleasure-seeking.... The work of Prometheus had advanced by another stride," wrote Robert Louis Stevenson, singing the praises of gaslight. "Mankind and its supper parties were no longer at the mercy of a few miles of sea-fog; sundown no longer emptied the promenade; and the day was lengthened out to every man's fancy. The city folk had stars of their own; biddable, domesticated stars. It is true that these were not so steady, nor yet so clear, as their originals; nor indeed was their lustre so elegant as that of the best wax candles. But then the gas stars, being nearer at hand, were more practically efficacious than Jupiter himself. . . . But the lamplighters took to their heels every evening, and ran with a good heart. It was pretty to see man thus emulating the punctuality of the heavens' orbs."[30] Gas lighting proved as popular in American cities as it was in foggy London, and by 1875, there were more than four hundred gas lighting companies in the United States.

As gas lamps replaced (or supplemented) the age-old light of oil and candles, and brought nocturnal lighting for the first time to many murky city streets, inventors and entrepreneurs struggled to come up with a workable version of the brilliant arc light Sir Humphry Davy had demonstrated to such acclaim in 1809. The British operated a few isolated and remote arc light lighthouses using large batteries, while a few daring and avant-garde members of the surviving French aristocracy tried them on their grounds. When engineers installed experimental arc lights near a château in Lyons, the local paper reported, "One could in fact have believed that the sun had risen. This illusion was so strong that birds, woken out of their sleep, began singing in the artificial daylight."[31] The arc light spectrum was indeed close to that of sunlight, and many commented on its evenness and steadiness, so unlike the flickering, swaying light produced by burning gas jets, oil lamps, or candles. But to become a commercial entity, the arc light needed a far simpler, better design and a practical electrical generator. While batteries served well enough for the minimal power needs of the telegraph and then the telephone, battery power was twenty times more expensive than that supplied by steam engines and certainly far too expensive a form of energy to compete with popular gas lighting.

Creating a better generator was a deeply baffling task. It would be almost thirty-five years from Faraday's epochal demonstration of his "electric machine" to the triumphant emergence of a truly practical dynamo. Many labored on the dynamo problem, but the man who ultimately prevailed "to make it useful" was Belgian engineer Zénobe-Théophile Gramme, who worked for a Parisian maker of electrical devices. By the early 1870s, Monsieur Gramme had not only designed a far more powerful direct current generator, but, equally important, had also invented the electric motor, which he showed was just a dynamo or generator running in reverse. Gramme incorporated a major advance introduced by Werner von Siemens: The bit of genius that propelled his generator ahead of all others was using electromagnets rather than regular magnets. A Gramme dynamo featured a ring of iron encircled by coils of wire that revolved in the plane of the lines of force between two electromagnets.

With the Gramme dynamo in hand, the time was ripe for the arc light. And in 1876, Russian military engineer Paul Jablochkoff (also living in Paris) finally came up with a commercial version—the Jablochkoff "candle"—that cast a gentler light than the earlier, glaring arc lights. The "candle" paired two tall, thin carbon sticks separated by a layer of kaolin cement, which served as both insulator and binder. "No mechanism was required for operation; once started the carbons continued to burn until consumed, lasting about two hours. Clusters of candles were arranged so that when one burned down, another was automatically started up."[32] Unlike earlier arc lights, Jablochkoff's could run for as long as sixteen hours, because as one pair sputtered out, another would take over.

However, Gramme's original dynamo design, where a device called a "commutator" directed the current to mimic the direct current produced by batteries, caused one carbon stick to burn twice as fast as the other, a serious flaw. Monsieur Gramme, who now had his own company, solved the problem by redesigning his dynamo to generate an alternating electric current that burned the candle pairs at equal rates. Unlike direct current (DC), where the electrons are spaced evenly as they flow steadily along the conductor, alternating current (AC) causes the electrons to cluster and advance and retreat in fits and starts. Despite all the advances, early arc lights were still more difficult to service than gaslights, and their dazzling glare still restricted use to large spaces—major squares and avenues, department stores, railroad stations, circuses, building sites, wharves, and factories. They required the erecting of special towering poles so the light would not impinge on the normal visual field, hurting people's eyes.

One visitor to Paris hailed the arc light for its "magnificent illumination," exclaiming, "The whole street, to the tops of the loftiest houses, is ablaze with a flood of beaming light which makes the streets seem like the scenes of some grand play at the opera."[33] Robert Louis Stevenson, however, was appalled when he first saw the arc lights of Paris, denouncing them as "horrible, unearthly, obnoxious to the human eye; a lamp for a nightmare! Such a light as this should shine only on murders and public crime, or along the corridors of lunatic asylums, a horror to heighten horror. To look at it only once is to fall in love with gas, which gives a warm domestic radiance."[34] The French authorities and many businesspeople felt differently, and by 1878 the half mile of the supremely elegant avenue de l'Opéra was ablaze with arc lights, as were other major Parisian venues, including the Magasins de Louvres and the Théâtre de Châtelet.

American scientists and businessmen were as interested in arc lights as their counterparts on the Continent, and when they learned of the Jablochkoff "candles" in Paris, the race was on to introduce something comparable in the United States. The Yankee businessmen who came up with the first workable arc light system might become the Vanderbilt of lighting, reaping fame and fortune brightening the American night. At the Centennial Exposition of 1876 in Philadelphia, most famous for its 1,400-horsepower Corliss steam engines and Thomas Edison's multiplex telegraph, electrical inventor Moses G. Farmer exhibited three of his own versions of the glaring arc lights. They were powered by the first dynamo designed by Americans, the work of the brilliant Farmer and William Wallace, his partner and proprietor of the nation's foremost brass and copper foundry in Ansonia, Connecticut. Within the year, a major arc light competitor emerged out in Cleveland: a young chemist named Charles F. Brush. He beat Wallace and Farmer to the market and by the fall of

1878 was installing his hissing, brilliant arc lights inside a Boston department store, Continental Clothing House.

Thomas Edison's longtime friend Professor George Barker of the University of Pennsylvania was certain this was a terrific and fertile field for Edison, and he tried to pique his interest by sending along numerous reports on this newest form of artificial illumination. When that had no effect, on Sunday, September 8, Barker escorted Edison to Wallace's large brass foundry. Under a cool gunmetal sky, Edison and Barker stepped off the train with a journalist in tow from Charles Dana's *New York Sun.* For the first time, Edison finally had a chance to really see and examine Wallace and Farmer's steam-run 8 horsepower electric dynamo, which they dubbed the "telemachon." This was the machine that lit up their whole line of eight arc lights at one time. Reported the *Sun,* "Edison was enraptured. He fairly gloated over it. . . . He ran from the instruments to the lights, and from the lights back to the instruments. He sprawled over a table with the SIMPLICITY OF A CHILD, and made all kinds of calculations. He estimated the power of the instrument and of the lights, the probable loss in transmission, the amount of coal the instruments would save in a day, a week, a month, a year, and the result of such saving on manufacturing."[35] The actual machines and their blazing light were having exactly the effect Professor Barker had hoped for. Edison was now afire with excitement. Ever the competitor, he turned to his host, William Wallace, and said, "I believe I can beat you making the electric light. I do not think you are working in the right direction."[36] William Wallace, who had been working on arc lights for several years and had his system up and going, was a good sport. He accepted the bet and shook hands on it.

Then Edison rushed back to quiet, bucolic Menlo Park, his research workshop in backwater New Jersey, to throw himself into creating a better and more practical electric light. He worked feverishly, thrilled at the possibilities of this new field. "It was all before me. I saw the thing had not gone so far but that I had a chance. I saw that what had been done had never been made practically useful. The intense light had not been subdivided so that it could be brought into private houses."[37] Edison always liked to go after "big things." In examining Wallace's lights, he had grasped both the immense possibilities of the dynamo and the limited nature of the blazing arc lights. The man who came up with the best arc light system might well make a fortune stealing away even that 10 percent of the gas lighting business—that of the streetlights. But the man who could subdivide the light—to take it indoors and tame it into a gentle glow—and power it with a dynamo, he would be the true Promethean, the blazing electrical pioneer, the hailed benefactor of humankind (and wealthy to boot). The race to illuminate with electricity the houses and offices of America—nay, of the entire world—and to power the machines of the new industrial order, was on.

R.F. Outca

CHAPTER 3

Thomas Edison: "The Wizard of Menlo Park"

On a mild September day in 1878, a reporter for the *New York Sun* headed to the Cortlandt Street dock, making his way through the cursing teamsters, their horses straining under great loads, past the oyster stalls, and into the Pennsylvania Railroad ferry terminal, there to board the railroad's ferry for Jersey City. Up on the boat's top deck, one could feel the harbor breeze and study the panoply of commercial vessels plying the Hudson River—the brigs and schooners, a three-masted clipper billowing toward the open seas, the workhorse barges coming down from the Erie Canal, and long side-wheel steamers. At times, it seemed the whole world, with all its woes and its wealth, was converging on Manhattan. Half a million immigrants surged in every year now. True, most kept moving, their eyes and hearts set on farms or striking it rich in the mines out west, but fifty thousand stayed here, cramming into the old tenements and filling the almshouses. The downtown slums had become sinkholes of cholera and typhoid. At night, paupers huddled in doorways near the steam grates. When the cold grew killing, those piteous beings retreated to the city's dank indoor sleeping dens. Yet the nation's wealth poured in, too. Every day the wharves were jammed with sailing vessels and steamships, more than ten thou-

Thomas Edison's laboratory, Menlo Park, New Jersey

sand a year coming and going. The magnificent buildings of Wall Street and the financial district reflected this great economic power, as did the sheer energy of the rushing, intent crowds. It was generally believed that the hard times following the Panic of 1873 were finally over.

On the New Jersey side of the river, hundreds who had debarked the Pennsylvania trains waited to board the giant ferry, all heading toward the cacophony of daytime New York. On the southbound accommodation train, it was just over twenty miles to Menlo Park, where Thomas A. Edison had moved from Newark two years earlier and established America's first invention factory. There Edison planned to develop "a minor invention every ten days and a big thing every six months or so."[1] Menlo Park had no real railway station, just a small wooden platform, and arriving visitors walked up a steep flight of crude steps to find themselves on the highest point between New York and Philadelphia, a pastoral place of fields edged by stands of woods, cows grazing in the distance, and a mellow sky that arched high overhead. Shortly after moving, Edison wrote a friend that his new lab was located at "Menlo Park, Western Div., Globe, Planet Earth, Middlesex County, four miles from Rahway, the prettiest spot in New Jersey, on the Penna. Railway, on a High Hill."[2]

Walking up the dirt road from the train tracks, one heard only birdsong, the sound of the wind, and mechanical rumblings from the handful of big plain buildings surrounded by a white picket fence. The two-story clapboard of grayish white was already famous as Edison's laboratory. Telegraph wires sprouted from the upper reaches, met up with tall wooden poles, and were carried off toward Manhattan. The laboratory was the center of this small universe, and Edison was its animating spirit, around whom all else revolved. Thomas Edison had announced he was becoming a full-time inventor in early 1869. The previous six years had been spent drifting from city to city as a crack itinerant Western Union operator, while he was continually devising improvements in telegraphy and soaking up technical books on telegraphy and electricity. Once committed to full-time inventing, Edison had done well enough with such clever items as an electric copying pen. But he really hit the jackpot in late 1874 when he sold rights to his quadruplex telegraph system to Western Union rival and Wall Street manipulator Jay Gould for $30,000.

This was heady success for a small-town boy from Port Huron, Michigan, whose father had muddled along in various grocery, real estate, and truck farming enterprises, while his mother took in boarders. Young Alva, as he was known, had received very little formal education, being taught mainly by his mother, who had briefly been a schoolteacher. His Michigan boyhood revolved largely around his many ingenious efforts to make mechanical things or brew new chemistry experiments, including one that produced "an explosion [that] . . . wrecked a corner of the building and burned [Edison] and some of the other boys."[3] When Edison, age thirteen, joined the Grand Trunk Railroad as a newsboy, he impressed his bosses as hardworking, entrepreneurial, and intent on self-improvement. He spent $2 (two days' pay) to join the new Detroit Public Library and proceeded to read his way through its shelves.

It was during these railroading years that Edison became partially deaf. Once as he was struggling to get aboard a moving train with his newspapers, a conductor helping him clamber on "took me by the ears and lifted me. I felt something snap inside my head, and my deafness started from that time."[4] Ever the optimist, Edison viewed his deafness as an advantage, a built-in buffer against outside distractions that helped him concentrate on whatever he was doing. By his teens, that was telegraphy. Edison's avid curiosity about all things mechanical had led him to befriend the local telegraph operators wherever he was. When Edison turned sixteen in 1863, his natural flair for banging out and receiving Morse code (honed by eighteen-hour bouts of practice) earned him a slot as a junior operator. The Civil War was on, and telegraphers were in great demand. And so Edison was launched in the world of telegraphy, invention, money, and getting ahead.

Edison's work with the telegraph, telephone, and the amazing talking phonograph had given him an excellent grasp of the current primitive state of electrical knowledge. Such was his reputation that he was on retainer to Western Union for $400 a month. He had invested much of his considerable earnings in Menlo Park, determined to have at hand everyone and anything he might need to create and work out the practical problems that interested him. As one interviewer noted, "The keynote of [Edison's] work is commercial utility. He asks himself when a new idea is suggested, 'Will this be valuable from the industrial point of view? Will it do something important better than existing methods?' "[5] Now Edison, that marvel of hard work, imagination, and enterprise, having seen Wallace and Farmer's dynamos and arc lights, was, in September of 1878, concentrated on the electric light. The previous spring, a reporter and artist from the *New York Daily Graphic* had made the pilgrimage to Menlo Park to "see Edison and his wonderful inventions." When they entered the laboratory to sketch scenes for their readers, they found a long open room humming with activity: "The first floor is occupied by scribes and bookkeepers in one end, and at the other some ten or twelve skillful workers in iron, who, at anvil and forge, lathe and drill, are noisily engaged in making patterns and models for the genius of the establishment. His iron ideas, in tangled shapes, are scattered and piled everywhere; turning lathes are thickly set on the floor and the room is filled with the screech of tortured metal.

"Upstairs we climb, to a room the size of the building, with twenty windows on sides and ends. It is walled with shelves of bottles like an apothecary shop, thousands of bottles of all sizes and colors. In the corner is a cabinet organ. On benches and tables are batteries of all descriptions, microscopes, magnifying glasses, crucibles, retorts, an ash-covered forge, and all the apparatus of a chemist."[6]

Menlo Park offered cheap real estate and blessed peace and quiet. Edison's right-hand men and many of his workers had moved out with him from Newark, where he had opened a lab and workshops in 1870. Preeminent were Charles Batchelor, a dark-bearded English mechanic who had learned unholy patience and exquisite motor skills in Manchester's textile mills, and John Kruesi, a master Swiss machinist who sported a huge drooping mustache over a thick black beard. Kruesi's task, at which he excelled, was to translate Edison's rough sketches into high-quality working models. When Edison was on to a problem, there was no day or night, just hours in which to work, as his long-suffering, neglected wife and two small children well knew. Though his family lived in a wooden house just a few hundred yards down a plank road, Edison could rarely pull himself away long enough to dine at home, instead fueling himself on yet another slice of pie, preferably apple. Most of the men roomed across the way at Mrs. Jordan's boardinghouse.

For forty years, scientists and inventors—American and English, French, Russian, Belgian—had been largely frustrated in their efforts to create a practical indoor electric light, some kind of enclosed glass globe that could safely and brightly glow. Edison himself had toyed briefly with both arc and incandescent lights in November 1877, but with little success. He would later recall, "The results of the carbon experiments, and also of the boron and silicon experiments, were not considered sufficiently satisfactory, when looked at in the commercial sense, to continue them at that time, and they were laid aside."[7] Now on this September Saturday—a mere *week* since his tour of Wallace's shop—Edison pronounced with characteristic hubris to the reporter from the *New York Sun,* who had come out to Menlo Park, that *he,* Edison, would be the one to succeed with the electric light (and more—far more!) where all others had failed. He, Edison, would be the Prometheus who would divine the secrets of this mysterious agency and light up America and the world. *He* had—in one inspired week—just invented the first practical incandescent light bulb, one where a wire inside the glass bulb glowed brilliantly as electricity flowed through it.

So on September 16, 1878, the *New York Sun* duly proclaimed: EDISON'S NEWEST MARVEL. SENDING CHEAP LIGHT, HEAT, AND POWER BY ELECTRICITY. As befitted one who as yet had not secured his patents, Edison was understandably vague on the details of his historic breakthrough, except to say, "I have obtained it [the light] through an entirely different process than that from which scientists have sought to secure it. They have all been working in the same groove. When it is known how I have accomplished my object everyone will wonder how they never thought of it. . . . I can produce a thousand—aye, ten thousand lights from one machine."[8] Edison's great but unrevealed breakthrough was, says biographer and Edison scholar Paul

Israel, "a thermal regulator to prevent the incandescing element of his lamp from melting."[9] Edison was proclaiming these breakthroughs—a workable light bulb and a whole electrical lighting network—with a large dollop of showmanship, intended to attract investors and scare off rivals.

Edison proclaimed to the *New York Sun* reporter, "I can light the entire lower part of New York city, using a 500 horse power engine. I propose to establish these light centres in Nassau street, whence wires can be run up town as far as the Cooper Institute, down to the Battery, and across to both rivers . . . the same wire that brings the light to you . . . will also bring power and heat . . . you may cook your food." Edison's vision had already vaulted beyond mere light bulbs into a glorious and immediate future of electrical grandiosity: Not only did he have in hand a workable incandescent light bulb (that would in short order make gas lighting obsolete), he would create an entire electrical power system. In a brief editorial, the *Sun*'s famous editor Charles Dana wryly allowed as how "if Edison is not deceiving himself, we are on the eve of surprising experiences."[10] Yet Dana, whose early years of poverty made him and his popular two-penny daily ardent champions of the workingman, did not view Edison's Promethean assertions as worthy of his lively front page, faithfully devoted to murder, mayhem, and disaster. Electricity would not displace detailed coverage of the yellow fever plague ravaging the South, or the riveting trial for the "murder of Mrs. Jesse Billing," or the previous day's FINDING A CORPSE IN A BARREL IN A SECLUDED RAVINE, or the three boys who drowned in a coal chute.

With Edison suddenly promising cheap and easy lighting within mere months, his attorney, Grosvenor P. Lowrey, leapt into action. Within the week, he informed Edison that he was rounding up investors for Edison's new venture. Developing practical electricity would be costly, far beyond the inventor's own resources. On October 3, 1878, Edison wrote, "Friend Lowrey: Go ahead. I shall agree to nothing, promise nothing and say nothing to any person, leaving the whole matter to you. All I want at present is to be provided with funds to push the light rapidly."[11] As reporters streamed down to Menlo Park and Edison insisted that he was on the verge of lighting up Manhattan, Lowrey established the Edison Electric Light Company on October 16, 1878, with 3,000 shares of stock. Edison was assigned 2,500 shares worth $250,000 for his electric light patents—those pending and in the future—leaving 500 shares worth $50,000. These were subscribed to by the initial investors, including Lowrey; three of his law partners; Western Union president Norvin Green; Drexel, Morgan partner Egisto Fabbri; capitalists Tracy Edson and James Banker; financier Robert L. Cutting Jr.; and last but not least, Hamilton McK. Twombly, son-in-law of the immensely wealthy William H. Vanderbilt.

While Lowrey was rounding up capital for Edison's light, the Wizard of Menlo Park was busily hosting reporters, and by mid-October he was actually demonstrating his much bruited new light bulb. The *New York Sun* reporter who returned again to Menlo Park to bear witness wrote reverently, "There was the light, clear, cold, and beautiful. The intense brightness was gone. There was nothing irritating to the eye. The mechanism was so simple and perfect that it explained itself. The strip of platinum that acted as burner did not burn. It was incandescent. It threw off a light pure and white. It was set in a gallows-like frame, but it glowed with the phosphorescent effulgence of the star Altaire. . . . It seemed perfect."[12] Edison could be equally rhapsodic about his aborning incandescent light bulb: "There will be neither blaze nor flame, no singeing or flickering; it will be whiter and steadier than any known lamp. It will give no obnoxious fumes nor smoke, will prove one of the healthiest lights possible, and will not blacken ceilings or furniture." Of course, what Edison didn't mention about his wondrous new light was that it lasted only an hour or two. It was still far from commercially viable.

In an age of great formality, when gentlemen wore fine Prince Albert suits, a proper stiff collar and cravat, and a shiny silk top hat when venturing forth in public, Edison preferred to play the unschooled hick at Menlo Park, affecting rumpled blue flannel workman's suits, silk neckerchiefs, a simple cloth skullcap, and solid boots. In truth, Edison was a voracious and penetrating reader, hungry for knowledge and possessing an amazing memory. His early deafness only made him more likely to lose himself in a book. At age twenty-one, while still working as a telegrapher for Western Union in Boston, Edison had avidly consumed all three volumes of British scientist Michael Faraday's *Experimental Researches in Electricity and Magnetism.* Faraday became an immediate hero, a poor London boy who rose to the top ranks of science on his brains and hard work. To Edison, Faraday had been living proof that the secrets of nature could be revealed through determined experiment and astute observation.

That November, Edison immersed himself in journals and books about the gas lighting industry, knowing that he had to understand what kind of system he would be challenging. Since the 1840s, the quarter of Americans who lived in reasonably big towns and cities had had access to gas lighting. The great majority of the nation, however, still lived out on farms and in villages and used cheap tallow candles, whale oil, or kerosene for light. Only in bigger cities was it economical for gas made from coal to be piped along under the streets, where it lit up street lamps, and from the street into stores, theaters, factories, and homes, traveling just as water did, through special pipes. Meters recorded usage. Of course, each gaslight had to be individually lighted and snuffed out and its glass globe cleaned. Each gas flame flickered and gave off, as it burned, small quantities of ammonia and sulfur, as well as carbon dioxide and water. Over time, these fumes visibly blackened not just the encasing glass globe, but a room's interior decor. Crowded, closed rooms lit by gas could quickly become deficient in oxygen and make people feel ill. Electricity had none of these drawbacks.

Beloved as Edison was by an awed and respectful public, his cocky ways and phenomenal early success had deeply irked his many scientific and inventing rivals, especially the gentlemen of academia. As the press breathlessly parroted Edison's overblown light bulb claims, the scientists responded with disdainful disbelief. In Britain, Professor Silvanus Thompson scoffed in a public lecture, "We have heard a great deal of late of Mr. Edison's discovery of a means of indefinitely dividing the light. I cannot tell you what his method may be, but this I can tell you, that any system depending on incandescence will fail." Another prominent English electrician, John T. Sprague, declared, "Neither Mr. Edison nor anyone else can override the well-known laws of Nature, and when he is made to say that the same wire which brings you light will also bring you power and heat, there is no difficulty in seeing that more is promised than can possibly be performed. The talk about cooking food by heat derived from electricity is absurd."[13] They were joined by their countryman and fellow electrical scientist William Preece, who jeered that "a subdivision of the electric light is an absolute *ignis fatuus.*" The Latin phrase meant literally "foolish fire," but it was an even worse insult because it referred to almost imaginary swamp gases.[14] But such was Edison's worldwide reputation that gas stocks in the United States and England plunged in value. British Parliament tried to reassure investors by appointing a committee to review Edison's claims. The conclusion: Edison's wild dreams might be "good enough for our transatlantic friends" but were "unworthy of the attention of practical or scientific men."[15]

Meanwhile, back at Menlo Park, where only the regular passing of Pennsylvania Railroad trains disturbed the bucolic tranquillity, Edison was settling into what he now appreciated was a far more difficult task than he had imagined. Swept up in his Promethean dream, Edison worked feverishly amid the batteries and bottles, seeking the ideal, long-burning filament, the properly shaped glass, the perfect atmosphere within that glass. Because he was inventing not just a light bulb, but a whole electrical power network to run those light bulbs, he was already thinking about the problem in terms of what made economic sense. Very early on, Edison had realized that—contrary to common electrical wisdom—he should be seeking to create incandescence with a very high-resistance material. Those inventors who had walked the light bulb path previously had all veered toward the path of low-resistance materials. But, always conscious of ultimate network costs, Edison concluded that the only way to diminish the horrific cost of copper wire for transmission was to run very low currents through thin copper wires. According to Ohm's law, first formulated by a German physicist in 1827 and as yet little understood or honored, the magnitude of the electrical current flowing in a conductor (amperes) was equal to the electromotive force (this being the pressure or volts) *divided* by the resistance to the current, the resistance being measured as ohms. So Edison calculated that if he was going to run a low current (of 1 or 2 amps) through his thin copper wires to save money, he would have to develop a high-resistance light bulb (200 ohms) operating at a relatively low voltage (110 volts). When he had demonstrated his breakthrough bulb in mid-October to the journalists, it featured a thin incandescent spiral of platinum inside an orange-size glass globe atop a thin neck. And, indeed, it gave a quite satisfactory light—but only for an hour or so.

The truth was, Edison was having trouble finding a reliable high-resistance filament. He was getting disenchanted with platinum, which took high heat well but was fragile and simply did not burn for all that long. So he plugged away, looking for something better. But just as he had made the important realization that high resistance was key, so too had he discovered that the greater the vacuum inside the light bulb, the longer and better the filament burned. Therefore much time and energy were applied to developing better vacuum pumps. By February of 1879, Edison was almost completely preoccupied with finding the perfect high-resistance material and the ideal vacuum. Originally, he assumed he would use William Wallace's direct current generator to light up his lamps. But just as he had concluded that in order to slash copper costs he would need low currents pulsating invisibly over his wires and a high-resistance filament, now he realized he must have more powerful generators to supply the necessary horsepower for the many thousands of light bulbs he envisioned radiating their cool, quiet glow in Manhattan's gloomy offices and brownstones. In his usual methodical way, Edison had ordered the five best existing dynamos and begun making improvements, homing in on how the armatures were wound and then the size and shape of the all-important magnets.

Grosvenor P. Lowrey, meanwhile, had been importuning Edison to let him escort his increasingly restive Wall Street investors out to Menlo Park to see firsthand his wonderful progress. After all, Edison felt free to brag to journalists that he was making this or that marvelous breakthrough. They also knew he had already spent prodigious sums—of *their* money. In January, Edison had written a friend, "The fund I have here is very rapidly exhausted as it is very expensive experimenting. I bought last week $3,000 worth of copper rods alone, and it will require $18,000 worth of copper to light the whole of Menlo park 1/2 mile radius." At the same time, rumors were circulating that Edison was hopelessly bogged down. So, on the raw, chill evening of Monday, March 26, Edison played host to the money men. Lowrey and the financiers trooped off the train and up to a trim new brick building. There Edison met them in the warmth of the elegant office and welcomed them into the upstairs library, all furnished—at Lowrey's insistence—in the finest cherrywood furniture. Wall Street millionaires and other important visitors required more than a shack.

Edison spoke for half an hour or so about progress on various fronts: a better filament—platinum plus iridium, tighter vacuum, improved DC dynamo. Then he led the way through the blustery raw evening to the nearby laboratory. It was a moonless night, very dark out, ideal for Edison to display his platinum bulbs in the pitch-black lab. The *New York Herald* reporter described twelve incandescent bulbs in the large machine shop doing the work of eighteen gas burners: "The light given was clear, white and steady, pleasant to the eye."[16] So the money men saw that Edison was making progress, and the journalist from the *Herald* duly reported, "All were much pleased with the result." Yet the hard truth was that even the improved platinum light bulb—despite Edison's assertions to reporters that it was ready for the world—was nowhere close to fully functional. Nor had Edison yet improved any of the DC dynamos enough to show one off as his own. While this show was calculated to convince the naysayers that Edison was on the verge of building his central generating station in downtown Manhattan and lighting up New York with electricity, he had a great deal more work to do.

By late April of 1879, the Menlo Park gang had all reason for genuine cheer, for they had at last devised a superior dynamo whose appearance earned the affectionate nickname the "long-legged Mary-Ann." Edison's machine had a pair of three-foot-tall iron poles (hence, the legs). What was really new was placing "the dynamo's armature between the poles of a powerful, oversized magnet . . . a concentrated source of Faraday's lines of magnetic force."[17] Edison's generator was far superior to existing electrical generators, more efficient, and capable of lighting many light bulbs. He did this by "making the internal resistance much smaller than the external load, rather than having equal internal and external resistance," as was the norm.[18] But *what* light bulb? For all of Edison's advances in understanding resistance and vacuum, practical success on the filament remained maddeningly elusive. All through the spring and summer, the Menlo Park gang laboriously tinkered in fits and starts with endless variations of the platinum-filament bulb. In August, the young German immigrant glassblower Ludwig Boehm joined the crew, installing his bellows and glassblowing table in a corner. A dandyish lad, he wore pince-nez and was quick to remind others that he had studied with the great German master Heinrich Geissler. At a time when the Menlo gang was starting to feel the wearying effects of months of frustration, Boehm's lively zither playing provided a pleasant diversion on warm evenings.

As October rolled around and the air grew brisk and the large ash tree outside the laboratory shed its leaves, Edison and Charles Batchelor began experimenting with baked carbon filaments, the first ones being made from kerosene lampblack that was rolled into reed-thin strips, carefully coiled, and then gently carbonized in a furnace. During the testing of these lamps, Batchelor was always assisted by young Francis Jehl, who was in charge of making sure the batteries were fresh and full of power. Jehl also had the slow and tedious ten-hour task of evacuating (with the unwieldy vacuum pump) as much air as possible from each new carbon-filament bulb tested. Then, on October 22, Batchelor wrote in the extremely detailed Menlo Park lab notebooks, "We made some very interesting experiments on straight carbons made from cotton thread."[19] A plain cotton thread had been baked in the special carbonizing oven, installed gingerly in the filament holders, and fitted into one of Boehm's handblown pear-shaped bulbs; then the bottom was closed and the air in the bulb was slowly, slowly pumped out by Jehl. When attached to the batteries and turned on, this lamp showed resistance above 100 ohms. Moreover, these thread-filament bulbs burned for two and three hours, a signal improvement over platinum. Batchelor pressed on, systematically testing eleven other fiber variations—"thread rubbed with tarred lampblack," "soft paper," "fine thread plaited together 6 strands," "cotton soaked in tar (boiling) & put on."

At 1:30 in the morning, Batchelor and Jehl, watched by Edison, began on the ninth fiber, a plain carbonized cotton-thread filament (in a horseshoe shape) set up in a vacuum glass bulb. They attached the batteries, and the bulb's soft incandescent glow lit up the dark laboratory, the bottles lining the shelves reflecting its gleam. As had many another experimental model, the bulb glowed bright. But this time, the lamp still shone hour after hour through that night. The morning came and went, and still the cotton-thread filament radiated its incandescent light. Lunchtime passed and the carbonized cotton fiber still glowed. At 4:00 P.M. the glass bulb cracked and the light went out. Fourteen and a half hours! Francis Upton, of the soulful dark eyes and graduate degrees, the one Edison man trained in math and physics, now began evaluating the electrical properties—the improved filament and superior vacuum—of this most promising experimental lamp. On November 4, Edison applied for the light bulb patent that would catapult him to even greater fame—the carbonized cotton-thread horseshoe-shaped filament burning inside a pear-shaped bulb largely voided of air.

Edison now settled in with his unlit cigar and his verdigris microscope, systematically examining hundreds of other prospective filament fibers. Those whose structure looked promising he passed over to the genial Charles Batchelor, whose great patience and wonderful dexterity helped the Edison lab work methodically through Chinese and Italian raw silk, horsehair, teak, spruce, boxwood, cork, celluloid, parchment, and New Zealand flax, to name but a few. But most memorable and most amusing were the hairs harvested for possible filaments from the "luxurious beards" of Swiss machinist John Kruesi and a Scotsman from Michigan. Recalls Jehl, "Bets were placed with much gusto by the supporters of the two men, and many arguments held over the rival merits of their beards."[20] Kruesi's carbonized beard hair filament flamed out first, making him the loser of the fiber "derby," a loss followed by much good-natured grumbling about unfair discrepancies in currents. Ultimately, carbonized cardboard emerged that fall as the best of the possible filaments—even better than the carbonized cotton—and production was begun.

The whole atmosphere at Menlo Park quickened during November and early December of 1879, for their official public display was to be New Year's Eve. They had their light bulb. They had their generator, the "long-legged Mary-Ann." Now they ordered steam engines to power them. While these were the major components of the system, Edison also had had to invent and manufacture dozens of other parts, including switches, fuses, distribution lines, regulators, and fixtures. Western Union agreed to send out men to help with the electrical wiring. The ever faithful Lowrey was kept informed. (Edison, former ace telegrapher, himself often personally tap-tap-tapped his telegraphic communications to Lowrey's Manhattan office.) Flush with success, Edison was gearing up for the all-important public lighting display, the radiant refutation of all his academic naysayers. More important, he knew he needed to woo back his surly Wall Street investors so they would reopen their wallets for the next phase, creation of a prototype of the New York electrical network. Rumors in the New York press were rampant, for after all, travelers in the Pennsylvania Railroad trains passing at night began to report brilliant lights gleaming through the windows of Menlo Park edifices. Edison's colleague Francis Upton wrote jubilantly to his father that December of 1879, "The light is still prosperous; I have had six burners in my [Menlo Park] house during the past week and illuminated my parlor for the benefit of a party of visitors from New York. The exhibition was a success. Mr. Edison's and my house were the only ones illuminated. There will be a great sensation when the light is made known to the world for it does so much more than anyone expects can be done."[21]

Edison, who had always been easily available to any reporter looking for a story, now put off all requests. He allowed access only to his favorite, Marshall Fox of the important and influential *New York Herald,* a Republican daily preeminent in foreign news, expensive to purchase at three cents, and consequently read only by the best people. But the condition was that Fox publish the article, which Upton would help edit, only when Edison gave the go-ahead. Yet in true journalistic fashion, Fox broke the story as soon as possible. Thus, on Sunday, December 21, 1879, readers of the *Herald* opened their papers to find a full-page story headlined EDISON'S LIGHT—THE GREAT INVENTOR'S TRIUMPH IN ELECTRIC ILLUMINATION—A SCRAP OF PAPER—IT MAKES A LIGHT, WITHOUT GAS OR FLAME, CHEAPER THAN OIL—SUCCESS IN A COTTON THREAD. "Edison's electric light," wrote Fox, "incredible as it may appear, is produced from a tiny strip of paper that a breath would blow away. Through this little strip of paper is passed an electric current, and the result is a bright, beautiful light, like the mellow sunset of an Italian autumn.... And this light, the inventor claims, can be produced cheaper than that from the cheapest oil."

Edison's investors remained chary. Egisto Fabbri, the Drexel, Morgan partner, had journeyed down to Menlo Park while Fox was hanging about preparing his article. He had seen Upton's and Edison's plain wooden houses "illuminated," as well as the laboratory, but Fabbri still harbored memories of previous overblown statements by the ever optimistic Edison. With the New York papers full of stories about the light and the New Year's Eve official debut looming, Fabbri wrote Edison on December 26, "I suggest to you the wisdom & the *business* necessity of giving the whole system of *indoor* & outdoor lighting *a full test of continuous work* for a week, day & night, *before* inviting the public to come and look for themselves.... Any disappointment would be extremely damaging and probably more so than may appear to you as a scientific man."[22]

In truth, it was too late. With the *Herald*'s article, the word was quite out. Each subsequent afternoon and evening, flocks of electricity sightseers crowded off specially scheduled Pennsylvania Railroad trains or pulled up in the crudest of farm wagons and the most luxurious of broughams, carriages equipped with coachmen and gleaming pairs. There, as the freezing December evening enveloped the snow-clad Jersey countryside, and clouds scudded across the black night sky, the visitors would head through the dark toward the bright laboratory, there to push through and gaze in awe at the magical display. The official public unveiling was December 31, 1879, New Year's Eve. And that evening, as the 1870s became the 1880s, three thousand people poured in to Menlo Park, ignoring the stormy weather, to see the miracle of incandescence.

Despite Fabbri's concerns about premature displays and humiliating failures, Edison's light was a smashing success, a fabulous vindication. He had, as promised, divided the electric light. Reported a breathless *New York Herald,* "The laboratory was brilliantly illuminated with twenty-five electric lamps, the office and counting room with eight, and twenty others were distributed in the street leading to the depot and in some adjoining houses. The entire system was explained in detail by Edison and his assistants, and the light subjected to a variety of tests.... Many had come in the expectation of seeing a dignified, elegantly dressed person, and were much surprised to find [Edison] a simple young man attired in the homeliest manner, using for his explanations not high sounding technical terms, but the plainest and simplest language." Inside the laboratory, important men sporting elegant evening wear, accompanied by women in fashionable silk gowns with short fur jackets and muffs, pressed through the crowds of country boys in checked suits and derbies to see the electricity turned on and off, to gaze in wonder at the glowing light captured in a pear-shaped glass bulb.

The triumphant display of the new electric light impressed the New York investors sufficiently that they parted with a further $57,568 to underwrite Edison's next phase. Now began the great push to transform Menlo Park's jury-rigged system that dazzled crowds of electrical tourists for a couple of weeks into a truly commercial network that would function reliably in the demanding hurly-burly of New York and still compete in price with gas. Edison's plan was to create a miniature lighting network in the still-frozen fields in and around his laboratory, there to test out his plan to generate electricity in a central station and then send it forth to the world via insulated copper wires buried just below the city streets in (still to be dug) sunken subways. Once those insulated copper wires reached a building, they would just be run through the existing gas pipes into existing gas lamps, where light bulbs would be attached. As the exhilaration of New Year's Eve subsided, the hard truth of Menlo Park in the early days of January 1880 was the vast task before them, for every single component required major improvements. To that end, Edison again doubled his laboratory workforce, bringing it up to sixty men.

Edison scholar Paul Israel notes that when Edison built his Menlo Park laboratory, it was the "largest private laboratory in the United States and certainly the largest devoted to invention." Edison's unique access to big corporate money—first through contracts with Western Union and then through Wall Street light bulb money—gave him an enormous advantage over his rivals. When Edison embarked on his electrical quest, Israel points out, he was very much a "traditional though highly ingenious inventor, working with two or three close assistants and a few skilled experimental machinists. . . . By the beginning of 1880, as he turned from basic research to the development of a commercial system . . . Edison had begun to resemble the modern director of research and development. . . . And like the modern research director, Edison depended on the support of corporate capital. While Edison the individual is celebrated as the inventor of the electric light, it is the less visible corporate organization of the laboratory and business enterprise that enabled him to succeed."[23] Edison was inventing not just the light bulb, but a new kind of relationship—however prickly and difficult—between corporate capital and scientific creativity.

The first order of business was perfecting the light bulb, whose brightly shining carbonized cardboard filament may have thrilled young and old, rube and city slicker, but simply was not reliable for use day in, day out. The filaments generally burned out after only three hundred hours or so, a big improvement but still not enough. Moreover, rival light bulb inventor William Sawyer, when he saw Edison's patent for the carbonized cardboard bulb, promptly filed an interference suit, pointing out that he had already filed a patent for a light bulb with a cardboard filament. So Edison had every reason to invent yet a better light bulb. "Now I believe that somewhere in God Almighty's workshop," Edison was said to have uttered, "there is a vegetable growth with geometrically parallel fibers suitable to our use. Look for it. Paper is manmade and not good for filaments."[24] Once again, the patient Batchelor sat hour after hour, night after night, at the lab table, testing all kinds of fibrous natural substances. Months passed with no definitive breakthroughs.

Finally, on April 21, 1880, the digging of the electrical "subways" began. Spring had come to the lovely Jersey countryside, and the heavy clay soil had finally warmed up. Workmen wielding plows and shovels began excavating a system of long, narrow ditches that fanned out from the direct current generating station, ran along Menlo Park's few muddy streets, and then headed out to the surrounding fields. For his next great public demonstration of his electric network, Edison planned to illuminate all of Menlo Park with four hundred incandescent lights arrayed along eight miles. (He had solved the problem of lights arrayed in a series—as they were in arc lighting—where one burned-out light would break the whole circuit, by rearranging the circuit into ladderlike parallels; this meant that the electricity could be routed along individual "rungs," thus circumventing a light bulb that had been switched off or had burned out.)

The subways carried the insulated copper wires cradled in narrow wooden conductor boxes, which were coated with tar to protect against moisture and decay and then enclosed with a top. Once the wires were laid, the conductor boxes were sealed and the trenches refilled. Day after day, the men dug and installed, as the days lengthened and the strange broiling heat of May gave way to a mercifully cool summer. By mid-July, the men had dug, installed, and covered five miles.

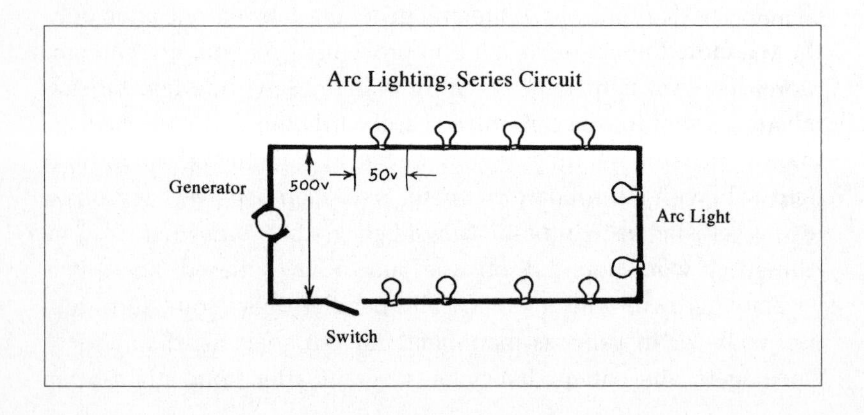

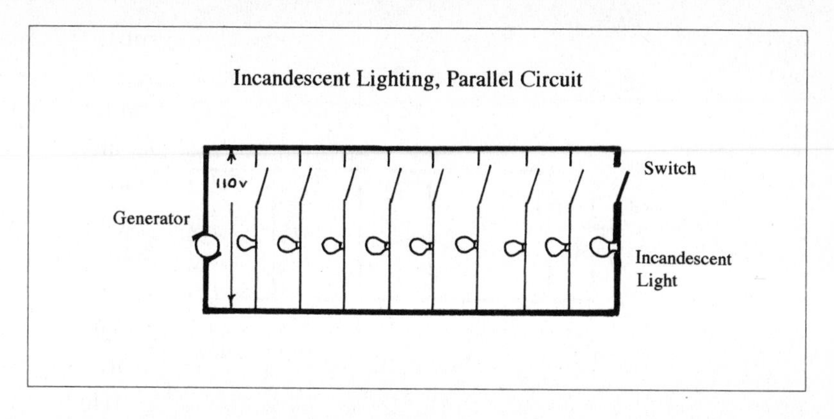

At this point, Francis Upton began testing the lines, only to find "some of the circuits are very badly insulated and all more or less defective." One of the newcomers to the lab wondered why in the world all this had been done by "inexperienced men" and "without being required to test a single circuit or wire until the entire work is finished."[25] Now all the ditches would have to be opened up. But the real problem was devising decent insulation, a top priority. Over the next few months, new insulation was twice reapplied and the wires twice reburied. Then Edison and his men would wait for the sky to darken and the next summer rain to come soak through the earth. And again, for a second and a third time, the insulation failed. Through it all, Edison was his usual sanguine self, wandering about chewing on his cigar, vest half-buttoned, solving this problem and that, conferring with Batchelor and Upton, keeping Lowrey informed. He slept a few hours here or there. Edison's goal was to have his Manhattan prototype ready for public display by the Christmas holidays.

Always hanging over Edison's head was the cost of copper, the biggest and most daunting expense for his central station plan. The cost of copper had inspired his invention of the high-resistance light bulb. But even that huge savings would not be enough if he was to match, much less undercut, the cost of gas lighting. In the summer of 1880, even as the "subways" were being laid out again and again, Edison had one of his profound breakthroughs, coming up with a "feeder and main" system of distribution that mimicked—under the

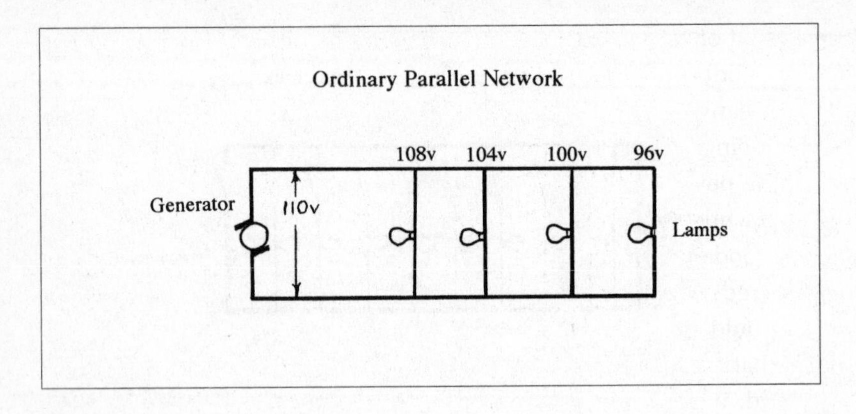

streets—the new parallel circuits to be used within buildings. This brilliant new approach would cut copper costs to *one-eighth* the earlier estimates.

Essentially, instead of one or two very thick (and costly) copper trunks carrying electricity forth and then branching off to each individual building, Edison proposed a network of much thinner multiple "feeder" copper wires coming from the central station DC dynamo and intersecting with many small mains that lit large clusters of lights, thereby eliminating the bulk of the copper. When this elegant and simple answer to problems of cost and maintaining pressure was demonstrated in England, someone asked the brilliant Glasgow physicist Sir William Thomson, knighted for his critical role in the practical success of the transatlantic cable, why no one else had

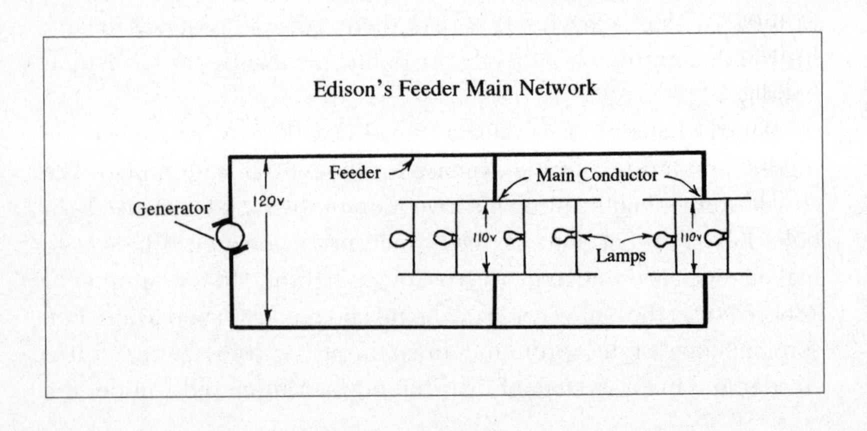

thought of it. He said, "The only answer I can think of is that no one else is Edison."[26]

Meanwhile, the search for the ideal filament had been slowly advancing. During that weirdly hot spring, the lab was fully focused on the possibilities of bast, the woody outer layer of flax or hemp. Meanwhile, Edison was busy installing his light bulb factory in an old wooden barn across the railroad tracks, for he would need many hundreds of light bulbs for his New Year's display. Soon thereafter, he would need thousands of light bulbs stockpiled for his future Manhattan customers. Then, on July 10, according to Edison lore and legend, while the ditch diggers were laboring away to finish their eight miles of trenches, Edison was sitting in the lab idly fanning himself against the still summer air with a bamboo fan. He looked at the fan, cut off a long thin piece for a filament, examined it under his microscope, and handed it over to be tested. The results seemed promising, and better-quality bamboo was obtained, carbonized, and tried. By August 2, the lab was fully focused on bamboo. Edison scholar Paul Israel debunks this legend as "plain wrong," for the lab notebooks show an order of bamboo coming in July 7. Moreover, Edison had conducted careful literature searches, and it was these, not the heat, that put him on to bamboo.

Throughout 1880, Edison's experimental inventor's laboratory had been steadily transformed into a production and testing facility for light bulbs, dynamos, conducting wires, and insulation. Each component of the new lighting network had to be designed, tested, redesigned, and retested. "Everything is so new that each step is in the dark," Edison said. "I have to make the dynamos, the lamps, the conductors, and attend to a thousand details the world never hears of." The truth was, this was completely new and complex technology, and no one really knew how long it would take to make it work or what it might ultimately cost. In Menlo Park, September had rolled around and vast flocks of birds at times darkened the broad country sky as they began heading south. Now, for the fourth time the workmen began to lay the wires in the eight miles of conduits. And at last, insulated with a compound that involved several layers of muslin and then "parafine, tar, Linseed oil and Asphaltum," the wires remained sound when the rain came.[27] On Monday, November 1, a blustery cold evening, the first Menlo Park streetlights began to glow, their subtle radiance coming from electricity generated by a "long-legged Mary-Ann" in Edison's central station and then flashed through the copper wires buried in the subways. Soon Edison's and Upton's houses were connected to the central station, as were the miles of lights running up the hamlet's central plank road and then far out into the golden autumn fields.

The very next day was Election Day, and by the next night, when it was known that Republican James Garfield had won the presidency by a hairbreadth, Edison celebrated his party's victory by lighting up a whole turnpike of lights near the passing railroad. President-elect Garfield was, like Edison, a self-made man who had overcome his family's penury. A tall, handsome, affable scholar, Garfield had been president of Hiram College in Ohio and a member of Congress when the Civil War broke out. He quickly organized a brigade for the Union Army, becoming a war hero at the battle of Chickamauga. Elected a U.S. senator by the Ohio state legislature, Garfield resigned his army commission at President Abraham Lincoln's request. By 1880 Garfield was a powerful and respected senator, but he was nominated as the Republican presidential candidate only on the thirty-third ballot. As Garfield prepared to enter the grand world stage, so did Edison. Each hoped to make an important mark on a fast-growing nation finally feeling prosperous again.

Edison, despite two arduous years rife with setbacks, mounting skepticism, and pointed sarcasm about his being far behind his own original announced schedule of mere months, was his ever blithe, hubristic self. Thus he was able to write to his European business associate Theodore Puskas in October 1880—before the Menlo Park system was even up and running smoothly—that he could "safely say that the Edison Electric Light Company of America will have one station established and in full working order lighting the lower portion of the City of New York before the first of May 1881."[28] Such was Edison's fame in those days that there were continual visitors wanting to meet the great man. So when the beautiful French tragedienne the "divine" Sarah Bernhardt performed in New York during a national tour, she, too, longed to meet "*le grand* Edison." Edison investor and director Robert L. Cutting was more than happy to escort the glamorous Madame Bernhardt out to Menlo Park, arranging a special train after an evening performance. At 2:00 A.M. on December 5, she stepped off the private car into the raw rural cold of Menlo Park, thrilled by the soft glow of the incandescent lights lining the plank road. Edison, who showed little interest in women (including his own wife), was smitten by this glorious, vivacious creature in her exquisite French gown with its voluminous, swishing skirts. "She was a terrific 'rubberneck,' " he would later remember. "She jumped all over the machinery, and I had one man especially to guard her dress. She wanted to know everything. She would speak in French, and Cutting would translate into English. She stayed there about an hour and a half."[29] In the comfortable library, Edison held her hand and explained the secrets of the phonograph. She recited favorite passages into it from Racine's *Phèdre* and was amazed to hear her voice. When the great inventor flashed the hundreds of outdoor lights on and off in the pitch dark of the early morning, on and off, on and off, she clapped with pure Gallic delight. Finally, she and Cutting had to return to New York. "*C'est grand, c'est magnifique!*" she exclaimed in that world-famous voice.[30]

Bernhardt was not the only important visitor Edison entertained that December. The moment was drawing near to conquer the Empire City, for Edison apparently envisioned few further delays. But before he could install an entire electrical system from scratch in crowded, noisy, dirty Manhattan, there were a few political details to master. The Edison investors knew the company needed to secure City Hall's permission to dig up Manhattan's streets, and with notoriously corrupt Tammany in firm control, one could not be certain what that might entail. The ever diplomatic Grosvenor Lowrey quickly arranged a posh and persuasive evening out in Menlo Park for the city's Tammany aldermen, hoping to dazzle them into quick and friendly action with lights and Lorenzo Delmonico's finest catered fare and copious champagne. Numerous reporters tagged along for the fun. *The New York Times* reported that as the group got off the train in "the bleak and uninviting place where Mr. Edison has chosen for his home," the hundreds of electric lights illuminating the plank road to the laboratory and surrounding fields "cast a soft and mellow light . . . beautiful to look upon." Edison, in a sealskin hat, awaited the politicos at the brick office building and "grasped the hand of each one as he passed and smiled with all the frankness of a pleased school-boy."[31] The inventor spoke briefly and then led the visitors through the cold, well-lit December night to the laboratory. There he proudly introduced the new bamboo-filament light bulb, which he said should last six months with normal use. He then turned off and on various rows of lamps. Next, with one turn of a wheel he put out all the 290 outdoor lights aglow in the snowy streets and nearby pastures. Then, with a turn of the handle, Edison brought those 290 globes back to glowing life. For men who knew only gaslights that required individual lighting and snuffing, this was something astonishing indeed!

After that radiant display, Lowrey and Edison escorted the Tammany crew back through the cold to the elegant brick building and Edison's upstairs library, where the men settled in comfortably. Edison explained why his electric light would be as cheap as gas when it became available in the fifty-one blocks bounded by the East River, Spruce, Wall, and Nassau Streets. He boasted a bit about his 250-some patents granted for electrical innovations. Then, to liven matters up, he displayed his ever popular phonograph. The reporter from the *New York Truth* noticed that "by this time the city fathers had begun to look quite dry and hungry, and as though refreshments would have looked much more palatable to them than the very scientific display."[32] Perhaps their hearts sank when Lowrey proposed they revisit the laboratory. Instead, there beckoned a table as long as the room, groaning under Delmonico's aromatic delicacies—turkey, duck, chicken salad, and ham, all to be washed down with the best wines and champagnes. Soon the aldermen were feeling much more jolly. The city's superintendent of gas, no less, toasted Edison on his success. "Gas," said the superintendent, "is dangerous. It is very easy for a man to go to his hotel, blow out the gas and wake up dead in the morning. There is no danger of a man blowing out the electric light."[33]

The *New York Post* observed in its story about the aldermen's visit, "There are now six different companies at work introducing electric lights in this city, the lights being known as the Brush, Maxim, Edison, Jablochkoff, Sawyer and Fuller (Gramme patents) lights." Most of these were arc-lighting companies, but rival inventor Hiram Maxim had boldly helped himself to Edison's incandescent light bulb (and to Ludwig Boehm, who had defected to the better-paying enemy) and had raced ahead, displaying *his* new incandescent lighting system in the Mercantile Safe Deposit Company's vaults and reading rooms. Maxim's system had been up and running for two months already. Such timely competition did nothing to soothe Edison's money men, who wondered what was keeping their man from getting his lights running.

The very night the aldermen trooped out to Menlo Park, the Brush Electric Company had had its New York debut. At 5:25 P.M., the generators at its central station had roared to life and illuminated seventeen powerful new electric arc lights, lighting up Broadway for the three-quarters of a mile from Union Square up to Delmonico's Restaurant at 26th Street. The *New York Evening Post* described the new arc lights as blazing "with a clear, sharp, bluish light resembling intense moonlight, with the same deep shadows that moonlight casts."[34] When the brilliant arc lights flashed to life in the cold night, the strolling crowds of fashionable holiday shoppers exclaimed and clapped in admiration. Suddenly, passing horses, streetcars, and omnibuses emerged from the usual gloom of gaslight, uncannily visible. One reporter was struck by the "artistic effects" created by this new light. "A pair of white horses attached to an elegant private carriage outside of Tiffany's was illuminated with a brilliancy which, contrasted with the deep black outline, formed a picture. The great white outlines of the marble stores, the mazes of wire overhead, the throng of moving vehicles."[35]

Just before the aldermen's visit, Lowrey had organized a new corporate structure, the Edison Electric Illuminating Company, comprising much the same directors—Western Union and Morgan people—as did the Edison Electric Light Company. Thomas Edison had been telling Lowrey and his fellow capitalists for months that he now needed not thousands of dollars, but millions to light up lower Manhattan. But the Wall Street investors were loath to wade any further into Edison's financial bog, where money swiftly disappeared, never to be seen again. Where was the lighting system that was to churn out all the profits? So, some years later Edison would explain, "We were confronted by a stupendous obstacle. Nowhere in the world could we obtain any of the items or devices necessary for the exploitation of the system. The directors of the Edison Electric Light Company would not go into manufacturing. Thus forced to the wall, I was forced to go into manufacturing myself." To one of his New York investors, Edison declared, "Since capital is timid, I will raise and supply it.... The issue is factories or death!"[36] To show he was not kidding, Edison had boldly established the light bulb factory out at Menlo Park, which by the end of the year was turning out several hundred bulbs daily. This was controlled and financed by Edison himself, who sold Edison Electric stock and borrowed wherever he could. Again, the nature of the new and little understood electrical science and its many unknowns were dictating these first groping corporate arrangements.

In February of 1881, Edison and his key staff members began shifting, at long last, into Manhattan, joining the already legendary cacophony of the Empire City, what one guidebook of the era described as "the intense activity and bustle alike visible and audible in all the conditions of its street-life. The crush of carriages, drays, trucks, and other vehicles, private and public, roaring and rattling over the stone-paved streets; the crowds of swiftly moving men walking as if not to lose a second of time, their faces preoccupied and eager; the sidewalks encumbered, without regard to the convenience of pedestrians, with boxes and bales of goods—in a word, the whole aspect of New York in its business portions is a true key to the character of its population, as the most energetic and restless of people."[37] By the end of February, Edison had signed a lease for the handsome former Bishop Mansion at 65 Fifth Avenue, an ornate four-story double brownstone in the city's most fashionable quarter, just below 14th Street.

At the highly visible headquarters for the new Edison Electric Illuminating Company, Edison quickly rigged up a steam engine and generator and by mid-April had equipped its tall-ceilinged rooms with numerous "electroliers" (electric chandeliers) and other attractive light fixtures. Illuminated every evening and long into the wee hours, 65 Fifth Avenue was the glorious and radiant new electrical reality, where Edison held court most nights. The back parlor served as his campaign headquarters, complete with a wall-size map of Manhattan and his designated first lighting district. Long before moving into New York, Edison had thoroughly canvased his prospective "first district" central station electric customers and determined that 1,500 were coal-gas customers using twenty thousand jets. All this was indicated on the map, along with the planned routes of the subways, switches, and so forth. For his new role as businessman, Edison moved up sartorially from his old blue flannel suit to a seedy Prince Albert frock coat. He was in his usual ebullient spirits: "We're up in the world now. I remember ten years ago—I had just come from Boston—I had to walk the streets of New York all night because I hadn't the price of a bed. And now think of it! I'm to occupy a whole house in Fifth Avenue."[38]

Edison certainly liked to promote this Horatio Alger story that predated those popular books, describing himself as arriving in New York an almost penniless youth with a great talent for wireless technology and machines. He happened, so he said, to be in the first days of a lowly job at the Gold Indicator Company, which supplied Wall Street via ticker tapes with the fast-changing price of gold, when the equipment ground to a halt. As the officers began to panic, Edison examined the silent machine and observed that the trouble was a broken contact spring. Amid the hysteria, he quickly fixed the problem and was duly promoted to the important position of technician. And so was launched his prosperous career as an inventor and improver of telegraphy equipment. The reality, says Edison scholar Paul Israel, was that while Edison certainly started life with few advantages, when he first came to Manhattan he was already well connected and had a respectable—not lowly—job as an engineer, and any money problems were short-lived.

When Edison returned to New York this time, he brought with him many of his main Menlo Park crew, each assigned new and greater responsibilities. In mid-February, John Kruesi, Edison's trusted Swiss mechanic, the man who could fabricate and make almost anything work, had opened the Edison Electric Tube Company at 65 Washington Street. Kruesi was the general in charge of Edison's toughest campaign—the manufacture and then physical installation (beneath some of Manhattan's busiest, filthiest thoroughfares) of fourteen miles of underground distribution cables and wires. He would command a big gang of Irish laborers, many of whom viewed electricity as some evil sprite. They would share the nighttime streets with the city's denizens of the dark, including the great army of rag pickers and their dog-pulled wooden carts, each licensed to root through the daily refuse for salvageable cloth. Kruesi's longtime Menlo Park assistant, Charles Dean, was put in charge of the all-important Edison Machine Works, located in an old ironworks building at 104 Goerck Street near the East River docks on the crowded and noisome Lower East Side. Here in this grimy setting would be perfected and manufactured the workhorses of the Edison system—the generators. Meanwhile, back at Menlo Park, Francis Upton, the scientist, was running the light bulb factory, now churning out a thousand lights a day. These three enterprises were all organized and financed by Edison or his closest associates.

Even as Edison himself settled into Manhattan, his other right-hand man, Charles Batchelor, sailed off to Paris to launch the Edison European branch. Business manager and sometime Edison promoter Edward Johnson headed to England to push the Edison light there. Both were to lay the groundwork for the electrical empire Edison had envisioned from the start. Edison was a famous commercial name already overseas, for he had conducted major European business with his previous inventions. He had existing partners and contacts, and now Batchelor and Johnson were to begin launching their famous boss's biggest enterprise yet—central electric stations, as well as isolated or stand-alone electric plants for individual factories or buildings. The Edison Electric Light Company of Europe had already been formed in January of 1880. So even as Edison labored away in Gotham, Charles Batchelor was hard at work in Paris organizing the all-important Edison system display for that summer's International Electrical Exhibition. Across the Channel, Johnson began building a demonstration central station that would light up the centrally located Holborn Viaduct.

Back in New York, down on Wall Street, Edison's investors were again questioning the necessity—and the ensuing huge expense—of burying the electric wires in "subways," something Edison had been determined to do from the start. By the 1880s, anyone lifting his or her gaze above street level in the commercial blocks of American cities could barely see the sky for the ugly maze of hundreds of electric wires strung higgledy-piggledy between towering wooden poles. The wires crisscrossed the streets and were festooned from windows and rooftops as if huge crazed spiders had run amok. A range of fast-expanding industries now depended upon electricity (most of it still battery produced)—including the telegraph, telephone, stock tickers, fire and burglar alarms, and certain small manufacturers. In any city, numerous companies vied to provide these various services, and where they found customers, they installed more poles—some towering a hundred feet and higher. Firms came and went, but their wires remained, deteriorating, fraying, dropping onto one another, and creating short circuits. However, all these early electric-based services operated on very low voltage direct current derived from large batteries. These wires might cause a shock but would not electrocute a hapless passerby.

All that changed with the coming of the new outdoor arc lighting in the 1880s. The extremely high voltage alternating current required to operate these lights—as high as 3,500 volts—made their outdoor wires potentially truly perilous. The Brush Electric Company had installed its first lights on Broadway between 14th and 34th Streets at the end of 1880, and their brilliant blue white light soon earned Broadway its sobriquet "the Great White Way." New York City then contracted with Brush to light more of Broadway and several squares.

Hotels, theaters, and other public spots installed arc lights. Brush built three central power stations and transmitted its high-power electricity—typically 2,000 to 3,000 volts—on wires strung among the existing low-voltage tangle. Edison wanted nothing to do with these mangled nests of live and abandoned wires and insisted that the Edison system, by burying its wires, would be both safe and reliable. The new Edison system operated on low-voltage direct current, which was efficient and economical only within a half-mile radius of the generator. Beyond that distance, the cost of copper wiring became prohibitive and the energy loss too great. However, Edison prided himself upon the low voltages of his system and believed its buried wires added a great margin of safety for the general public and his customers.

In late April 1881, the Edison Electric Illuminating Company finally received city permission to begin digging its subways. In the meantime, the company had already wired about fifty first-district homes and office buildings and promised that current would be flowing by the fall, providing them lovely electric light just as the winter gloom closed in. The city's permission came with one deeply worrying caveat: Five city inspectors—to be paid $25 a week by Edison—would monitor progress. Edison envisioned all kinds of trouble aimed mainly at extracting bribes. But in true Tammany "do no work" fashion, the inspectors appeared only on Saturday afternoon to collect their pay. Kruesi quickly broke ground with his Irish street crews, working largely at night when the city's much maligned street-cleaning crews spread out to remove the two to three million pounds of equine manure left behind each day by the city's 150,000 horses. Kruesi soon found that digging down two feet was far more time-consuming and arduous than expected. Edison and Kruesi personally had to install the connector boxes located every twenty feet. Worse yet, the suppliers of the copper wiring and the iron pipes (the latter substituting for the original wooden boxes) had stopped delivering. June was slow going, as it rained every day but one. Then, on July 2, the Edison men organizing the evening's subway work heard shocking news: President James Garfield, waiting at the Baltimore and Potomac station for a train to the cool of the New Jersey seaside, had been shot twice in the back by an angry job seeker. Still alive, the president was carefully conveyed to the stifling heat of the White House. The nation prayed that Garfield would survive this terrible attack.

At the time, Edison was scouting the worst slum streets of his first district, looking for a cheap but capacious building to house the heart and soul of his system—the central generators. That August, with the summer heat exacerbating the usual stench of horse piss and manure, great piles of garbage, and sour beer and sawdust from the ubiquitous bucket shops, Edison purchased 255-57 Pearl Street for $65,000. From this squalid block, Edison's electricity would eventually gently hum forth a half mile in each direction, lighting up the all-important financial district centered on Wall Street and much of newspaper row. "The Pearl Street Station," Edison later said, "was the biggest and most responsible thing I had ever undertaken. It was a gigantic problem, with many ramifications.... All our apparatus, devices and parts were home-devised and home-made. Our men were completely new and without central station experience. What might happen on turning a big current into the conductors under the streets of New York no one could say."[39] The Edison dynamos—powered by coal-fired steam engines—produced an initial alternating current electricity that was then gathered from the machine by "commutators" and brushes and turned into a direct current. One of the perennial problems with these early generators was that the constant friction against the commutators and brushes meant regular replacement. Every step of the way, myriad technical problems arose that had to be resolved. Again and again, the starting date was postponed.

Summer ended and still President Garfield clung to life, a bullet lodged next to his spinal cord. From the start, reported the *New-York Daily Tribune,* his physicians had told him "he had one chance in a hundred of living. 'Then we will take that chance,' he said. All that mortal man could make of so slender a chance was made. His courage never faltered.... For seventy-nine days the agonizing struggle was prolonged."[40] On September 6, it was decided to move the president to Elburon, New Jersey, where he had been heading when shot. It was, therefore, no great surprise when on the night of September 19 New Yorkers heard church bells start to toll out a doleful dirge. The president was dead. Vice President Chester A. Arthur was hurriedly sworn in to the nation's highest office at his Manhattan town house. Englishwoman Iza Hardy, then visiting New York, wrote, "The heart of the nation beat with one regret; the word on every lip was, 'The President is dead.' " A week of mourning followed. "From the highest to the lowest, from Fifth Avenue mansion to the squatter's shanty, each home hung out its sign of sorrow. . . . The star-spangled banner, generally looped with crape, floated in all sizes and of all materials from a thousand windows; from the little ten-cent paper flag to the imposing patriotic bunting waving from side to side of the road."[41] Garfield's assassin, Charles J. Guiteau, was arraigned October 14, the trial beginning a month later. By the end of the year, as the trial dragged on, the *New-York Daily Tribune* described it as a "vulgar peep show" where Guiteau, determined to prove he was innocent by reason of insanity, dominated the courtroom with "his drivel, abuse, malevolence, and smudge." The editors denounced the circus atmosphere as a "disgrace to the country" allowed by a "judge whose backbone seems to be made of tissue paper."[42]

Edison was deep into his own travails, as he struggled to get all the components of his central station up and functioning. The New York press had become far less friendly. On December 2, 1881, *The New York Times* described in a brief, page eight article that the Edison company "have laid a considerable quantity of wire, but so far as lighting up the downtown district is concerned, they are as far away from that as ever." Winter, snow, and frozen streets brought the digging to a halt. More months passed with little progress. In an article entitled "The Edison Dark Lanterns," *The New York Times* reported, "Much grumbling has been done lately by businessmen and residents of the district bounded by Nassau, Wall, South, and Spruce streets because there seems no prospect of the Edison Electric Light Company putting in the lights they promised to have burning by last November." A grumpy Edison official acknowledged that only half of the fourteen miles of subways were in place, in part due to sporadic delivery by suppliers of iron and copper and in part because of suspension of work while the ground was frozen. With the advent of spring, they were working at top pace again. When pressed repeatedly for a completion date, the Edison officer replied, "All we can say now about the prospects of lighting up is that we are doing our best to get the wire laid, immediately after which we will be able to light the lamps." Of course, it was in this month that J. Pierpont Morgan's Italianate brownstone on Madison Avenue was first lit up, much to his delight. But his was an isolated electrical plant, not part of the central station. He lived too far north for that service. On June 30, Guiteau, "this most despicable of assassins," was hanged, and he turned his awful end at the gallows into a final frenzied display of shrieks and tears. By late August 1882, as the city steamed in the summer heat, Kruesi at long last led his Irish crews through the final and fourteenth mile.

Very quietly, the Edison Electric Illuminating Company began testing its system. No formal announcement was made, but Thomas A. Edison had begun running his generating and distribution system, hooking up various customers, and testing the lights. The truth was, Edison had momentarily exhausted his penchant for ballyhoo. For four years, he had worked as hard as ever he had at any one project, and he was understandably nervous that it would actually perform as promised. Edison was now thirty-five years old, and while his face still looked as youthful as ever, his thatch of brown hair had turned gray in the years since he had blithely and innocently promised to light up all of lower Manhattan with his platinum bulb and William Wallace's dynamo. The New York press found out that the Edison network was being tested when horses passing along Nassau Street at Fulton "became suddenly electrified, gave a sudden jump, and with a snort, ran off as speedily as possible." At first the Edison people could not believe their system was responsible, but a steam-heating company also digging had indeed broken their iron pipe and shorted the wires. Such minor electrical disasters were unnerving, for Edison knew that no one could truly say what might happen when electricity went coursing forth from Pearl Street. They certainly hoped it wouldn't regularly escape to shock the unwary.

On September 4, 1882, a pleasant, warm day, a slightly chastened Edison, attired for the occasion in a better frock coat and a white, high-crowned derby hat, spent the morning and early afternoon repeatedly checking all aspects of the Pearl Street operation, abandoning his collar early on. Just before departing for Wall Street, where he was going (at long last) to formally launch the Edison Electric Light Company's service, the great inventor synchronized his watch with Pearl Street employee John Lieb. Now, as the long-awaited event neared, Edison walked into the Morgan offices with Edward Johnson, John Kruesi, and a few others. J. Pierpont Morgan and many of the directors of the board also gathered in the burnished Drexel, Morgan offices. Four years of hard work of the most original, difficult sort and almost $500,000 had brought them to this crucial moment. Edison had been operating under a pall of skepticism for some time. And he knew better than any man present just how many small things could go wrong over at Pearl Street or in his electrical subways to bollix up his company's formal electrical debut. To break the palpable tension, Johnson joked to Edison, "One hundred dollars they don't go on!"

"Taken!" said Edison. He looked at his pocket watch. It was three o'clock. The moment of truth was upon them. Over on Pearl Street, John Lieb stood on his tiptoes and threw the main circuit breaker. Blocks away in Morgan's office, Edison closed the switch next to him.

"They're on!" cried the directors. It was a wondrous vindication, for all around them some one hundred incandescent bulbs had glowed softly to life. Three hundred more glowed in nearby offices, delivering an energy visibly superior to flickering, odorous gaslight. It was not until darkness fell, wrote *The New York Times,* whose Edison lights also came on that day, that "the electric light really made itself known and showed how bright and steady it is. . . . There was a very slight amount of heat from each lamp, but not nearly as much as from a gas burner. . . . The light was soft, mellow, and grateful to the eye . . . without a particle of flicker to make the head ache . . . the decision was unanimous in favor of the Edison electric lamp as against gas." In the coming months, another two thousand lights in additional buildings were lighted up. It was not happenstance that the

Edison Electric Light Company's first customers included such influential entities as his financial backers, Drexel, Morgan & Co., the Park Bank, and *The New York Times.* (The *New York Herald* had an isolated stand-alone plant.)

Understandably, Edison gloried in his amazing accomplishment of creating the first true incandescent electric light network. To a *Sun* reporter, he said that day, "I have accomplished all I promised." And he had indeed brought the incandescent electric light to New York City. But from the start, Edison had seen New York City as just the beginning. Already he had his corporate generals readying for new conquests. Before him the great inventor saw only more glory and great fortune, which to Edison translated into utter freedom to exercise his prodigious gifts as an inventor. He explained, "My one ambition is to be able to work without regard to the expense. . . . I want none of the rich man's usual toys. I want no horses or yachts—I have no time for them. What I want is a perfect workshop." While only the most perspicacious yet understood the potential of electricity, farseeing capitalists were already envisioning the day the clearly superior electric light would displace the vast and lucrative manufactured gas industry, worth $400 million just in the United States. Edison's electric light would span the world, and he and his backers would be even more famous and very rich. Pearl Street was just the tiny beginning of a great and lucrative radiant empire.

CHAPTER 4

Nikola Tesla: "Our Parisian"

In April of 1882, a tall, slender young Serbian engineer of twenty-six named Nikola Tesla descended from a train amid the vast, sooty clangor of the Gare de l'Est and emerged into the resplendence of belle epoque Paris. The dreamy and romantic Tesla was enchanted by the magnificent grandeur of late-nineteenth-century Paris. For days after he arrived, he just wandered. He strolled the broad, expansive boulevards created by Baron Haussmann, lined with fashionable cafés and shaded with fragrant flowering chestnut trees; he admired the formal city parks with their splashing fountains and geometric designs of clipped greenery, he peered into the ancient and lovely churches set back in the warrens of old neighborhoods, lively with street markets pungent with every fish and cheese imaginable. Tesla lingered by the silvery Seine, spanned by one sculptured bridge after another.

Paris at night, he found, was perhaps more wondrous. Kilometer after kilometer of gaslit avenues glowed in the darkness, lined by luminous shop windows and department stores. After dark, animated crowds swirled about like moths in their commercial penumbrae. The Paris Opera in those soft spring evenings was a gilded chimera, its new electric lighting casting a moneyed sheen on the

---

Nikola Tesla, "our Parisian"

ladies' diamonds and the gentlemen's silk top hats and opera cloaks. The city's traditional gaslight, now joined by the new arc lights and the fledgling incandescence, combined to produce a distinctly modern metropolis, one of haunting nocturnal moods. Mirrored Parisian cafés and theaters, lit up luxuriantly, came to scintillating life, wonderfully alluring and atmospheric. "I can never forget the deep impression that magic city produced on my mind," Tesla said decades later.[1] But Nikola Tesla had come to the legendary City of Light not to gawk like a country rube, but to work as a junior engineer in Ivry-sur-Seine at Thomas Edison's Société Industrielle, headed by Charles Batchelor, veteran of Menlo Park and engineer to the newly established Compagnie Continentale Edison.

Those first few days of bedazzlement over Parisian glamour soon gave way to a regimen and rhythm of hard work. For Nikola Tesla the junior engineer, however dreamy and eccentric he might be, was truly possessed by only one great passion—the mystery of all things electric. So he rented rooms on the edge of the ancient and picturesque Latin Quarter with its many students and professors and settled into a strenuous schedule that began at 5:00 A.M. "Every morning regardless of the weather," Tesla explained, "I would go from the Boulevard St. Marcel, where I resided, to a bathing house on the Seine; plunge into the water, loop the circuit [swim laps] twenty-seven times and then walk an hour to reach Ivry, where the Company's factory was located. There I would have a wood chopper's breakfast at half past seven o'clock and then eagerly await the lunch hour, in the meantime cracking hard nuts for the Manager of the Works, Mr. Charles Batchelor, who was an intimate friend and assistant of Edison."[2]

It was a considerable coup for a young man whose whole soul resonated to the little-known mysteries of electricity: He was here in the expanding Edison empire, for Thomas Edison was the great practical man in the field, the one who had shown the highly skeptical scientists you could indeed subdivide the electric light and take it indoors. And Tesla's boss, Batchelor, had been present at the creation of the very incandescent light bulb that was launching a new luminous epoch. Charles Batchelor, after his years in Newark and then

Menlo Park, knew very well what hard work was, and he had arrived in Paris determined to conquer and electrify all of Europe, starting from scratch. Yet this was a monumental ambition, and at times, even the usually genial Batchelor felt deeply burdened, as this irascible note to Edison shows: "My job here is no fool of a job, what with lamps, dynamos, chandeliers, and all the extras. I am just in up to my neck; then I have so much outside work of such a responsible nature and involving so much money that I wear a hat about three sizes larger than when I left New York."[3]

Young Tesla, in contrast, was but a novice, a very junior engineer at the Société Industrielle. But he was also one who quickly showed he was a reliable troubleshooter capable of solving most electrical tangles. Tesla spoke good, formal, heavily accented English and was also fluent in numerous other languages, notably French and German. But observant colleagues concluded that however talented Tesla was as an engineer, he was also a decidedly odd fellow. Always fastidious in appearance, his black hair waving back gently, his mustache neatly trimmed, the tall, slender Tesla was prey to strange habits and phobias. He (silently) counted each step he took as he made his early morning walk down to the Ivry factory. Every activity ideally had to be divisible by three (hence the twenty-seven laps each morning in the Seine). Before eating or drinking anything, he felt obliged to calculate its cubic contents. He deeply disliked shaking hands with anyone. He had a "violent aversion against the earrings of women," pearls above all. "I would not touch the hair of other people except, perhaps, at the point of a revolver." The mere sight of a peach brought on a fever.[4] Moreover, Tesla could (and happily did) recite long swaths of Serbian poetry from heart.

Tesla's very presence in the noisy, busy Ivry Edison factory showed the deep incursions already made by the new modern industrial order on long-held traditions and life patterns. All the men in his deeply conservative Serbian family had always been destined for the church or the army, honored professions in their strategic small world ill situated between the decaying Ottoman empire and the crumbling European monarchies. "I was," Tesla conceded readily, "intended from my very birth for the clerical profession and this thought constantly oppressed me."[5] His eminent father was an Eastern Orthodox minister. His highly inventive mother had devised and fashioned many handy household items and tools. She was also a master weaver who spun much of her own thread. "When she was past sixty," wrote Tesla, "her fingers were still nimble enough to tie three knots in an eyelash."[6] But Tesla seemed destined only for electricity. All his life he recalled this formative episode at age three with his beloved cat, Macak. "It was dusk of the evening and I felt impelled to stroke Macak's back. Macak's back was a sheet of light and my hand produced a shower of sparks loud enough to be heard all over the place." What was this? the young boy wondered to his father. " 'Well,' [his father] finally remarked, 'this is nothing but electricity, the same thing you see on the trees in a storm.' My mother seemed alarmed. 'Stop playing with the cat,' she said, 'he might start a fire.' I was thinking abstractedly. Is nature a giant cat? If so, who strokes its back? It can only be God, I concluded. . . . Day after day I asked myself what is electricity and found no answer."[7]

During high school, Tesla, a prodigy in math and physics, fell even more deeply and irrevocably in thrall to the still nascent science of electricity. He alarmed his professors with his voracious and exhausting appetite for work, especially if it had to do with electricity. "It is impossible for me to convey an adequate idea of the intensity of feeling I experienced in witnessing [my physics teacher's] exhibitions of these mysterious phenomena. Every impression produced a thousand echoes in my mind. I wanted to know more of this wonderful force; I longed for experiment and investigation."[8] But hovering always was the impending burden of the family priesthood. When the adolescent Tesla was stricken with cholera and teetered on the brink of death, his anxious father agreed that he could study to become an electrical engineer. So Tesla, the only son, recovered and was free to begin his studies in Graz, Austria.

In 1877, during his second year at Graz, Tesla walked into his favorite physics class and saw sitting upon the wooden table a fascinating-looking machine, an assemblage of magnet and metal. Just in from the fabled city of Paris, it featured a large, standing, horseshoe-shaped laminated field magnet that stood over and around a hollow cylinder encased in tightly wrapped wire—the armature. This was the new dynamo invented by Belgian Zénobe-Théophile Gramme. It was causing a great stir in Western Europe and the United States because here, at last, was a dynamo that could generate enough electricity—when powered by a steam engine—to run the new sun-bright arc lights in factories and on city streets. But equally thrilling, the Gramme machine, when run in reverse, could also serve as a motor. If machines, too, could be moved by electricity, the implications of this new motive power would be enormous.

Raptly watching his teacher's demonstration of this Gramme machine with its magical combination of two long known materials—magnet and metal—Tesla could little dream how this clever machine would change his whole life trajectory. When the Gramme machine was run as a motor, its commutator brushes (segments of copper that rotated with the armature and caused the current to flow in one direction only), noticed Tesla, were "sparking badly, and I observed that it might be possible to operate a motor without these appliances. But [my professor] declared it could not be done and did me the honour of delivering a lecture on the subject, at the conclusion he remarked, 'Mr. Tesla may accomplish great things, but he certainly will never do this. It would be equivalent to converting a steadily pulling force, like that of gravity, into a rotary effort. It is a perpetual motion scheme, an impossible idea.' "[9]

Initially deeply embarrassed by such a public rebuke, Tesla the dreamer could not resist, however, thinking about the pointlessness of the sparking commutators. And indeed, these were the glaring weak points of the first new experimental electric motors. The gauze brushes that rubbed against the commutators were essential for picking up the naturally alternating electric current and sending it back into the motor as a nice, tame direct current. At the same time, commutators were also expensive to maintain for obvious reasons—whenever the motor ran, the commutator brushes were wearing down and generating sparks. So young Tesla was soon immersed almost daily in an electrical reverie, his blue gray eyes ablaze, his mind cogitating various designs of motors that did not involve sparking brushes, reviewing again and again how he might combine motors and generators. "The images I saw were to me real and perfectly tangible."

In the fall of 1880, as Thomas Edison was laboring away at Menlo Park to get his central power network ready for his second New Year's Eve demonstration, far across the Atlantic, Nikola Tesla, who had managed to get himself kicked out of high school for gambling and then done some youthful drifting, was entering the university in the ancient city of Prague, Bohemia, at age twenty-four. He stayed in Prague but a year, for his father died, forcing Tesla, now twenty-five, to get a job. He moved to Budapest, Hungary, the thriving commercial capital of the Austro-Hungarian empire, where he worked for a new telephone company run by a family friend, one Ferenc Puskas, brother of Thomas Edison's European friend and representative Theodore Puskas. Through all these intervening years, troubles, and various relocations, Tesla had never stopped wrestling in his head with how to design a motor that did not awkwardly scoop up electric current with a commutator and brush.

At the phone company, Tesla's intense and voracious work pace brought on an excruciating breakdown. The glorious city of Budapest with its showplace parks along the Danube River, famous castle, and lively cafés was reduced to tiny, tormenting sounds. "I could hear the ticking of a watch with three rooms between me and the timepiece. A fly alighting on a table in the room would cause a dull thud in my ear. A carriage passing at a distance of a few miles fairly shook my whole body. The whistle of a locomotive twenty or thirty miles away made the bench or chair on which I sat, vibrate so strongly the pain was unbearable."[10] The doctor was mystified and offered no hope of recovery. Yet even as the world hurtled and smashed upon him, Tesla felt that all this hypersensitivity was slowly unleashing from his unconscious the motor design that he had been seeking for almost five years.

One of Tesla's closest college friends, Anthony Szigety, had also moved to Budapest to work at the new telephone company. He suggested to the debilitated Nikola that he begin to exercise to regain his health. With exercise and fresh air, Tesla began to recover. One chilly February late afternoon in 1882, the athletic Szigety persuaded Tesla to wander forth to a city park as the sun was setting lushly. Tesla, as was his dreamy wont, began reciting poetry, Goethe's *Faust,* to celebrate the blazing sky before them:

> *The glow retreats, done is the day of toil;*
> *It yonder hastes, new fields of life exploring;*
> *Ah, that no wing can lift me from the soil,*
> *Upon its track to follow, follow soaring . . .*

"As I uttered these inspiring words the idea came like a flash of lightning and in an instant the truth was revealed." Tesla had been swaying and waving his arms gracefully as he declaimed, as if he were about to soar aloft. Now, tall, emaciated from his illness, he stood stock still. Szigety was worried that his friend had been stricken again and tried to steer him to a bench. Instead, Tesla swooped down and snatched a big twig. "I drew with a stick in the sand. . . . The images I saw were wonderfully sharp and clear and had the solidity of metal and stone, so much that I told him, 'See my motor here; watch me reverse it.' I cannot begin to describe my emotions."[11] Decades later, Tesla passionately relived this exultant electrical epiphany with his first biographer, science editor John J. O'Neill, recalling how he, Tesla, rapturously gestured to his simple designs in the dirt and declared to Szigety, "Isn't it beautiful? Isn't it sublime? Isn't it simple? I have solved the problem. Now I can die happy. But I must live, I must return to work and build the motor so I can give it to the world. No more will men be slaves to hard tasks. My motor will set them free, it will do the work of the world."[12]

While Tesla came of an educated clan, he had grown up among farmers and laborers and knew well the relentless, bone-wearying drudgery that always had been (and still was) the daily lot of most of humanity. If a field had to be plowed, or sown, or harvested, people would do each backbreaking step, helped perhaps by their animals. If a well had to be dug, men with shovels would dig it, meter by meter.

If a stout tree had to be felled, men would saw arduously back and forth until finally it toppled. If water had to be fetched and carried, women or children would haul it in heavy, sloshing bucketfuls. If clothes had to be washed, that dirty laundry would all be scrubbed by hand.

The steam engine had already wrought a great revolution in transportation and manufacturing. Steam engines were powering new factories, allowing textile mills to spin out cloth in great undulating waves, and railroads spanned whole continents, reducing perilous journeys of many months to mere days. Now, Tesla saw his AC induction motor similarly obliterating life's tiresome daily chores and burdens in a thousand ways.

Szigety was also an electrical engineer, and he gradually grasped that Tesla had—astonishingly—figured out the motor he had been obsessing over for almost five years. Tesla had at long last extricated his tormenting electrical visions and given them substance. He had dispensed with the awkward commutators and brushes to produce power from an almost magical rotating magnetic field. There in the Budapest park, Nikola Tesla had finally figured out how to design a motor that operated on the undulating electrical rhythms of alternating current. The swift back-and-forth reversals of AC as they advanced along their conductors (versus the steady, forward-only advance of DC electrons) would prove to be completely pivotal in the ultimate development of Tesla's polyphase vision, but in Budapest, Tesla knew only that he'd figured out his AC motor. Tesla and Szigety stayed up all night rejoicing in Tesla's brilliant and completely original new motor design. Tesla's first biographer, the science editor of the *New York Herald Tribune,* explained Tesla's advance thus: "Up to this time everyone who tried to make an alternating current motor used a single circuit. . . . What Tesla did was to use two circuits, each one carrying the same frequency of alternating current, but in which the current waves were out of step with each other. This was the equivalent of adding to an engine a second cylinder. . . . [These currents created] a rotating magnetic field . . . [that] possessed the property of transferring wirelessly through space, by means of its lines of force, energy."

In short, Tesla planned to so position his currents that as the first waned, the next would kick in, creating an invisible whirling magnetic field, a beautifully simple AC induction motor, what some would later call a "wheel of electricity," with almost no wearing parts. However, all this was still firmly in Tesla's head—down to the most minute details, for he never worked from blueprints, only his own prodigious, three-dimensional memory. Of course, no one but Szigety and Tesla knew of his great breakthrough. Nor, as Tesla would soon find out, would many people—even fellow electricians—in the early 1880s be able to appreciate the brilliance and originality of his motor. Why, they would wonder, should anyone care much about an AC induction motor that ran off an AC generator—just as arc lights did—when geniuses like Edison were conquering the world with central stations generating DC? And Edison had motors aplenty to offer customers who wanted them to power their factories. Tesla was about to discover the perils and frustrations of being ahead of his time.

For the next few weeks, the always dreamy Tesla drifted into an ecstatic reverie. "For a while I gave myself up entirely to the intense enjoyment of picturing machines and devising new forms. . . . The pieces of apparatus I conceived were to me absolutely real and tangible in every detail. . . . In less than two months I evolved virtually all the types of motors and modifications of the system."[13] About this time, Ferenc Puskas, the family friend and patron, sold the telephone business where Tesla had been working and returned to Paris. Tesla, his brain afire with whirling electric motors, soon followed. For the other Puskas brother, Tivadar or Theodore, Thomas Edison's European promoter and business partner, had agreed to introduce Nikola Tesla to Charles Batchelor, Edison's right-hand man and the main force behind European Edison. Theodore Puskas had ably represented Edison's telephone and phonograph patents in Europe, and Edison had kept him informed from the start about the glorious prospects of the incandescent light and the central station system.

So it was that in 1884 Nikola Tesla arose each morning in Paris, swam his twenty-seven laps in the pool, and then strode down (presumably counting every step) to Ivry-sur-Seine, where Batchelor had established a large factory to produce dynamos and the other elements needed to install isolated stand-alone plants or central stations. Young engineers like Tesla learned all the intricacies of the various Edison machines and distribution systems, preparatory to fanning out across Europe to bring incandescence and electrical power to the Old World. The ever reliable Batchelor had been entrusted with a formidable first task upon reaching Paris: to work with Puskas to install and promote the Edison lighting exhibit in all its technical glory at the Paris Electrical Exposition of 1881. To the amazement of the cosmopolitan crowds of visitors attending the vast exhibit hall dominated by a working lighthouse, Edison's gigantic and powerful new jumbo generators lit up an astonishing five hundred incandescent lamps of 16 candlepower each.

Even the most cosmopolitan were puzzled by electricity. Unlike a steam engine, its power was obscure, invisible. Wrote one Frenchman, "We are not yet in the habit of observing machines that function without apparent cause. Their occult workings baffle us. The secret of their existence escapes us."[14] Edison's electrical triumph was such that he swept all the top honors for his dazzling luminous display, leaving his competitors like the Englishmen Joseph Swan and Lane Fox, and fellow American Hiram Maxim, sharing a paltry second place. To Edison, struggling back in New York to get his central station working, this had been a soothing balm.

Perhaps sweeter yet was the complete and public capitulation of English scientist William Preece. The longtime Edison critic had scoffed time and again at Edison's assertion that he had subdivided the light. After visiting the Paris exhibit, Preece wrote, "Mr. Edison's system has been worked out in detail, with a thoroughness and mastery of the subject that can extract nothing but eulogy from his bitterest opponents. Many unkind things have been said of Mr. Edison and his promises; perhaps no one has been severer in this direction than myself. It is some gratification for me to be able to announce my belief that he has at last solved the problem he set himself to solve."[15] As if that were not triumph enough, the French conferred the Légion d'Honneur on Edison—which, along with all the other accolades, imbued his new Paris company with great prestige and credibility.

At the Edison factory in Ivry, Tesla had met his first Americans, who "fairly fell in love with me because of my proficiency in Billiards!" Of course, the naive and enthusiastic Tesla was soon explaining his wonderful alternating current induction motor and full system to these new colleagues and bosses, assuming that they, of all people, would appreciate it. At a time when the understanding of electricity was still quite primitive, polyphase alternating current was a quantum leap and difficult to grasp. Tesla's exuberant and idealistic plan to liberate the world from drudgery was not at all obvious even to those working at Ivry. What they knew and understood was direct current electricity, where the electrons flowed only in one direction and created little magnetic field. There was no way to increase the voltage (or driving pressure) of that electricity. Really, the only way to increase the amount of direct current flowing out into an Edison power grid was to provide heavier (and highly expensive) copper wires to carry those greater numbers of electrons to their ultimate destination. The greater the current, the greater the heat on the wires. Hence the need for sturdy, low-resistance copper. Moreover, as those heavy currents traveled over the wires, they lost some of their energy to expended heat. Since the *amount* of delivered electric power is a function of voltage (or pressure) times current (flowing electrons), direct current electricity systems—with their low voltages—needed to be high-current systems. It was this immutable Ohm's law of electricity and the high cost of copper that kept the Edison systems confined to small, high-density locales. It also meant that no matter where you might come in accidental contact with an Edison DC system, you could not get badly shocked. The overall power was just too low.

One of the Paris Edison men, D. Cunningham, foreman of the mechanical department, proposed to the young and hardworking Serbian that they form a company and finance it by selling stock to develop Tesla's fabulous invention. "The proposal," Tesla would later write, "seemed to me comical in the extreme. I did not have the

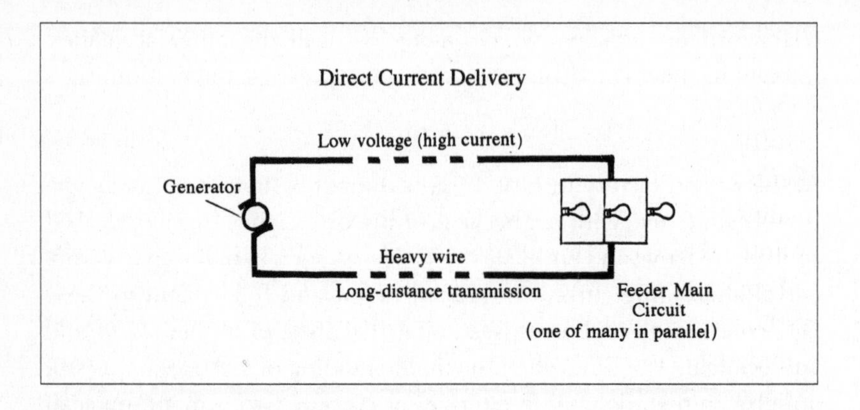

faintest conception of what he meant, except that it was an American way of doing things. Nothing came of it, however, and during the next few months I had to travel from one place to another in France and Germany to cure the ills of the power plants."[16] That fall, as the Edison Electric Light Company in New York was officially launching its Pearl Street Station power network, Tesla again returned to Paris. After spending some months working closely with the Edison machines, he now proposed to his Paris bosses some improvements in the standard dynamos. "My success was complete and the delighted directors accorded me the privilege of developing automatic regulators which were much desired."[17]

Although Tesla freely shared his vision of a new and wondrous AC induction motor, none of the harried Edison managers seemed interested. Theirs was a DC system. They had good motors to offer their customers. What possible need did they have for an AC system or an AC motor? They had many other matters on their minds. Establishing the European Edison companies was fraught with the predictable difficulties of creating any new enterprise, but it was exacerbated by the scope and ambition of the company, the sheer newness of the product, and the technical and managerial inexperience of virtually all involved. As another electrical manager of the time explained, "People generally did not at all appreciate the need or value of electricity. They had to be educated to its use.... Suitable manufacturing methods as well as adequate ways of distributing the manufactured product had to be devised.... Customers did not exist, they had to be created."[18] Despite the great triumph of the Paris Expo, Parisians showed little serious interest in a central station. Batchelor's almost daily letters and cables back to New York are a litany of insufficient capital, shipments that needed to be speeded up, fickle clients of isolated plants, problematic and erratic machines, and poor-quality supplies. Many of these difficulties were resolved fairly easily, but others were major embarrassments that threatened grave financial consequences. Into the latter category fell the disastrous case of the Edison lighting plant at the new railroad station in the important commercial city of Strasbourg, in Alsace-Lorraine, a region held by Germany since the war of 1870.

In early 1883, the Strasbourg city fathers held a dedication ceremony to showcase the new railroad station's light, an important event graced by the august presence of the aged but powerful Emperor William I of Germany. When the switch was turned on to light the station's 1,200 Edison bulbs and flood the grand space with incandescent radiance, there was instead a terrifying explosion that blew out a wall. The German government, which had placed the order, was outraged at this violent fiasco and refused to accept the plant, much less pay for it. "On account of my knowledge of the German language and past experience," wrote Tesla, "I was entrusted with the difficult task of straightening out matters and early in 1883, I went to Strasbourg on that mission. . . . The practical work, correspondence, and conferences with officials kept me preoccupied day and night."[19] By summer, the railroad station plant was largely repaired and negotiations launched with the German bureaucracy for payment. Tesla now had the time to do what he had been chafing to do—create a prototype of his AC induction motor. Having grasped the fundamental advantage of AC over DC, Tesla had long since designed in his head a system that would put it to work. "As soon as I was able to manage, I undertook construction of a simple motor in a mechanical shop opposite the railroad station, having brought with me from Paris some material for that purpose. . . . [That summer] I finally had the satisfaction of seeing the rotation affected by alternating currents of different phase, and without sliding contacts or commutator, as I had conceived a year before. It was an exquisite pleasure but not to compare with the delirium of joy following the first revelation."[20]

Having been further exposed to the American notions of stock companies and working capital, Tesla tried through some of his new and influential friends in Strasbourg, including the former mayor, to persuade a number of local wealthy men to invest in his revolutionary electric generator. To his mortification, no one evinced the smallest iota of interest, despite his being able to demonstrate his working prototype. Certainly there was a place for electricity in train stations, factories, and the homes of the adventurous rich, but if the world-famous, much honored American Edison was already installing electricity all over the Continent, why would anyone need the untested ideas of an obscure poetry-reciting junior engineer from Croatia? Who would risk their money on such an unknown? Having failed to sway any Strasbourg capitalists, Tesla was anxious to return to the cosmopolitan wealth of Paris and try his prospects there, now that he had his working model. But the Germans' punctilious observation of every rule of rank and protocol held him hostage many months more in Strasbourg, as the completion of the railroad job stretched on and on through the fall and winter of 1883 and into the New Year. It was not until the spring of 1884 that the Germans finally accepted and paid for the Strasbourg Hoffbahn plant.

Nikola Tesla then returned to Paris (after a year away), just as the Paris Salon was opening at the Palais de l'Industrie. For several years, the great hall full of paintings had been lighted with arc lights at night, a decision that had outraged some artists but had quadrupled the number of strolling art lovers (or the merely inquisitive) to seven hundred thousand, for many more had evenings free for such diversions than had days. In such small ways were the social fruits of electricity already apparent. So Tesla reveled in the beauties and energy of Paris as he strode down to Ivry-sur-Seine, happily anticipating "liberal compensation" for his dynamo improvements and his successful resolution of the Strasbourg assignment. With a thick wad of bonus francs, he could finally finance a full working model of his AC system, including his beloved motor and the necessary polyphase AC generator to run it properly. Then he would attract some venturesome French capitalists. Instead, the slightly older and wiser Tesla found his Edison bosses all passing the buck as he sought his bonus. "It dawned on me that my reward was a castle in Spain," a mirage.[21]

Considering the Edison companies' constant scrounging for more capital and the rather perilous and very nascent state of their American and European operations, it is little wonder that no manager felt inclined to dole out cash rewards. In its 1883 annual report, the Edison Electric Light Company had reported only losses. Moreover, there is no indication in the vast Edison archive of business correspondence that Tesla's Paris bosses valued Tesla's contributions as highly as did the young man himself. Seeing that Tesla was restless and ambitious, Edison officials urged him instead to go to New York and work with the master himself. Charles Batchelor had already sailed home, having overseen the installation of more than a hundred isolated plants all over Europe, in textile mills, factories, hotels, theaters, stores, steamships, shipyards, and railroad stations. But only three central stations had been sold and installed, in the cities of Milan, Rotterdam, and St. Petersburg.

Tesla's young life thus far was marked by a series of dramatic events—several youthful near misses with death, his passionate embrace of electricity as a life's work, and the revelation of the induction motor and the AC polyphase system. He was especially prone in these Paris years to certain intense hallucinations, notably encompassing flashes of light. "They were my strangest and [most] inexplicable experience," Tesla would later say. "They usually occurred when I found myself in a dangerous or a distressing situation or when I was greatly exhilarated. In some instances I have seen the air around me filled with tongues of living flame. Their intensity, instead of diminishing, increased with time and seemingly attained a maximum when I was about twenty-five years old."[22] (Tesla also liked to describe his birth as occurring at the stroke of midnight during a severe electrical storm, obvious portents of his electrical future.) Nikola Tesla's departure from Europe and arrival in the New World (what he termed the "Land of the Golden Promise") featured further dramas. When he arrived at the Paris railroad station, "I discovered my money and tickets were gone," not to mention his valise. With the train pulling out in great billows of steam, the frantic Tesla decided to go anyway and ran down the platform and scrambled on. The steamship company allowed him to board at the last minute only when no one else arrived to claim his berth. "I managed to embark for New York with the remnants of my belongings, some poems and articles I had written, and a package of calculations relating to solutions of an unsolvable integral and my flying machine. During the voyage I sat most of the time at the stern of the ship watching for an opportunity to save somebody from a watery grave."[23]

Nikola Tesla debarked at the port of New York on June 6, 1884, a fair, pleasant Friday, as did thousands of Germans, Irish, Scandinavians, Italians, and Russian Jews, all, like the young electrician, full of American dreams—peasants yearning for farms; young men heading to mines, mills, factories; husbands and fathers hoping to amass quick savings to return home to buy land or a business; wives and mothers envisioning happier, more prosperous futures for themselves and their offspring. There in the echoing spaces of Manhattan's Castle Garden immigration depot, a gloomy former fort and onetime theater on a pile of rocks down by the Battery, the multitude of nationalities mingled, their many languages a Tower of Babel as they shuffled through the immigration lines and into an America fast becoming the most prosperous place on earth. For it was in these years between the end of the Civil War and the turn of the century that the United States would become urban, industrial, and exceedingly rich, its gross national product soaring from $9.1 billion to $37.1 billion, its per capita income tripling. And the three compelling reasons for this rising prosperity were there on view at Castle Garden. The first element was swift population growth. More than seven million hopeful souls passed through Castle Garden alone between 1855 and 1890, when Ellis Island opened as the new federally run immigration depot for New York. These foreign immigrants surged by the thousands off steamships and quickly set out via America's far-reaching railroads, seeking their destiny in the great booming cities, remote territories, and new states. And there lay the second reason for burgeoning American wealth—its excellent system of transportation and communication. The fast-expanding American railroad network knit together its vast and disparate spaces, making commerce easily possible, as did the indispensable telegraph and the new telephone.

The people and the transport in turn made possible efficient exploitation of the nation's phenomenal natural blessings—its vast coal and ore deposits, its rich farmlands and forests. Great fortunes and great enterprises were coalescing as this new American world of industry and cities arose. Men like William H. Vanderbilt were now richer than the queen of England. Millions were similarly striving to become rich—in railroads, steel, oil, lumber, coal, gold, silver, sugar, department stores, or such new consumer products as cigarettes, ready-made clothes, soap, biscuits, and colas. But in all those daily arrivals in that early summer of 1884, it is hard to imagine any immigrant whose dreams and visions equaled those of the obscure and impecunious Serb Nikola Tesla, who stepped onto the pedestrian precincts of crowded, dirty lower Manhattan still swathed in reveries about bringing cheap and abundant power and light to the whole world.

So Tesla set forth confidently in the New World, with all of four cents in the pocket of his one suit, the address of a friend, and the prospect of a job with Edison. New York City of the mid-1880s was not Paris. The Battery offered a broad green turf, wide paths, and shady trees. But as soon as one walked north from there, one was hit by the city's frenetic commercial pace and palpable atmosphere of moneymaking and money getting. There were no sweeping boulevards, grandiose palaces, or formal public gardens. The downtown districts were a hodgepodge of large warehouses, commercial establishments, and shabby tenements, bisected by the four north-south lines of the elevated trains, whose noisy steam engines rained down soot and ash on the unwary from on high. Once summer settled in, the streets were hot and dusty and stank of horses, for equine teams pulled vehicles of every sort, the huge colorful old-style city omnibuses, the new streetcars gliding more quickly on sunken rail tracks, all manner of big teamster carts, and, for those who could afford it, the personal carriages. The immediate crowded blocks around Wall Street were fine and very imposing, built up with noble and costly ten-story edifices. Newsboys everywhere hawked the city's bestselling newspapers, the *Sun* and the *Herald*, and the new fast-rising favorite, the *World*. And jutting up high over the warm and smelly commercial streets towered a veritable forest of wooden poles and their tangle of electrical wires, drooping and draping from pole to rooftop to window and on to other poles. In some commercial blocks, the festoons of hundreds of crisscrossing wires almost obscured the early-summer sky.

Nikola Tesla was walking north and taking in all these jangling sights and sounds when he noticed a foreman in a shop standing over an electrical motor and looking most exasperated. When Tesla entered the doorway, the man explained the machine was broken. Tesla offered his services, eventually got the motor working, and was promptly offered a job. He declined, saying he already had prospects of one. The grateful shop owner paid him $20, a munificent sum when a worker's daily wage was generally $1. A skilled, educated, and experienced engineer like Tesla made $18 a week. So Tesla had his first taste of the riches of the New World. He continued on, located his friend, and secured a night's rest. The next day he headed toward Fifth Avenue, where, one guide said, were "concentrated the wealth and aristocracy of the city . . . all the [brownstone] blocks are massive and palatial . . . [consisting of] the huge club-houses, the expensive libraries, the fine picture galleries, and richly furnished drawing rooms of this region of merchant-princes."[24] There, each late afternoon, one could marvel further at the conspicuous display of wealth, in this instance the passing parade of some of the world's finest horseflesh and most luxurious vehicles en route to and from Central Park. There, fashionable New Yorkers liked to show off their fine equipages (and themselves) as they made their way amid the park's artfully placed carriage paths.

Nikola Tesla fit in well with the well-dressed, attractive throngs on Fifth Avenue. His education and his year in Paris had given him a certain polish, so he confidently entered Edison headquarters at 65 Fifth Avenue. The impressive brownstone was luxuriously fitted out with beautiful electrical chandeliers and lamps to entice the rich, while its topmost floor had been given over to housing for some of the unmarried employees. On yet another floor, Edward Johnson had started a night school for fledgling electricians, for the company desperately needed trained men. Edison, wearing his shabby Prince Albert coat, was often to be found squiring around the more exalted visitors or holding court in his comfortable back office, smoking cigars. "The meeting with Edison was a memorable event in my life," Tesla later wrote. "I was amazed at this wonderful man who, without early advantages and scientific training, had accomplished so much. I had studied a dozen languages, delved in literature and art, and had spent my best years in libraries reading all sorts of stuff that fell into my hands, from Newton's 'Principia' to the novels of Paul de Kock, and felt that most of my life had been squandered. But it did not take long before I recognized that it was the best thing I could have done."[25]

Edison was not quite as unkempt in appearance now that he was a New Yorker, but he remained utterly practical, plainspoken, world famous, a great kidder, and at age thirty-seven a seasoned veteran of the corporate rough-and-tumble on both sides of the Atlantic. Before him he saw Tesla, erudite, unknown, tall, slender, formal in dress and manner, a dreamy young man a decade his junior who expressed himself in a flowery, heavy-accented style and came across as an utter naïf. He quickly nicknamed him "our Parisian." Tesla recalled, "I was thrilled to the marrow by meeting Edison, who began my American education right then and there. I wanted to have my shoes shined, something I considered below my dignity. Edison said, 'You will shine the shoes yourself and like it.' He impressed me tremendously. I shined my shoes and liked it."[26]

Tesla soon proved his worth, for the passenger steamship SS *Oregon* was stuck in dock over on the East River, unable to depart on schedule because Edison's on-board lighting system had broken down. "The predicament was a serious one," Tesla recalled, "and Edison was much annoyed. In the evening I took the necessary instruments with me and went aboard the vessel, where I stayed for the night. The dynamos were in bad condition, having several short circuits and breaks, but with the assistance of the crew, I succeeded in putting them in good shape. At five o'clock in the morning, when passing along Fifth Avenue on my way to the shop, I met Edison with Batchelor and a few others, as they were returning home to retire.

'Here is our Parisian running around at night,' he said. When I told him I was coming from the Oregon and had repaired both machines, he looked at me in silence and walked away without another word. But when he had gone some distance I heard him remark, 'Batchelor, this is a good man.' "[27]

In the summer of 1884, the Edison Electric Illuminating Company in New York was still expanding, adding new customers, and always looking to supply electricity reliably and more profitably. They were now lighting up such important institutions as the New York Stock Exchange, the *New York Commercial Advertiser,* the New Haven Steamboat Company's offices and large pier, Brown Brothers & Company on Wall Street, and the North British & Mercantile Insurance Company on William Street. Both Pearl Street in Manhattan and Menlo Park in New Jersey served as demonstrations of central stations.

But Edison's direct current central station system, with its half-mile limitation in any direction, was proving a tough sell in less compact cities and towns. After all, it was far harder to convince the hundreds of businesses necessary to make the network of a central station profitable that they needed electric light than it was a single wealthy homeowner like J. P. Morgan or a single factory owner, who could more easily see the advantages. By the end of 1884, only eighteen central stations were installed in the United States. In contrast, isolated stand-alone lighting plants, which required only the enthusiasm of a factory owner or a hotelier, were a popular product. By the fall of that year, there were 378 such plants all over the country. Once again, as in Paris, Tesla was working on improving dynamos and troubleshooting. But with each passing week, he was "more and more anxious about the [AC induction motor] invention and was making up my mind to place it before Edison."

Edison, who was thoroughly fed up with the listless management of the main company controlled by his Wall Street directors, was very much preoccupied that fall and winter with winning back control and restructuring the Edison companies to better push his central stations and generally advance their products. On August 9, his much neglected wife, having been ill on and off for several years, died out at Menlo Park, leaving Edison with three children. Edison threw himself even harder into his work. But in some late night dinner or other encounter, apparently he and Tesla did discuss his young employee's proposed system for alternating current. Tesla pointed out that a central station based on alternating current dynamos could liberate electricity from the one-mile shackle of Edison's DC plants. And if his, Tesla's, induction motor was developed, it could fill the big, looming gap of an AC system that aspired to go beyond lighting. Moreover, his AC induction motor would surely be superior to those operating on DC. Edison, said Tesla, responded "very bluntly that he was not interested in alternating current; there was no future to it and anyone who dabbled in that field was wasting his time; and besides, it was a deadly current whereas direct current was safe."[28]

The arc light companies all operated their blazing lights with high-voltage alternating current, and certainly there had been unfortunate accidents where electrical workers had been severely shocked and even killed by inadvertent contact with the AC apparatus. Edison denounced AC as far too dangerous for domestic use, a stance seconded by such other leaders in the field as Glasgow's Sir William Thomson, a scientist whose giant reputation gave his opinions great weight. Perhaps just as relevant, Edison or his company had not invented or developed the alternating current generators used for powering arc lights. Edison took enormous and justifiable pride in having invented—although often with the help of his subordinates—every aspect of his low-voltage electrical system. A person touching any part of Edison's DC system—from the dynamo to the wires to the bulbs—would receive only a mild shock. He wanted nothing to do with AC.

Nikola Tesla had arrived in the United States just in time for the lively every-four-years blood sport of presidential politics, American style. Corruption, which had sullied many aspects of local and national life, was the central issue. Senator James G. Blaine, the Republican candidate, was seen as deeply in thrall to the unbeloved and bullying railroads, which blatantly bought whom and what they wanted. The Democratic candidate was reformist New York governor (and former mayor of Buffalo) Grover Cleveland, known as "Grover the Good" for his fight against Tammany corruption. Democrats portrayed their man as a rare paragon of honesty in an age when dirty dealing in business and politics was rife. Then on July 21 the Republican *Buffalo Evening Telegraph* broke the scandal that the unmarried Cleveland had fathered a bastard! Wired Cleveland to the party, "Whatever you say, tell the truth." The truth was that it *might* have been his child, so he had provided support. The Republicans now taunted the Democrats with the gibe, "Ma! Ma! Where's my pa?" while the Democrats taunted back, "Gone to the White House, ha! ha! ha!" Also running was New Jersey lawyer Belva Lockwood of the Equal Rights ticket, even though no American woman could vote.

The race between the tall, fat, bland-looking Cleveland and the gray-whiskered Blaine was neck and neck when Blaine, the candidate of the new industrial titans, descended upon Manhattan at the end of October. New York, with its legions of rich men and newspapers that were read nationwide, had become deeply influential. In the afternoon, candidate Blaine met with important Protestant ministers at the vast marbled elegance of the Fifth Avenue Hotel. One minister denounced the Democrats as "the party whose antecedents are rum, Romanism, and rebellion." A tired Blaine did not demur. Perhaps he was thinking of that evening's festivities at Gotham's most famous and delectably high-toned restaurant, Delmonico's, renowned for its luscious lobster Newburg and the ever showy and scrumptious house dessert, baked Alaska. Even Edison, that man of the people, preferred Delmonico's above all other restaurants. And on the night of October 29, 180 of America's richest and most prominent men were gathering there to fete their Republican candidate, whose main campaign plank was excluding imports through high tariffs, thus leaving the vast and lucrative American market completely to his hosts' industrial and consumer blandishments.

Unfortunately for Blaine, the scrappy Joseph Pulitzer had the previous year bought the fading *New York World* and was consciously refashioning it into the must-read paper of the new upwardly mobile workingman and -woman, people who were becoming deeply concerned at the rise of a plutocracy that seemed determined to control everything through convoluted monopolies, money power, and virtual ownership of amenable politicians. After taking over, Pulitzer assembled the paper's existing staff and announced, "Gentlemen, you realize that a change has taken place in the *World.* Heretofore you have all been living in the parlor and taking baths every day. Now I wish you to understand that, in future, you are all walking down the Bowery."[29] Pulitzer proceeded boldly, abandoning the prim, reserved one-column format that indicated a story's importance only by multiple decks of subheads. Instead, he splashed big news across many columns with blaring headlines and huge illustrations.

So, to the great indignation of the rich Republicans, the October 30 *New York World* devoted the entire front page to the Delmonico's dinner, declaring in a giant banner headline: A ROYAL FEAST OF BELSHAZZAR BLAINE AND THE MONEY KINGS. The well-fed rich—recognizable caricatures of William H. Vanderbilt, Andrew Carnegie, Jay Gould, and so forth—were sipping Monopoly Soup and ladling Lobby Pudding while ignoring a poor wraithlike couple begging for scraps. The *World* was also careful to report on the "rum, Romanism, and rebellion" remark, which outraged Irish Catholics, many of whom had previously been Blaine supporters. Blaine lost in New York State by only 1,149 votes and thereby forfeited the whole election. New York City was not only the nation's financial center, it also wielded enormous political power as the nation grappled with its new industrial identity. Grover Cleveland became the first Democrat elected since the Civil War, and his ascension was a sign of a nation restive over rising plutocracy.

Through all this no-holds-barred political combat over the nation's future, Nikola Tesla labored diligently away at his electrical duties for Thomas Edison. "During this period I designed twenty-four different types of standard machines with short cores and uniform pattern, which replaced the old ones. The Manager had promised me fifty thousand dollars on the completion of this task." In the spring of 1885, Tesla sought this big bonus. He told his first biographer, John J. O'Neill, that Edison reneged, explaining that the offer had been a big joke. "Tesla," said Edison, "you don't understand our American humor."[30] It is exceedingly hard to imagine that the Edison companies, still being in dire need of cash and capital themselves, would promise that kind of huge money—as much as they had gotten in hard cash to launch the initial company—to a salaried employee. Moreover, Tesla's most recent biographer, Marc Seifer, points out that Tesla could not even get a $7-a-week raise. When another employee approached Batchelor on Tesla's behalf (Tesla believing he deserved to have his salary raised from $18 a week to $25), Batchelor curtly refused, saying, "The woods are full of men like [Tesla]. I can get any number of them I want for $18 a week."[31] Whether a joke or a betrayal of a promise, Tesla was sufficiently outraged to quit.

Tesla had lasted less than a year as Edison's employee in New York. In truth, he and Edison were like oil and water, each amused and annoyed by the other. A great dandy by nature, Tesla prided himself on being dapper and fashionable and abhorred Edison's slovenly indifference: "If he had not later married a woman of exceptional intelligence, who made it the one object of her life to preserve him, he would have died many years ago from consequences of sheer neglect."[32] Far worse, believed Tesla, was Edison's approach to science: "If Edison had a needle to find in a haystack, he would proceed at once with the diligence of the bee to examine straw after straw until he found the object of his search. . . . His method was inefficient in the extreme, for an immense ground had to be covered to get anything at all unless blind chance intervened and, at first, I was almost a sorry witness of such doings, knowing that a little theory and calculation would have saved him 90 percent of his labor."[33] Edison, in turn, dismissed Tesla as a "poet of science" whose ideas were "magnificent but utterly impractical."

Tesla was catching on to the American way of operating, and now he took an absolutely practical tack. He needed to make an independent name for himself with something very salable and useful if he was going to interest investors in his far more visionary AC induction motor. So he set himself a straightforward and mundane task—designing improved arc lights that did not flicker and a better kind of generator to run them. In mid-March of 1885, Tesla began meeting with patent attorney Lemuel Serrell and his patent artist, who instructed him on preparing and submitting his first arc light patents. Serrell also introduced Tesla to two Rahway, New Jersey, businessmen who eagerly backed the new Tesla Electric Light & Manufacturing Company. The company's first project was municipal arc lighting for Rahway, the investors' hometown. For the next year Tesla labored away producing and installing his system, which lighted certain major streets along with a few factories. Tesla did all this to establish his bona fides. The *Electrical Review* was sufficiently impressed with his advances to feature his new system on its front page of August 16, 1886. In subsequent advertisements in the same journal, the Tesla Electric Light & Manufacturing Company announced it was "now prepared to furnish the most perfect automatic, self-regulating system of electric arc lighting yet produced." There was "no flickering or hissing of the lamps" under this "entirely new system of automatic regulation resulting in absolute safety and a great saving of power."[34]

In Rahway, New Jersey, over the next few months, the Tesla Electric Light & Manufacturing Company finished successfully installing the powerful moonlight white arc lights around the city. In the course of this inaugural business venture, Tesla had received his first American patents—a commutator for a dynamo electric machine on January 26, 1886, followed by an electric arc lamp on February 9, 1886, and a regulator for a dynamo electric machine on March 2. In fall of 1886, Tesla proposed once again to his partners that they widen their horizons, think pioneering thoughts (rather than prosaic practical arc light ones), and get to work developing his world-changing alternating current motor and electrical system. "The delay of my cherished plans was agonizing," Tesla recalled years later.[35] Not only were these shortsighted New Jersey gentlemen not interested, but they elected to cheat Tesla, ever the naïf when it came to business, out of his patents and oust him from the very company he had founded. So in that autumn, poor Nikola Tesla, the romantic and dreamer, was reeling from "the hardest blow I ever received. Through some local influences, I was forced out of the company, losing not only all my interest but also my reputation as engineer and inventor."[36] As he put it later, "I was free, but with no other possession than a beautifully engraved certificate of stock of hypothetical value."[37] This elegant, erudite electrician and immigrant suddenly found himself as penniless as he had been when he'd stepped off the boat more than two years earlier. Moreover, he had no immediate prospects and was far too proud to seek help or work from his former Edison colleagues.

It is unlikely that Nikola Tesla harbored any illusions about the bleak desperation of being down-and-out in Manhattan. The Edison Machine Works, where Tesla had worked, sat amid some of New York's vilest slums, windowless warrens where summer heat baked in the stench of overflowing outdoor privies and where bone-chilling winters drove the cold deep into the stained walls. Was it any wonder that disease and death awaited so many there, or that other familiar fate, the oblivion of vice and crime? In 1886, an arctic winter and a sour economic spell had reduced more families than usual to beggary. Clashes between labor and capital escalated venomously. Fast-swelling union ranks challenged the harsh new industrial order. Employers yielded as little as possible. Nonetheless, opined the censorious editors of the *New-York Daily Tribune,* "organized labor has astonished the country with exhibitions of its political power," starting with blockaded railroads, striking tanners, curriers, carpet weavers, coal miners, and the Chicago meat packers. In Manhattan, fifteen thousand striking streetcar drivers and conductors fought street battles with police and fill-ins until they extracted a twelve-hour day for a $2 wage. All through 1886, "business in every quarter of the country has been disordered in consequence of these and hundreds of similar uprisings."[38] On May 4, anarchist bomb throwers killed seven policemen and injured dozens of strikers in Chicago's Haymarket Square, sending shock waves of fear undulating through the whole nation. That violence curdled the better classes' already lukewarm support for the American labor movement.

Into this harsh economic climate ventured an unemployed Nikola Tesla, who found little need for his specialized electrical talents. "My high education in various branches of science, mechanics and literature seemed to me like a mockery," he would say later.[39] The newspapers were filled with hundreds of ads placed by men seeking work—as coachmen, butlers, private waiters, valets, and that all-around position, "useful man," one who could tend a furnace or assist in the stable. In contrast, the "Help Wanted" column might have a handful of prospects, a family at Park Row looking for someone to care for their horse, cow, and garden. Or there was the Madison Avenue household in need of a man to run the furnace, wash windows, and make himself generally useful. With his money gone and his family far away, Tesla wandered the streets of Manhattan, the plush restaurants and luxurious shops a mocking reminder of his gnawing penury. "There were many days when [I] did not know where my next meal was coming from. But I was never afraid to work, I went to where some men were digging a ditch . . . [and] said I wanted to work. The boss looked at my good clothes and white hands and he laughed to the others . . . but he said, 'All right. Spit on your hands. Get in the ditch.' And I worked harder than anybody. At the end of the day I had $2."[40]

In this year of unrest and anarchism, the new electrical industry had its own share of woes, though working conditions were idyllic compared to the dark, fiery, and often fatal dangers of the coal mines and steel mills. Thomas Edison's own businesses had so expanded that he no longer personally knew all his men. At the lamp factory out in New Jersey, eighty highly skilled filament sealers formed a union and "became very insolent," said Edison, "knowing that it was very impossible to manufacture lamps without them." When they objected to the proposed firing of one of their members, Edison quickly designed thirty machines to automate their work. Then he fired the man as planned. "The union went out," said Edison, following up with the punch line: "It has been out ever since."[41]

In March of 1886, a committee at the Edison Machine Works on Goerck Street parlayed with their immediate boss, Charles Batchelor, demanding the right to form a union and seek better pay and conditions. The Edison management, which paid average wages, was amenable to reducing the workday from ten hours to nine, but they wanted no truck with unions. The shop was "to be run just as the managers decided & no interference whatever to be tolerated." Nor would they accept the end of piecework. Batchelor explained to reporters, "If [a worker] loafs or gets drunk he loses his own time, not ours."[42] On May 19, the Edison workers struck. Shortly thereafter, Thomas Edison moved the machine works upstate to the quiet town of Schenectady, an easy ride north and then west on the New York Central Railroad. He wanted, he said, "to get away from the embarrassment of the strikes and the communists to a place where our men are settled in their own homes."[43]

As for Nikola Tesla, "I lived through a year of terrible heartaches and bitter tears, my sufferings being intensified by material want."[44] Scraping by in the cold winter, Tesla was reduced many a day to working in a New York labor gang. Presumably, he was painfully aware as he swung his pick or shoveled his dirt that others might be pressing ahead with AC systems. As the days began to warm, and the daylight became mellow and lingered longer, fate smiled again on this dreamer. In the early spring of 1887, one of Tesla's foremen realized that this hardworking Serb really was no ordinary laborer. He arranged to introduce the down-on-his-luck electrician to a high-level engineer at the Western Union Telegraph Company. The engineer, Alfred S. Brown, was impressed by Tesla's fervid description of his AC motor that would power the world, and he in turn introduced Tesla to Charles F. Peck, a distinguished lawyer and investor. Peck was informed enough to know that no one had yet managed to design a commercial industrial AC motor that actually worked. Why should he believe great deeds would come from this former junior Edison engineer and failed arc light entrepreneur whose English was elegant but thickly accented? He declined even to watch some tests.

Tesla racked his brain, wondering how to impress this bland attorney. "I had an inspiration," he recalled many years later. He asked Peck, "Do you remember the 'Egg of Columbus'?" Tesla was referring to a banquet where the explorer challenged all those skeptical of his quest to show how they would balance an egg on its end. After they tried in vain, Columbus took the egg and, cracking the shell slightly by a gentle tap, made it stand upright. This small feat led to an audience with Isabella, the queen of Spain, who pledged her support. Peck was now intrigued. Was Tesla, he inquired, planning to balance an egg on its end? And why? "Well, what if I could make an egg stand on the pointed end without cracking the shell?" Tesla saw he had his man's interest. Peck said, "If you could do this, we would admit you had gone Columbus one better." Tesla pressed him on whether this would make him a potential patron. "We have no crown jewels to pawn, but there are a few ducats in our buckskins," Peck conceded. "And we might be able to help you to an extent."[45]

With that Tesla hurried forth through the tumult of the downtown streets clogged with wagons hauling barrels and every kind of freight to find a hard-boiled egg and a blacksmith. The demonstration took place the very next day. "A rotating field magnet was fastened under the top board of a wooden table and Mr. Tesla provided a copper-plated egg and several brass balls and pivoted iron discs for convincing his prospective associates. He placed the egg on the table and to their astonishment it stood on end, but when they found it was *rapidly spinning* their stupefaction was complete. The brass balls and pivoted iron discs in turn were sent spinning rapidly by the rotating field, to the amazement of the spectators. No sooner had they regained their composure than they delighted Tesla with the question, 'Do you need any money?' "[46]

Nikola Tesla's life was taking another one of its dramatic turns. After the betrayal of his Rahway partners, his winter of pauperism and hard labor, Nikola Tesla was once again launched on his long deferred electrical dream. Very shortly, he, Peck, and the Western Union engineer Alfred S. Brown had formed the Tesla Electric Company, and Tesla was happily setting up his first laboratory at 89 Liberty Street, a busy thoroughfare a few blocks in from the jumbled wealth of the Hudson River wharves and ferry lines and north of the moneyed blocks of Wall Street. While the yells of the truckers driving huge teams to and from the docks drifted in through his windows all that summer and fall, Tesla labored to put to paper the designs he needed to patent. Finally he would be able to build the whole AC system he had dreamed of for so long, especially his many induction motors. Finally he would free the world from drudge labor.

CHAPTER 5

George Westinghouse: "He Is Ubiquitous"

The bold and dynamic Pittsburgh inventor and entrepreneur George Westinghouse had spent much of 1883 and 1884 living in New York, watching closely as Gotham gradually came alive electrically, a metropolis nightly more scintillating with artificial light. Now each wintry eve, when a lavender darkness enveloped the city, the dazzling arc lamps bathed the major avenues in their brilliant blue light. Broadway had become a new nighttime stage for stylish crowds of promenaders shopping and strolling, the new electric light gleaming off the curved blackness of men's derbies and high hats, casting a soft sheen on the women's fur-trimmed velveteen cloaks, the jet black of their beads and feathered bonnets. Farther south, music halls and theaters sparkled brilliantly with electric lights, while Edison's incandescent light lit up newspaper offices on Park Row, the best hotels, and many of the financial edifices around Wall Street. All up Fifth Avenue and over on Madison the brownstone mansions of the merchant princes and the new industrialists were aglow each dusk with electric candlepower. For those who cared to notice, the electrical future was quietly taking shape.

And George Westinghouse, one of the era's industrial titans, was the sort who cared to notice. A forceful, well-built man of six feet,

A young George Westinghouse

Westinghouse was an imposing presence with his thick mane of chestnut hair, vigorous sideburns, and huge walrus handlebar mustache. By 1884, at just thirty-seven, he had already assembled a formidable empire and fortune in the free-wheeling world of railroads, the most important and ruthless corporate force in America. His first railroad-related inventions, a "car replacer" that got derailed trains quickly back on the tracks and a long-lasting steel "frog" that prevented derailing at junctions, met a sorry fate, licensed at first by railroads only to be quickly "improved" so the roads could claim their own patents. Then, in 1869, Westinghouse, all of twenty-two, introduced his most momentous invention to date, a revolutionary air brake that allowed the engineer of a passenger train for the first time to quickly and safely stop all the cars. Westinghouse had had a struggle finding anyone willing to back this novel and expensive venture. And so, far, far wiser, when he finally did introduce his air brake, he staunchly refused to give the railroads licenses, saying only *he* would manufacture them in his small Pittsburgh works.

As the young Westinghouse improved his brakes and shored up his patents, the railroads tried to circumvent and eliminate this upstart. Wrote one railroad manager to another, "Do you use Westinghouse and can you make any improvement upon his apparatus without his permission and cooperation?"[1] When Westinghouse felt the railroads, however powerful, were treading on his turf, he intervened forcefully, threatening patent lawsuits, usually in person. "Westinghouse stopped by," passed along the railroad manager, "and warned that if we try the vacuum [brake], even experimentally, he will bring suit."[2] Having seen his first patents expropriated by the railroads and his first company consequently dwindle away, Westinghouse assumed a lifelong ferocity when it came to his products and patents.

Just as Edison became familiar with electricity through telegraphy, Westinghouse learned about the "subtle agency" through its use in railroad signals. Having swiftly established the Westinghouse Air Brake Company as preeminent in the United States, Westinghouse, who combined hard-driving ambition with a famously winsome, persuasive charm, set out to conquer England. He took with him his new wife, the refined and cultured Marguerite Erskine, whom he had met on a train and soon married in her hometown of Brooklyn after a whirlwind courtship. In England, he eventually sold air brakes but also discovered railroad signaling. In 1881, he began buying up promising patents, most important one that controlled electric circuits set off by trains, thereby activating signals. By combining these with his own improvements and inventions, Westinghouse soon dominated this new field through yet another Westinghouse entity, the Union Switch and Signal Company, organized in 1882.

The oil lamps used in the signaling system were problematic, and existing electric companies were unhelpful with solutions. George's brother Herman, another mechanically inclined businessman, "had become acquainted with a live wire, Mr. William Stanley, Jr. and through him the Electric Company began to develop."[3] The immediate problem was the signal lamps, but they were something of an excuse. For if George Westinghouse was going into electricity, he would do it as he did everything, in a very big way. This suited William Stanley. Tall, slender, his middle-parted thin hair plastered neatly, and his luxuriant mustache ending in long, strange wisps, Stanley was indeed a "live wire." After a dutiful semester at Yale, he bailed out to work with things mechanical, writing his parents, "Have had enough of this, am going to New York."[4] In 1884, after quick tours with the Swan Incandescent Electric Light Company and Hiram Maxim and two successful sojourns in his own small-business ventures, Stanley agreed to come work for the formidable George Westinghouse, ready for bigger ventures.

The truth was, when Westinghouse surveyed the state of the electrical art as embodied by Edison and his competitors, it did not much excite him, except as a sure way to develop a large enterprise. The physical limitations of Edison's direct current central station were more than evident: The future foretold an insatiable demand for small direct current central stations serving mile-square areas and individual isolated plants, such as had been installed at J. P. Morgan's mansion and in many factories and offices. Why shouldn't George Westinghouse make and install electrical plants as well as Edison or Brush or Swan or Weston or Thomson-Houston? The world was brimming with ambitious young electricians for hire. George Westinghouse had had no trouble luring away one of Edison's promising young mechanical engineers, H. M. Byllesby, just by offering a bigger salary. Preparatory to entering the electrical field with his own DC plants, Westinghouse paid Swan Incandescent Electric Light Company $50,000 to retrieve two patents they held from his new employee William Stanley. These patents covered a self-regulating DC dynamo and a carbonized silk-filament light bulb. William Stanley's initial March 1884 contract paid him a handsome $5,000-a-year salary, with the stipulation that the company would own any of his patented inventions it manufactured and sold. Stanley would receive 10 percent of profits. The young inventor and electrician concentrated on setting up a commercial light bulb facility in Pittsburgh and developing a DC system for Westinghouse. The new Westinghouse system had its marketing debut at the 1884 Philadelphia Electric Exhibit. Reported *Electrical World* that September, "The company are now prepared to do business. Their display comprises electric motors as well. An ingenious arrangement of the lamps is shown by which when one goes out another is switched into circuit, and by which also a bell announcing the occurrence can be rung at any chosen place."[5]

From the start, George Westinghouse had machines in his blood, spending his formative years in Schenectady, New York, a small village overlooking that rich commercial waterway the Erie Canal. There young Westinghouse worked amid the pounding screech and whine of the family machine shop, which successfully manufactured threshing machines of his father's own design. When the War Between the States broke out, George, a mere stripling of fifteen, immediately ran off to serve, as had two older brothers. His father pulled him home. Two years later, as the fighting and dying dragged bloodily on, George, seventeen, was allowed to enlist, serving first with the cavalry and then with the navy. When the Civil War came to its weary end, George Westinghouse briefly attended Union College before returning gratefully to the family machine shop. That fall, at age nineteen, he patented his first invention, a rotary engine. Years later Westinghouse would say, "My early greatest capital was the experience and skill acquired from the opportunity given me, when I was young, to work with all kinds of machinery, coupled later with the lessons in that discipline to which a soldier is required to submit."[6]

So, in the beginning of 1885, as George Westinghouse, industrialist-entrepreneur of Pittsburgh, marshaled his inventive faculties and his bright young electrical lieutenants, he was very much a man to contend with, the founder of four successful companies in the United States and overseas. Unlike Edison, who preferred to use only his own patented work, Westinghouse already had long and reasonably happy experience with purchasing other inventors' better ideas and improving them in his own shops. Of course, Westinghouse in 1885 was nowhere near as famous as the flamboyant Edison. By this time, Edison was a completely lionized and beloved figure, generally a favorite with reporters, the disheveled genius who minced no words, lived off apple pie, and embodied everything big and bold and can-do about a self-confident young nation. To the press, Westinghouse was just another successful Pittsburgh inventor and manufacturer, and worse yet, one who usually refused requests for interviews and stories. "If my face becomes too familiar to the public, every bore or crazy schemer will insist on buttonholing me," he explained.[7] Even when he did grant an interview, Westinghouse rarely made for good copy. He was pretty dull stuff.

In private, however, he was intensely compelling, forthright, blunt, often charismatic. Wrote one biographer of Westinghouse, "With his soft voice, his kind eyes, and his gentle smile, he could charm a bird out of a tree. It is related that in a knotty negotiation it was suggested to the late Jacob H. Schiff, then the head of a great banking house, that he should meet Westinghouse. 'No,' said the astute old Jew, 'I do not wish to see Mr. Westinghouse; he would persuade me.' "[8] This was Westinghouse's charming side. He could also be blunt to the point of offense. But Westinghouse's public persona was reserved and serious. While he had a solid reputation for making novel technologies work in the real world, his major inventions—

railroad air brakes and automated signaling systems—however much they had improved the safety and productivity of the nation's most important industry, lacked the glamour and glitz of Edison's riveting discoveries, the talking phonograph and the incandescent light. The mesmerizing French actress Sarah Bernhardt had never begged to meet "*le grand* Westinghouse."

Like Edison, Westinghouse was much liked and admired by his workers. One young Westinghouse apprentice would always remember the following telling episode at Westinghouse Electric Company's Garrison Alley Works:

"One day several of us were back of the 'Iron clad' building at the side of Duquesne Way, which was an unpaved quagmire. A young foreigner was wheeling copper ingots unloaded from a freight car on the other side of Duquesne Way. An iron slab served as a run-way for the wheel barrow. The wheel slipped off one side into the soft mud. Our crowd enjoyed the predicament and jeeringly gave advice to the helpless lad.

"Mr. Westinghouse appeared, in his long-tailed coat and high hat. He removed his gloves, took hold of the wheel and lifted it onto the slab. He said nothing. It made a lasting impression on me."[9] Here was a boss whose first impulse was to help, to set things right, and at the same time to drive home a powerful but unspoken lesson. They were all working together, from top to bottom.

George Westinghouse, like Edison, thought money was important only as a form of "stored energy" to use as he wished in his work and expand his businesses. He was interested not in being rich, but in helping the world. He strove incessantly to deliver better, more reliable products. But he had another goal also. "My ambition is to give as many persons as possible an opportunity to earn money by their own efforts," he once explained, "and this has been the reason why I have tried to build up corporations which are large employers of labor, and to pay living wages, larger even than other manufacturers pay, or than the open labor market necessitates."[10] After his first trip to England, he instituted a half-day off on Saturdays, beginning in June 1871, the first local firm to do so. Westinghouse companies would be pioneers in worker safety, disability benefits, and pensions.

George Westinghouse had his great electrical epiphany in the spring of 1885 while reading the English journal *Engineering.* For Westinghouse, his *coup de foudre* was reading the description of an alternating current system on display in London at the Inventions Exhibition. This AC system used something entirely novel—a "secondary generator" (soon to be known as a transformer)—to step down higher AC voltages to those low enough to run individual incandescent lights. While others saw it as having limited application, Westinghouse immediately realized that here was something potentially revolutionary, a new way to economically transfer electricity not just to individual light bulbs, but over long distances. Right now, a DC central station had to be located in the middle of its service area. What if you could dispense with coal and steam and run electrical generators with hydropower at faraway waterfalls, then use high-voltage AC to send that electricity a great distance? This transformer ought to be able then to step down the high-voltage AC safely *before* it entered the factory or office building or house.

It so happened that one of Westinghouse's young employees, Guido Pantaleoni, was in Italy because his physician father had just died. Westinghouse cabled him to locate the inventors of the "secondary generator," Lucien Gaulard and John Gibbs, to see what he thought about their work. If it was all that Westinghouse hoped, they should secure options on the patents. Pantaleoni did not have to travel far, it turned out, for Gaulard was in Turin, where he was demonstrating the ability of their AC system to transmit electricity long distances from a waterfall-powered dynamo and then step it down for each individual light. In fact, Gaulard and Gibbs had been developing and exhibiting their AC system for a couple of years, but not until that spring of 1885 did it catch Westinghouse's eager attention.

Over in Europe, Pantaleoni, a junior engineer, felt very uneasy rendering judgment for his formidable boss about the commercial feasibility of what he saw: "a fifty-mile circuit that lighted the exhibition buildings, the Turin railway station, and stations at Veneria

Reale and at Lanzo, a small village in the Savoy Alps."[11] So the young man sought the advice of the far more seasoned German electrical firm of Siemens and Halske. Years later Pantaleoni would relate, "Werner von Siemens, whom I had known, assured me there was nothing whatsoever in alternating current, that it was pure humbug."[12] But knowing how excited Westinghouse was, Pantaleoni sought other advice from a Budapest firm, the Ganz Company, which strongly advised the young man to pursue the Gaulard-Gibbs system, for they themselves were busy improving upon it (some might say infringing upon their patent). Westinghouse bought an option on the American patent rights and arranged for delivery of one of the duo's transformers, along with a Siemens generator designed to run arc lights on alternating current.

In the United States, as sultry spring turned to summer, the whole nation was riveted by the courageous and highly public dying of revered Civil War hero Ulysses S. Grant. Stricken with a fatal throat cancer, the once mighty warrior was confined to his sickbed at the family's East 66th Street brownstone in New York. There he labored determinedly through June and then July's suffocating heat to complete his memoir. A man of probity himself, Grant had served two inglorious terms as a president whose administration had been marred by corruption. Then the ex-president found himself unwittingly enmeshed in a mortifying $14 million Wall Street banking swindle, only to be bailed out by the kindness of the nation's richest man, William H. Vanderbilt. Grant hoped his memoir would sell enough to erase his humiliating debts. His dignity during his slow and painful decline, and the ex-president's determination to salvage his honor, to assert the importance of simple honesty, had restored the general to the nation's deepest affections. As the July heat pressed in, the Grant family sought out the cool mountain air near upstate Saratoga, where the ever frailer soldier, wrapped in shawls, painstakingly revised his page proofs with the help of his editor and publisher, Samuel Clemens. On July 23, within days of finishing, Grant died peacefully. In New York City, the morning air reverberated with a solemn clangor of hundreds of church bells, punctuated by the newsboys yelling: "Extra! Extra!" Black-bordered editions featured page after page sanctifying the humble beginnings and heroic deeds of the great general of the Grand Army of the Republic, the man who had saved the Union. Within hours, the city's thousands of American flags were at half-mast and buildings draped in deepest black. Like Edison and Westinghouse, Grant was a small-town boy who had risen to great deeds from very ordinary circumstances.

Grant's New York funeral was an epic panoply of public grief, such as would be forever remembered by all who were present. On the blue, warm dawn of August 8, like an incoming tide, a vast sea of one and a half million Americans of every rank and class flowed in, engulfing the six-mile funeral route from Manhattan's City Hall all the way north to Riverside Park, where Grant would be buried. All were there to honor Grant and pay homage to the nation's very survival. "By 9 o'clock every balcony, window, and door commanding a view of the line of march was teeming; the roofs and cornices swarmed," reported *The New York Times.* "There was not an accessible point, however high and dangerous, but had its observer: men climbed the telegraph poles and clung to the wires; boys were high in the trees; carriages and wagons thronged the crossings where the police would allow them . . . the statues in the squares were black with climbers, and even the lamp post granted many a foothold."[13] At 10:00 A.M., as the air began to shimmer with heat, Major General Winfield Scott Hancock cantered forth on his magnificent black charger, followed by his staff. The funeral cortege set out on its slow and somber procession north. As far as the eye could see, behind General Hancock flowed a mighty river of military display: "massed regiments in their brilliant uniforms, their guns glistening in the sun, their colors draped, and their slow steps keeping time to the music of the many dirges for the dead." Hour after hour, as the heat intensified, the columns of soldiers, sailors, and marines swept by in all their finest military precision. Then followed volunteers and militia from many states, even the old Confederacy, marching, marching in columns to band music or muffled drums. And, finally, came the remains of the great dead hero Ulysses S. Grant, his purple-draped catafalque solemn and humble atop a black wagon drawn by twenty-four horses, each majestically caparisoned with purple, each horse led by a young black man in broadcloth and silk hat.

All through the silent, respectful sea of mourners, men and boys swept off their hats, while women dabbed their eyes and tears trickled silently down. Next rumbled past an elegant multitude of fine carriages, ferrying President Grover Cleveland and a wide array of the nation's political and diplomatic luminaries. The great state funeral had lasted four hours, but only at the very end came the most poignant of all the military sights: the aging, marching veterans of the Civil War—eighteen thousand strong—who had fought a full two decades ago, whether under Grant or Lee. On this day they were one. The crowds were shrouded in a respectful silence rare in that cacophonous city. Grant's funeral served as yet another reminder that the old, rural, preindustrial America was passing. The North with its railroads and superior communications had prevailed. No longer were great landowners the most wealthy and powerful. Now it was the eastern financiers and industrialists who were ascendant, the men who ruled the nation. Grant's classic memoir became a runaway bestseller, salvaging his honor and his widow's finances.

So when Pantaleoni sailed back into Manhattan from the Continent, he returned to a nation just out of mourning. Not long behind him on another ocean steamer followed Reginald Belfield, an English employee of Gaulard-Gibbs who was bringing with him their "secondary transformers." Wrote Belfield, "I arrived in America on the 22nd of November 1885, where I immediately saw Mr. George Westinghouse and Mr. Herman Westinghouse."[14] One can safely assume that once ensconced in Pittsburgh, Belfield found his first encounter with the gloomy industrial boomtown as astounding as did all other newcomers. Located in the far western hills of Pennsylvania, where the swift-flowing Allegheny and the Monongahela Rivers merged to form the broad and busy Ohio River, Pittsburgh had the advantage of being a pivotal river port. Pittsburgh also served as a major terminus for the powerful Pennsylvania Railroad (whose arrogance toward the locals had led the 1877 labor rioters to burn and loot the road's depot, yards, and trains). Equally important, the surrounding hills and valleys contained some of the world's richest coalfields, which cheaply fed the city's hundreds of filthy, belching factories and mills. When English writer Anthony Trollope passed through hardworking Pittsburgh in 1862, he found it "the blackest place . . . I ever saw." Six years later, about the same time that George Westinghouse relocated his works to Pittsburgh because all that coal made casting steel frogs cheap, author James Parton decried a city so ugly and grimy that "every street appears to end in a huge black cloud . . . [it was] smoke, smoke, smoke—everywhere smoke." In short, it was "Hell with the lid taken off."[15]

In 1882, a very proud Andrew Carnegie brought his idol, Darwinist Herbert Spencer, to view what the steelman cherished as Pittsburgh's utopian might. Spencer's verdict as his train steamed away from the stinky black haze: "Six months residence here would justify suicide."[16] An observer arriving in the same year as Reginald Belfield wrote, "Pittsburg [as it was spelled from 1890 to 1911] of to-day looks in the distance like a huge volcano, continually belching forth smoke and flames. By day a great pall rests over the city, obscuring the sun, and by night the glow and flash of the almost numberless iron-mills which fill the valley and cover the hill-sides, light the sky with a fiery glare. This great workshop of the modern Cyclops is one of the most important manufacturing centers of the country, and embodies our most valuable interests in iron and steel manufactures. . . . Though the suburbs of the city are beautiful and contain many charming residences, the aspect of the city itself is grimy and gloomy, in spite of the noble business blocks and open, spacious streets."[17]

Those who could afford it, like the Westinghouses, lived amid the flowering hedgerows and dappled lawns of suburban Homewood, six miles east of downtown on the Pennsylvania Railroad line. And it was there that Belfield was taken on that first day in late November 1885 to live as a guest of the family. The white brick Westinghouse villa, George's surprise birthday present to Marguerite in 1871, featured a square tower, a fashionable mansard roof, and twenty acres of surrounding gardens and lawns. It also had its own railway siding for Westinghouse's private railcar. But when Belfield came that late fall, the summer flowers had long since fallen before the first frosts. The villa was misnamed Solitude, for the Westinghouses loved to entertain and always had company around. "Hospitality," wrote one Westinghouse biographer of George, "was his greatest diversion and in this he was ably assisted by Mrs. Westinghouse. It was their normal life to have several guests in the house, and to have a dinner-party every night. The varied company included distinguished men of many lands."[18]

A famously genial and charming host, Westinghouse would frequently call his wife during the day from the office to say he would be bringing two, four, or even ten home to dine. The usual guests were business associates and their wives, but those Pittsburgh locals often arrived to find a distinguished sprinkling of visiting scientists, railroad executives, and foreign notables who had happened to be in town touring the big Westinghouse works. Marguerite, reported the local *Social Mirror,* "lives in greater style, entertains more splendidly, and wears more gorgeous, varied, elegant toilets, has more and finer diamonds than any woman in Pittsburgh."[19] The table always glittered with beautiful Sèvres and Dresden china, crystal, and solid silver and gold silverware, but the food was very simple and healthful. Westinghouse always made his own salad dressing and heard and said nothing while concentrating completely on this private dining ritual. If the dinner discussion became technical, Isaac Watson, the black butler, would discreetly hand his employer a pencil and pad. When Westinghouse had made whatever notes and sketches he wished, Watson would retrieve it and put it on a table in the library. Westinghouse might repair later with a few others there or to his billiards room, where papers could be spread out on the green baize, diagrams studied, and plans drawn.

Reginald Belfield, when he stepped off onto the Pennsylvania Railroad stop a few hundred yards from Solitude, may have noticed a towering wooden structure back behind the lovely villa. It was, strangely enough, a large natural gas well. Westinghouse had become interested in natural gas the previous year and had sunk an initial well, but as Pantaleoni would later recount, it was disparaged by the president of the local gas company as nice enough but *little.*

"If there was a thing that would cause Mr. W. to fight it was the thought that anything he was in was small and insignificant. Within a day or two a new rig was put in . . . with a terrific roar the tools were thrown out and smashed the hot house, as the most phenomenal well of Natural Gas was 'in.' For days you could not hear your own voice in the Homewood residence. . . . [Westinghouse] alone had a never disappearing smile on his beaming face, while the Fire Department was pumping water over the house to prevent it from catching fire from the burning gas. The Philadelphia Company (natural gas) was born."[20]

Just as Westinghouse was getting enthralled with electricity, he had started this new natural gas venture and invented an entire delivery system (for which he secured thirty-six patents). This patented system depended on the natural high pressure of the released gas pushing through narrow pipes for several miles, whereupon it was lowered to a pressure safe for use in homes and factories by the gradual widening of the pipes as the gas flowed into the factory or house. And so, historian Steven W. Usselman points out, one sees a certain theme emerging in Westinghouse's early enterprises: "First, they involved transmission over distance." With air brakes, compressed air was transmitted; with railroad signals, electricity; and with natural gas, the gas itself through pipes. But Usselman's second observation is, "Many of the technologies that George Westinghouse pursued involved a crucial linking mechanism that served to connect the long-distance transmission lines with the rest of the system. Often these devices incorporated feedback mechanisms that regulated the system."[21] It was little wonder that Westinghouse was finding himself drawn to the field of electrical power, for it meshed with his longtime inclinations and experience.

However, even as Westinghouse eagerly awaited the arrival of Belfield and the Gaulard-Gibbs apparatus in the fall of 1885, his leading electrician, William Stanley, had become unwell (perhaps the unhealthful Pittsburgh smog?). He had expressed discontent even back in the spring, having written his new wife, "I think that I will surely get up and leave Pittsburgh for good."[22] Stanley had been through many jobs and situations before coming to Westinghouse, and his early employer, the inventor Hiram Maxim, had said that Stanley may have been "very tall and thin, but what he lacked in bulk he made up in speed. Nothing went fast enough for him."[23] Now, once more Stanley was chafing. He preferred inventing on his own. By summer, Stanley had decided to relocate with his wife to the far more healthful and bucolic Berkshire Hills of Massachusetts, where he had vacationed as a child with his grandparents. He would, however, continue to work for Westinghouse. "My health gave out," Stanley would say later, "and there seemed to be grave questions as to my ability to withstand Pittsburgh and its work."[24]

But now, at last, in the final days of November, Reginald Belfield had arrived and gotten right down to business at Westinghouse's grimy Garrison Alley Works in Pittsburgh, there unpacking all that he had brought from its wooden crate. It was not an encouraging sight. "The Gaulard and Gibbs apparatus from England [the prototransformer] was sent in a very unsatisfactory condition," he would later write. "The work in the various parts was very faulty; portions not properly soldered only held together by the flux. The impression this gave was so bad that Mr. Pantaleoni was going to telegraph to London canceling the arrangement that had been made. Here Mr. Westinghouse stepped in and with his well-known sympathy towards a new invention, gave me another chance, for which I was most grateful, so that I had to practically reconstruct the entire apparatus and build most of it anew. This made a great deal of hard work but the result was satisfactory, inasmuch as Mr. Westinghouse determined to take up the apparatus. During this period I was staying at Solitude, and had occasion every night to discuss the matter with him, and the future development of the system."[25]

The first transformer had actually been created by Michael Faraday, just as the first generator had. In a transformer, the "primary" is the copper wire that is fed electric current, generating a magnetic field. The "secondary" is the copper wire that intercepts the primary's field of force. The two interact, self-inducing voltage. In an effective transformer, the two coils of copper wire must be coupled perfectly to create high self-induction. However, no one really understood the full potential of Faraday's prototransformer back in 1831, any more than most electricians did in 1885. Faraday had coiled many turns of an insulated wire around a bar of soft iron and then taken a second insulated wire and coiled that around the first. When alternating current was sent through the first coil, the second coil would also begin to generate current. The voltage of the insulated wire was purely a function of how many times it was coiled around the iron bar. The greater the number of wrapped coils, the higher the voltage would rise. Send that same alternating current into another iron bar with fewer coils wrapped around it and the voltage would diminish. A very rudimentary transformer. William Stanley said later in life, "I have a very personal affection for a transformer. It is such a complete and simple solution for a difficult problem."[26] So simple, few appreciated it. Westinghouse and his engineers did. "It is to be remembered," Stanley said later, "that at this time, that is, in 1885, there were no alternating current machines built in America. [There were European imports.] The only transformers or induction coils that I knew of were three or four or more of the Gaulard type that had been imported from England."[27] Moreover, transformers worked *only* with alternating current—for the swift oscillations of AC electrons created a changing magnetic field around them as they coursed along that was *the* key to the transformer's induction. If useful electricity was to travel far distances, that electricity could only be AC.

The Gaulard-Gibbs transformer that Reginald Belfield rebuilt at the Westinghouse plant in late November, he explained, "consisted of a bundle of iron wire forming the magnetic circuit surrounded by a large number of copper discs, having a hole in their centers, one to each turn of both primary and secondary, and each soldered to its neighbor, this multitude of soldered joints, each a source of trouble, was most impractical and very expensive to make." Historian Harold C. Passer notes that an English technical journal of the time dismissed Gaulard-Gibbs and their transformer because more famous electricians who had worked in the field already would not "have allowed this subject to dwindle away to mere nothingness had there been a chance of bringing the matter to a successful issue."[28] The two inventors themselves did not really understand the potential of their system, for they were using the transformer merely to step down voltages to individual light bulbs.

George Westinghouse, as was his wont, thoroughly enjoyed examining the whole Gaulard-Gibbs apparatus and then disassembling it. He discussed it at great length with Reginald Belfield all during December at Solitude. Belfield later recalled, "Those who knew Mr. Westinghouse will fully realize the great energy he threw into this question. . . . In an astonishingly short time [by the end of the year] the absolutely uncommercial [Gaulard-Gibbs] secondary generator was converted into the modern transformer."[29] A whole new design emerged that could be cheaply produced by machines: H-shaped iron plates that could be machine stamped now formed the core. The horizontal part of the H would pass through copper wire coils that could be machine wound and would serve as the primary and secondary. The ends were closed by means of I-shaped plates. Voilà! George Westinghouse had developed the modern transformer. His greatest innovation was organizing the transformers in parallel, with the number of parallels allowing great increases in voltage. Nonetheless, Pantaleoni reports that his fellow electricians at Westinghouse were still firmly against high-voltage alternating current and using the new transformer to increase voltages as they left the generator and to step it back down as it entered buildings. "The opposition by ALL the electric part of the Westinghouse organization was such that it was only Mr. George Westinghouse's personal will that put it through."[30] William Stanley, who was busy installing his first Westinghouse DC system that fall, seems to have been the only other enthusiast for AC aside from his boss. No one besides Westinghouse understood the tremendous breakthrough represented by the AC transformer: a machine that could take high voltages that had traveled long distances and step them down for safe use in entire factories or homes.

Undeterred by the chorus of naysayers at his own company, Westinghouse renegotiated William Stanley's contract, and Stanley now prepared his new Great Barrington laboratory on Main Street for AC work. Reginald Belfield would come north to the snow-covered rolling hills to assist. In preparation for the wintry cold, Belfield bought an "American ready-made cloth coat" at Kaufman's

Department Store in Pittsburgh just before Christmas. The coat came with a complimentary Waterbury watch. When Westinghouse saw it, the ever curious magnate had to take it apart at once. Recalled Belfield, "Mr. Westinghouse was not satisfied with taking the watch to pieces once, but he took it to pieces many times and put it together again, and it speaks volumes for his mechanical ingenuity that the watch never suffered for the treatment."[31]

Now that George Westinghouse had his transformer, there was the real and thrilling prospect of leading an electrical revolution, something far beyond direct current incandescence. On January 8, 1886, Westinghouse entered the fray officially, incorporating his fifth company, the Westinghouse Electric Company, with stock worth $1 million. The Pittsburgh magnate assumed the presidency, initially holding 18,000 of the 20,000 shares valued at $50 each. In the coming months, Westinghouse sold almost 8,400 shares to finance his new venture. William Stanley received 2,000 shares of Westinghouse Electric stock, a salary of $4,000 a year, and $600 a month for laboratory expenses.[32] Any Stanley inventions developed by Westinghouse for commercial use were the company's. The next major move was to dispatch Guido Pantaleoni and a Westinghouse attorney to Europe to buy the American patent rights to the Gaulard-Gibbs transformer for $50,000. Pantaleoni would find that others in Europe were hot on the same technical trail.

By early January of 1886, Belfield was in Great Barrington, busy constructing H-shaped transformers in Stanley's old barn. Once these pioneering commercial transformers were ready, Belfield (presumably in his warm Kaufman's coat) braved the February cold to string four thousand feet of heavy copper electrical wire onto ceramic insulators fastened to the huge bare elms of Barrington's Main Street. Concerned about competitors seeing the six transformers, Stanley encased each one in a wooden box installed out of sight in the basements of those buildings to be lighted. A steam engine was readied to provide power to the generator. Stanley then left town for a fortnight's vacation and rest.

Upon his return in early March, Stanley was chagrined to find that the Edison people had gotten the jump on him, becoming the first to dazzle Great Barrington with the miracle of incandescent light. The Edison light was showcased in a local mansion, where the company had installed a conventional direct current isolated plant. On March 10, 1886, the *Berkshire Courier* reported breathlessly on the Edison light in an article headlined A BRILLIANT SPECTACLE. The story noted, "During the last few evenings the Hopkins premises have been brilliantly illuminated upon the balconies and piazzas of the homestead, while the interior of the new home was brilliantly lighted, as were the surrounding grounds." In a forthright commentary on the very real fear of this brand-new technology, the newspaper also reported, "A fire engine . . . stands ready to promptly extinguish any fires that may break out upon the works."[33]

A week later William Stanley, twenty-eight, fired up the coal-powered 25-horsepower steam engine in the old barn by his Main Street laboratory. As the Siemens AC generator came to life, it began transmitting 500 volts of AC electricity out on the strung copper wires, which hummed along under the stately elms and into the basement of the store owned by his cousin R. I. Taylor, there to be stepped down through the transformer to 100 volts, which ran up through the interior wires to the incandescent bulbs. On March 17, the *Berkshire Courier* reported, "Last evening the interior and exterior of R. I. Taylor's store was lighted by three 150 candle power electric lights of the Stanley system. Two of the lights in the store made it as light as noon-day. A large number of businessmen were present to witness the effect, and were unanimous in their praise thereof."[34] On that same day, Stanley reported back to Westinghouse in a letter that he had run a successful test: "All the converters [transformers] are under lock and key so that no one knows anything about them. . . . I might say a great deal about the system, but briefly, it is all right."[35] Within the week Stanley had also connected up a drugstore, another general store, and a doctor's office. Hundreds of local citizens came out for a gala Saturday night viewing, strolling in the brisk March evening under the barely budding elms to see for themselves the much ballyhooed incandescent light.

Said Stanley, "My townspeople, though very skeptical as to the dangers to be encountered when going near the lights, rejoiced with me." The newspaper marveled how inside the stores the electric lights were "so powerful and so perfectly white, that green and blue can be readily distinguished though they cannot by gas light."[36] By the end of the month Stanley had several dozen new customers, including a handful of local doctors, the billiards parlor, the post office, a stove store, a shoe store, and L. B. Brusie's restaurant. The Edison people were not sitting back, but they still had only half a dozen customers. They were limited by distance, as Stanley was not. Rarely had such a major technological breakthrough—one with such vast implications for the future of electricity—been introduced so quietly and discreetly.

Stanley, of course, had stayed absolutely mum on the nature of the new AC system, so the local newspaper had made no distinction between his enterprise and the rival installations by the Edison people. Only a handful of people in America could appreciate the monumental nature of this pathbreaking new lighting system, among them the Pittsburgh visitors who arrived by train on April 6. George Westinghouse had eagerly traveled over from his factories with his brother Henry, along with Pantaleoni and inventor-engineer Franklin Pope, to see firsthand William Stanley and Reginald Belfield's great electrical achievement. Here, for the historic first time in America, high-voltage electricity had been generated, transmitted, and then

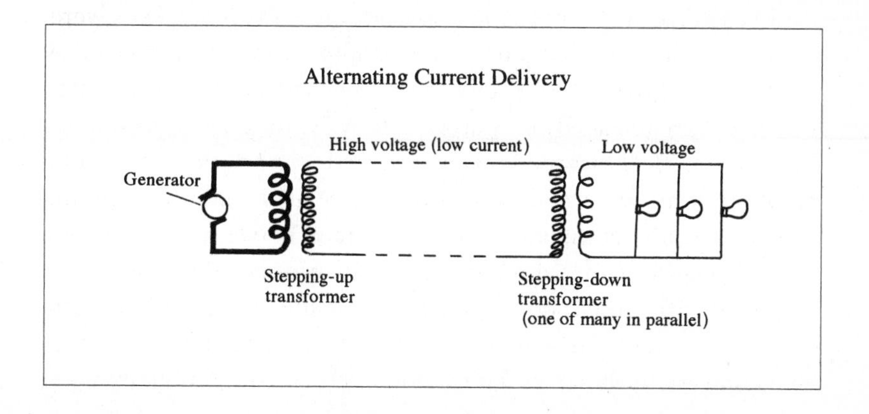

stepped down via newly designed transformers to a safe level for domestic consumption. No longer would electric central power stations, with all their coal and noise and smoke, have to be situated in walking distance of urban customers. Generating stations could be located far out of cities, nearer to their fuel sources, and their electricity dispatched quietly in great wire grids, humming across fields and rivers to deliver clean, quiet power. Several weeks later, the whole Westinghouse group returned yet again to admire and study the further expanded AC system at work. Stanley had designed a more reliable generator to replace the Siemens, which was really meant to run arc lights. Now every aspect of this new system had to be rigorously tested and improved, preparatory to invading the market and changing the world. Westinghouse directed that a new version of Stanley's AC system be installed back in Pittsburgh at the Union Switch and Signal Company and an electrical line strung three miles out to East Liberty. Transformers were established at each end, and throughout the summer the system was run and tested repeatedly. By fall, Westinghouse was ready to begin seeking his own place in the new electrical firmament, as he offered the world its first commercial AC incandescent system. Unlike Edison, who launched his every new invention with maximum public ballyhoo, George Westinghouse introduced *his* lighting revolution very quietly. No press account heralded the tremendous breakthrough of AC, the bursting of the DC shackles. The AC revolution began quietly and almost invisibly.

George Westinghouse's inaugural customer in his low-key electrical revolution was the Buffalo emporium of Adam, Meldrum & Anderson, a gigantic four-story Italianate palace of manifold consumer delights located on that upstate New York city's prestigious Main Street. Buffalo was growing rich as the transit point for immigrants flooding west on the railroads and for all manner of crops and goods—notably bumper midwestern grain harvests and gigantic herds of livestock—heading back east. The broad Buffalo waterfront was lined with monumental grain elevators, and there the mountainous piles of grain were bought, priced, and stored before being funneled onward to New York City and markets abroad. A wealthy Great

Lakes port, terminus of the Erie Canal, and national railway crossroads, Buffalo was poised for commercial glory.

Right after Thanksgiving, a front-page ad in the November 27, 1886, *Buffalo Commercial Advertiser* proclaimed that Adam, Meldrum & Anderson (despite month-long problems and delays) was now showcasing its incomparable selection of black cashmeres, carpets, draperies, bed and horse blankets, cloaks and shawls, dolls, and so forth with 498 Stanley lights run by the Westinghouse system. "There is no odor, no heat, no matches, no danger. We were the first business house in the city to adopt the plan of lighting our stores by incandescence. . . . The appearance is brilliant in the extreme. The light is steady and colorless. Shades can be perfectly matched. Come and see the grandest invention of the nineteenth century."[37] Two evenings later, on a Monday, the mammoth store was open—not to retail any of its lovely lace handkerchiefs, gloves, silk umbrellas, finest black silks, or eiderdown quilts, but purely to show off the Westinghouse lights. "No goods were sold," reported the *Commercial Advertiser,* "and the store was so thronged with visitors that it was difficult to get about." The well-dressed crowds streamed up and down the four floors to ooh and aah over the lights, exclaiming how akin it was to pure sunlight, how truly you could see the colors in the India shawls and the weave in the draperies. The Adam, Meldrum & Anderson store was just one of numerous spaces in Buffalo now luminous with electricity, including arc lights and rival incandescent plants. Consequently, "The store was visited last evening by quite a number of gentlemen interested in the different lighting companies of the city, and others, and all expressed great admiration at this Westinghouse light." Again, as in Great Barrington, the newspapers did not mention (or presumably know) that this new system marked a profound and momentous change in the commercial delivery of electricity. Nor did Westinghouse make any great public declarations. But AC's virtues were swiftly appreciated by those who could not be served by DC. Westinghouse soon had orders from twenty-seven new customers in numerous locales.

Back in Manhattan at 65 Fifth Avenue, Thomas Edison smoked his cigar and stewed, deeply incensed to learn that the Westinghouse Electric Company was invading the field of incandescent lighting, turf pioneered by Edison. The Brush Electric Company specialized in arc lights, but Edison disdained that product as an interim solution doomed to extinction. Edward Weston had developed an incandescent light bulb, but as far as Edison was concerned, Weston had simply stolen *his* light bulb and come up with a workable but inferior DC system. He was no great threat. Thomson-Houston was just another patent pirate, though they certainly hustled. Edison's patent lawyers were after them in the courts. But Westinghouse was altogether another matter, a formidable rival with immense achievements and access to major capital. He was not a man to scoff at, deride, or dismiss. The Pittsburgh industrialist was reputed to be a real fighter who, once decided, pursued a course of action full bore. And while Westinghouse flourished in the golden age of the robber barons, he very definitely was not one himself. In an era of flagrant venality and corruption, he remained a tremendous idealist and genuine democrat, completely dedicated to making the world a better place through engineering and machines. Whatever new project he launched, he was looking to be the best. Like Edison, he thoroughly enjoyed working in the noisy, dirty shops among his men, inquiring about the state of their projects, and infusing everyone with his own zest.

Understandably, then, Edison was concerned when he heard that a man like George Westinghouse was scaling his, Thomas Edison's, electrical Olympus. Moreover, the Pittsburgh magnate was entering the electrical arena equipped with something new, something potentially revolutionary. Edison was feeling deeply bitter about this new competition. He wrote darkly to Johnson: "Just as certain as death Westinghouse will kill a customer within six months after he puts in a [AC] system of any size. He has got a new thing and it will require a great deal of experimenting to get it working practically. It will never be free from danger." Edison was utterly perplexed by how the whole AC business could be practical, writing in the same report, "I cannot for the life of me see how alternating current high-pressure mains—which in large cities can never stop—could be repaired. Whereas the main of the direct current would not produce death if a man received a shock."[38] Here were the first angry rumblings in the coming hostilities that would come to be known as the War of the Electric Currents, a war between two titans whose technology was so potent, it was already changing the world.

CHAPTER 6

Edison Declares War

As most of Manhattan slept on Monday, March 12, 1888, a howling rainstorm engulfed the city. Its freezing downpour inflicted sodden misery on the armies of the poor huddled atop steam grates and curled up in the nooks and crannies of piers and alleys. The deluge coarsened into sleet, then thickened into driving snow. All night, the storm roared on. By dawn, its sixty-mile-an-hour winds were shoving the deep snow into drifts as high as second-story windows. When middle-class New Yorkers arose Tuesday morning, they were indignant to find they had no morning newspapers at their snowbound doors, no fresh milk from the milkman, no hot rolls from the baker. Instead, there was only an icy tornado of white screeching up the avenues. The temperature was a bone-numbing five degrees. Everywhere the great spiderwebs of electric wires sagged and collapsed, broken by the weight of the ice.

Reported *The New York Times,* "Before the day had well advanced, every horse car and elevated railroad train in this city had stopped running; the streets were almost impassable to men or horses by reason of the huge masses of drifting snow; the electric wires—telegraph and telephone—connecting spots in this city or opening communication with places outside were nearly all broken; hardly a train was sent out from the city or came into it during the

New Street looking to Wall Street, New York City, during the blizzard of 1888 entire day; the mails were stopped, and every variety of business dependent on motion or locomotion was stopped.... Probably if it had not been for the blizzard the people of this city might have gone on for an indefinite time enduring the nuisance of electric wires dangling from poles." The newspaper once again urged the electric wires all be buried, citing the very specific hazards: "The city is liable to be put into darkness and the consequent perils. There is also the danger of conflagration through the failure of the fire alarm wires."[1] Nowhere was mentioned the possibility of people being electrocuted. The blizzard of 1888, with its twenty-two inches, stood for sixty years as the city's worst snowstorm.

It was just about a month later that the great public fear of electrical "death by wire" began. On the cold, clear Saturday night of April 15, 1888, a youth in high spirits was heading along East Broadway toward Catherine Street, skipping and leaping. The snow was long gone, and most of the thick webs of electric wires had not been buried but were once again restrung on the forests of cedar poles. That crisp April evening, carriages were rolling by, their shiny cabs and sleek horses gleaming in the bright street arc lights. Passersby noticed the bounding boy seize one of the broken telegraph wires dangling down from high above. Wire in hand, he playfully skipped around and around the towering telegraph pole, when amid a spurting shower of sparks, the boy staggered back and fell reeling on the dirty sidewalk. A crowd quickly gathered and the ambulance corps soon galloped up to take the boy, later identified as Moses (or Meyer) Streiffer, to the Chambers Street Hospital. But he was dead.

The *New-York Daily Tribune* editorialized, "It is almost a pity that it wasn't a millionaire or other leading citizen that was killed by the electric light wire on Sunday morning. If it had been, the community would have been startled, and its indignation might have brought the wires underground. But it was only a poor boy peddler—a little fellow fifteen years old, a Rumanian, a stranger in this great city, selling collar buttons and pocket combs from a modest tray to help support his mother and eight brothers and sisters. A wire had been swinging for months from a pole near where the boy took his stand, he happened to touch it, gave 'a sort of a quack,' the policeman said, and was dead." The U.S. Illuminating Company, which ran its arc lights off high-voltage wires strung along the avenues, was charged with neglect for allowing dangerous wires to dangle.[2]

In the spring of 1888, "death by wire" became, for the first time, a great preoccupation of the New York papers. "Overhead cables suddenly leaped into prominence not only as eyesores but as a public peril," notes Westinghouse biographer Francis Leupp. "Leading newspapers which till then had confined their discussion to the expediency of exchanging gas for electricity, began, with astonishing unanimity, to make a display of every happening that could be used to excite animosity in the popular mind toward the [high-voltage] alternating current."[3] The newspapers' rising indignation mirrored that of Thomas Edison, long a darling of the New York press. Suddenly alert to every casualty and fatality caused by high-voltage electrical wires, the press now gave full attention to the scandalous, tangled state of Manhattan's overhead electrical wires and the dangers posed by high-voltage arc light wires.

Less than a month after the peddler boy Streiffer died, there was another electrical victim. On Friday, May 11, a lovely spring day, a Brush Electric Company worker was up cutting away old wires on a second-story cornice at 616-18 Broadway, high above the thick port traffic below on West Houston. An employee inside one of the buildings "saw smoke curling in the window and heard a spluttering sound. He found Murray dead and one of the electric wires partly cut through and the insulating material burning at the point where it was cut."[4] The lineman for the arc light company had failed to wear his heavy gloves and was electrocuted. Rescuers trying to pull the body in through the window were promptly shocked. Finally they obtained some rubber sheets and, after wrapping the corpse in those, pulled the poor charred fellow off the cornice and took him over to the Tenth Precinct Station House. For six years, arc lights and their wires had been viewed as an eyesore and a nuisance in Manhattan. But now, the citizenry would begin to associate high-voltage electricity with danger and death, just as did Thomas Edison.

Ever since George Westinghouse had turned on the lights in the Buffalo emporium of Adam, Meldrum & Anderson, Thomas Edison had been plagued by a steady stream of bad electrical news. In late 1886, Edison had jauntily written off his new rival, George Westinghouse, claiming, "None of his plans worries me in the least," albeit conceding that the "only thing that disturbs me is that Westinghouse is a great man for flooding the country with agents and travelers. He is ubiquitous and will form numerous companies before we know anything about it."[5] Indeed, with each passing month, Westinghouse was proving a more and more formidable threat to Edison dominance.

The hard-charging Westinghouse, after but one year in business, had constructed or had under contract 68 AC central stations. He had come from nowhere to loom as Edison's biggest competitor. Thomson-Houston, which had flourished primarily as an arc light company, had also taken up installing AC central stations just that spring, using Westinghouse transformers. *They* had already up or under contract 22 AC central stations. At the end of 1887, from the seat of his eight-year-old empire in Manhattan, Edison had built or had under contract 121 DC central stations in places as far-flung as Birmingham, Alabama, and Grand Rapids, Michigan. The Edison Electric Light Company's executives, in their 1887 annual report issued in late October, put a brave face on the AC situation, denouncing this electrical alternative as "from a commercial standpoint, having no merit in itself and, being of high pressure . . . notoriously destructive of both life and property."[6]

Edison was especially galled that Westinghouse had installed a big AC central station in New Orleans. Edison already had a plant in that steamy Louisiana port and viewed it as his exclusive incandescent turf. The 1887 Edison annual report chortled over the many Westinghouse woes there—constant breakdowns, transformers ruined by lightning, no AC motors for those who wanted to use their electricity for more than lighting. The report quoted W. T. Mottram, their Edison manager, as saying he was "thoroughly convinced that the Edison system is unassailable . . . they cannot compete with us or do us any permanent harm, and that a steady conservative policy will win the battle."[7] Brave words, these, and fine for the public, but meanwhile Edward Johnson was tartly rebuking Edison that without AC "we will do no small town business, or even much headway in cities of minor size."[8] What were the Edison salesmen to say to city fathers who needed electricity for homes and factories more than half a mile away from the DC central station?

Out in the far-flung marketplace of new central stations, the Edison camp was encountering continual and embittering setbacks. If a town bought a DC plant, it might well serve only half those who wanted electric lights. Those who lived more than half a mile away had to contemplate buying individual isolated plants or a whole other central station. In stark contrast, one AC station could serve the whole place and could be expanded as necessary to accommodate continued growth. So Thomas Edison's rancor toward Westinghouse continued to fester and grow with each passing month of 1887.

It was not that Edison did not have ready access to his own AC system. Even as Westinghouse was installing his first AC plant in late 1886, Edison's own people had been giving thorough consideration to buying a European-developed AC system that was causing a sensation on the Continent. Francis Jehl, a longtime Edison employee, had been in Europe on business in early 1885 when he was asked to travel to Budapest, with its famous castle and cafés, to look at the Ganz Company's new AC system. It was dubbed ZBD for its Hungarian inventors, Charles Zipernowsky, Otto Titus Blathy, and Max Deri. Jehl later recalled that in Budapest, "at the exhibition grounds I found 1,000 or more incandescent lights operating from an alternator running at a pressure of 1,300 volts. It was an 'eye opener,' for the pressure was reduced to that required by the lamps by means of poleless induction coils for which the Hungarian inventors called or coined the word transformers."[9] Jehl, who had once pumped the air out of the early incandescent bulbs at Menlo Park, described all the young European electricians as abuzz over AC.

The first ZBD customer had been none other than the Edison Company in Milan, run by John Lieb, former chief electrician of the Pearl Street Station. Lieb, who went down in history as the man who stood on his tiptoes to first pull the Pearl Street switch, had successfully launched the Milan branch. Wishing to light up a new customer, the Theater dal Verm, which was beyond the reach of the existing Edison DC mains, he had used the new ZBD system. So all through

1886 Lieb sang the wonders of AC, pleading with Edison to buy the ZBD patents. One of the ZBD inventors, Dr. Otto Titus Blathy of the Ganz Company, was equally anxious to hook up with Edison, for the Edison imprimatur would be the ultimate electrical accolade, a surefire guarantee of instant luminous glory and dominance. When Edison waffled and stalled, the good Hungarian doctor sailed to New York to personally lobby the great inventor, arriving in September of 1886. Edison still agreed to nothing, but he dispatched to Paris Francis Upton, who now ran the lamp company and had once supplied the math and physics back in the Menlo Park days. As Westinghouse had struggled to get its first AC system installed and functioning in the fancy Buffalo emporium, Upton had arrived in Paris, inspected the ZBD system, and strongly recommended the Edison Company buy an option on the American rights for $20,000.

When Edison had consulted Siemens and Halske of Berlin, *they* criticized the ZBD system as expensive, troublesome, and dangerous. They had, of course, given similar advice to Pantaleoni when he was considering the Gaulard and Gibbs apparatus for George Westinghouse. Pressed on all sides by his top men, Edison had reluctantly taken the option on the ZBD patent, well aware that Westinghouse was already launched. Yet he stubbornly and proudly refused even to consider using this AC system. Historians W. Bernard Carlson and A. J. Millard believe that Edison genuinely "feared that poorly designed and installed a.c. systems would impede the broad adoption of electric power."[10] One suspects a further cause—stubborn pride of authorship. Every aspect of the Edison DC system had been created from scratch by Edison or his colleagues. It is easy to imagine him obstinately balking at incorporating the inventions of others, especially if he could convince himself that their technology was dangerous.

The year 1887 was thoroughly trying to Edison. Not only was Westinghouse challenging Edison with his AC systems, the price of copper was also suddenly on the rise. This was worrisome for all electricians, for copper, with its peerless qualities of malleability and conductivity, was a major and critical component of their business. In December of 1887, the *Journal of Engineering and Mining* had explained the sharp rise—from ten cents a pound to sixteen cents a pound—as the work of "a powerful combination in Europe who have managed the whole business in a very skillful manner."[11] And indeed, across the wintry Atlantic in the business precincts of Paris could be found the mastermind of the copper corner: a bald-headed little executive named Hyacinth Secretan, who had built up the Société des Metaux into Europe's largest manufacturer of gleaming brass and copper goods. Closely attuned at all times to the copper market, Secretan traced the metal's every flutter and dip. In late 1886, when one of his tremendous orders for copper ingots had briefly elevated prices, Monsieur Secretan concluded the moment was propitious to quietly corner copper, drawing on lessons learned during a previous failed attempt to manipulate tin.

So the foppish Secretan began contracting with the principal copper mines of the world for their whole production (subject, of course, to restrictions) at thirteen cents a pound, a very nice profit for them. He then offered this copper on the world market at steadily escalating prices, hitting sixteen or seventeen cents a pound by late 1887, a very nice profit for him. Monsieur Secretan had little trouble signing up a host of European financial luminaries for his copper syndicate, including the Rothschilds, Credit Lyonnaise, and the Comptoir d'Escompt, France's second-largest bank. Seeing swirling around them the proliferation of the copper-reliant electric light, trolley cars, and the telegraph, all these financiers expected to ride the copper corner to even greater fortunes. The world's appetite for this metal could only grow.

No one knew better than Thomas Edison the critical importance of copper's cost. Not long after his fateful visit to William Wallace's hissing arc lights aroused a great ambition to "get ahead of the other fellows" and light the world with incandescent light, Edison had seen what all others had been blind to. Drastically reducing copper use was *the* key to creating a practical incandescent lighting system, one that could compete financially with the coal-gas illuminating American cities. And Edison had, through his genius, lowered the amount of copper needed to a tiny fraction of original estimates. But in the winter of 1887, the copper problem returned to harry Edison with a vengeance. As the *Electrical Engineer* noted by mid-February of 1888, "If the advance in the price of copper proves to be more than temporary in its effect, one of its incidental results will be to handicap seriously the low potential system of electrical distribution [Edison's DC], in their efforts to compete commercially with the high potential systems of more recent introduction [Westinghouse's AC]."[12]

By early March of 1888, a month later, the *Journal of Engineering and Mining* was reporting, "All the electrolitic copper in this country is now firmly in the grasp of the syndicate. There appears, in fact, nothing to prevent prices from being advanced to any figure the syndicate may wish."[13] This unfortunate and ominous turn of events was a real blow to the Edison Electric Light Company. For instance, in the spring of 1887, the company had been putting together a bid for a Minneapolis central station powering 21,700 lights. They estimated the feeders at 254,000 pounds of copper and the main at 51,680 pounds. At seventeen cents a pound, copper costs would total $51,965. Each one-cent rise in copper pushed costs up $3,056. A three-cent rise—for copper prices were escalating steadily—would add $9,000 to the almost $52,000 price tag for copper. In painful contrast, the new Westinghouse AC central plants required a third as much copper.

Just as he was being squeezed by Westinghouse on one side and the rising price of copper on the other, Edison was presented with an irresistible opportunity for wreaking some sub-rosa vengeance on his new enemy. In early November of 1887, Edison, America's most revered electrician, had received a beautifully penned letter from Dr. Alfred Southwick, a dour Buffalo dentist and one of three members of the New York State Death Commission. The commission's task was to find a civilized alternative to hanging for state prisoners condemned to execution. The committee chairman was the rich New York attorney and philanthropist Elbridge T. Gerry, best known for activism in the prevention of cruelty to animals. The third member was Matthew Hale, an Albany politician. After a series of repellently botched state hangings, New York governor David Hill had wondered if this "Dark Age" system might yield to something of a "less barbarous manner."[14] All the newspapers had described in revolting detail the state hangman's scandalous and repeated incompetence, whereby benighted criminals dangled in agony on too loose ropes, gasping until slow-motion strangulation finally silenced their death rattle. Or there was the equally gruesome but opposite problem: a noose so overtaut that it bloodily severed head from body before horrified official witnesses. In his letter to Edison, Dr. Southwick sought the inventor's opinion on using electricity for electrocuting condemned criminals and also wondered if he could suggest "the necessary strength of current to produce death with certainty in all cases and under all circumstances."[15] Edison wrote back to Dr. Southwick declining to get involved, saying he opposed capital punishment. That was in November.

But Dr. Southwick was as hard pushing as his hometown of Buffalo. He was quite adamant that clean and modern electrocution should prevail (having once seen a man keel over neatly dead in an electrical accident), and he wrote the nation's foremost electrician again in early December. Appealing to Edison's sense of civic duty, he pleaded, "Science and civilization demand some more humane method than the rope. The rope is a relic of barbarism and should be relegated to the past." These three commissioners of death—Dr. Southwick, politician Hale, and philanthropist Gerry—had made a thorough study of the history of lethality, conducting a survey of the state's judges, sheriffs, prosecutors, and physicians on the issue, and Dr. Southwick was pleased to report that eighty-seven of the two hundred responding favored electrocution. What the Buffalo dentist needed now was Edison and his enormous prestige "as an electrician" to persuade the legislature.

Tellingly, on December 9, 1887, Thomas Edison changed his mind and wrote Dr. Southwick again. But this time, the world's most famous electrical wizard was full of very definite and very damaging opinions. The quickest, most painless death, he asserted, "can be accomplished by the use of electricity, and the most suitable apparatus for the purpose is that class of dynamo-electric machine which employs intermittent currents. The most effective of those are known as 'alternating machines,' manufactured principally in this country by Geo. Westinghouse.... The passage of the current from these machines through the human body even by the slightest contacts, produces instantaneous death."[16] Was it the steadily rising price of copper? Had Westinghouse just won yet another central station contract from the Edison camp? Was it that the gas companies were proving harder to dislodge than anticipated, with their dropping prices and the much brighter gas mantle for their lights? We can only speculate. Whatever the reason for this hostile act, Edison's endorsement did the trick, and in mid-1888 the New York State Legislature would establish electrocution—much to the outrage of (the rest of) the electrical fraternity—as the new means of capital punishment starting January 1, 1889. Edison had just quietly (and secretly) planted something of a legislative land mine intended to damage his AC rivals.

By February of 1888, Thomas Edison was no longer content to vent his rancor with secret attacks. Using the vehicle of the Edison Electric Light Company, he lashed out publicly, issuing what surely stands as America's longest and most splenetic howl of corporate outrage. The eighty-four-page Edison diatribe, jacketed in angry scarlet and emblazoned with the title *WARNING!,* served as the official public salvo in one of the most unusual and caustic battles in American corporate history. Edison, with his DC system, was making his first open attack against Westinghouse and AC in the War of the Electric Currents. Thomas Edison, who had long (and reasonably) assumed that the electrical future was securely his—with all its glory and potential for riches—suddenly saw the famously tough, reckless, and industrially wealthy Westinghouse boldly swooping in from Pittsburgh to steal away his hard-earned prize. Edison would not sit back quietly and let what he saw as a dangerous system imperil not just his company, but the whole marvelous field of electricity.

What had triggered this furious verbal assault from the Wizard of Menlo Park? Why did he launch the War of the Electric Currents then? Edison, not the introspective sort, never did say. But we know that until 1885 Thomas Edison had been too busy (and fully confident of market dominance) to even bother suing the many lesser companies infringing on his 1879 light bulb patent. He had dismissed and largely ignored his competitors as shameless imitators, "patent pirates" who stole his ideas and inventions but who posed little genuine threat. But by 1885, as other companies began hurting his business, Edison had finally unmuzzled his top-dollar lawyers. Certainly part of his rising bitterness toward the Pittsburgh magnate was fueled by his anger at all infringers, for Westinghouse *was* among those making free use of an Edison-style bulb. Yet historian Harold Passer explains that Edison competitors "seriously questioned the validity of the Edison lamp patents. The United States company [controlled by Westinghouse], for example, considered its patent position much stronger than Edison's because it owned the incandescent-lamp patents of Farmer, Maxim, and Weston. Both Farmer and Maxim had worked in the incandescent lighting long before Edison."[17] Westinghouse, ever a fighter, had further goaded Edison by filing a counter light bulb patent suit. Thus, however much the Edison attorneys might huff and puff in the courts, there was still at this point, says Passer, "reasonable doubt that the basic Edison [light bulb] patent would be sustained. It is probable that few manufacturers and users of incandescent lamps considered it a serious business risk to make and use these lamps without permission from the Edison company."[18]

But angry as Edison was about light bulb infringers, George Westinghouse's biggest fault was daring to trespass at all on Edison's electrical terrain. When Thomas Edison first heard that the Pittsburgh industrialist was eyeing electricity, he famously snapped, "Tell him to stick to air brakes."[19] Not only had Westinghouse defied Edison, he was selling something original, a power system that was new, not just a second-rate copy of the Edison system. While the first half of *WARNING!* was dedicated to excoriating the light bulb infringers, much of the second half assailed Westinghouse. The whole AC system was "the most uneconomical yet offered to the public," insisted the Edison people, once you factored in the greater efficiency of DC generators, the reliable track record of the more tested DC systems, the lack of a meter to measure AC use, and the absence of any AC motor.[20] The DC motor remained Edison's great trump card. Those who preferred AC were dismissed as "Cheap Johns" and "the Apostles of Parsimony," shysters foisting inferior equipment on the unsuspecting.[21] Edison had no interest in acknowledging the great strength of the AC system—its ability to serve large areas and expand as needed.

But Edison reserved his greatest fury for AC's sheer dangers. Decreed the *WARNING!* booklet, "It is a matter of fact that any system employing high pressure, i.e. 500 to 2,000 units [volts], jeopardizes life."[22] The Edison people warned that if a transformer failed to step down the current, the whole building served would be a possible death chamber reverberating with high-voltage electricity. Thomas Edison had always prided himself on his system's safety: "There is no danger to life, health, or person, in the current generated by any of the Edison dynamos" and "the wires at any part of the system, and even the poles of the generator itself, may be grasped by the naked hand without the slightest effect."[23] No other electrical company had invested such time and energy in devising safe insulation for its wires and careful placement in the ground far away from the public as had Edison. In contrast, Edison detailed in his *WARNING!* pamphlet the gruesome deaths by high-voltage AC of numerous AC workers. DC was a gentle, friendly current. AC was a stone killer. Edison suggested the AC people were criminally indifferent to safety just to save a buck and get ahead.

Edison's rancorous corporate diatribe culminated by rallying the electrical troops to rise up against the infidels of AC: "All electricians who believe in the future of electricity ought to unite in a war of extermination against cheapness in applied electricity, wherever they see that it involves inefficiency and danger." Edison humbly volunteered to serve as the moral compass in this holy war. Those, like George Westinghouse, who dared to forge ahead on an alternate and alternating path, however shimmering and promising, were now the official enemy, the besmirchers of the sacred ways. George Westinghouse is, they sneered, "the inventor of the vaunted system of distribution which is to-day recognized by every thoroughly-read electrician as only an *ignis fatuus,* in following which the Pittsburg company have at every step sunk deeper in the quagmire of disappointment."[24] The first open, public shots had now been fired toward Pittsburgh in the War of the Electric Currents.

In the spring of 1888, even as the papers began to track every instance of "electrocution by wire" and the Edison Electric Light Company had made known its ire with AC, the small world of New York electricians began to buzz with rumors about the former Edison man Nikola Tesla. He was said to have reappeared and to be up to something big down on Liberty Street, filing a steady stream of patents related to an AC system. And indeed Nikola Tesla *had* been wonderfully productive, churning out one new AC machine after another. To handle all the work, he had summoned from Europe his old school chum and fellow engineer Anthony Szigety, who had sailed into the choppy waters of New York Harbor on May 10, 1887, to the rousing sight of Auguste Bartholdi's long-aborning Statue of Liberty. For six months now, the monumental goddess had been gleaming forth through each gray dawn, her torch of enlightenment radiantly lit from within by electricity and her bronze robes and grave face aglow from the thousands of candlepower at her feet. This luminous figure of welcome had been dedicated the previous winter with festive fanfare, after working-class readers of the *New York World* had sent in their pennies and dimes, showing the nation that ordinary people—not millionaires—would be the ones to finally erect the majestic sculpture on her Bedloe's Island pedestal. And so, emerging from Castle Garden to see the greensward of Battery Park, crisscrossed by shaded gravel paths and so wonderfully cooled by the harbor breezes, Szigety was reunited with his old friend Tesla. The two were soon putting in intense all-night hours constructing numerous variants on the AC induction motor Tesla had drawn for him in the sands of the Budapest park.

Years later, Tesla's first biographer, science writer John O'Neill, would recall how proud Tesla had been about the integrity of his vision. "When the machines were physically constructed not one of them failed to operate as he had anticipated.... Years had elapsed since he evolved the designs. In the meantime he had not committed a line to paper—yet he had remembered perfectly every last detail."[25] The rest of 1887 was a frenzy of creativity and secret construction as Tesla and his helpers turned out all the necessary components for three complete AC systems—single-phase alternating current, two-phase, and three-phase. He designed and built copper and iron models for each system—a dynamo (without the commutator!) that generated the electric current, an induction motor with its rotating magnetic core (again no commutators) to produce power, and then transformers to step up and step down that power. At the end of a whirlwind six months, Tesla had in his laboratory a whole system based on polyphase AC. On October 12, 1887, Tesla submitted an omnibus patent application, but the patent office requested it be broken down further. In November and December, Tesla filed for the first of what would eventually be forty patents covering the whole range of his AC system with its revolutionary induction motor. He was perfecting his AC system even as his old boss was denouncing AC to the world.

The ambitious young editor of *Electrical World,* Thomas Commerford Martin, a personable and ambitious bald English immigrant who sported a giant mustachio, stopped by Tesla's lab. He quickly grasped that this little-known but highly charming Serb was going to be the next electrical titan, a visionary whose radiant dreams rivaled Edison's. As a journalist, Martin savored the further drama that the unknown Tesla's electrical dreams clashed with those of the world-famous Edison—AC versus DC. Tesla, just thirty-one, was as much a true humanist as ever, seeking to ease the hard labor of the whole world with his spectacular induction motor and alternating current system. What AC had lacked up until now was a workable motor that it could power (although many a well-placed inventor was struggling to solve that puzzle and cover himself in glory). Now here was Nikola Tesla, a little-known electrician of minor accomplishment, seizing that prize.

Thomas Commerford Martin appreciated immediately the epochal nature of Tesla's AC motor and polyphase system and began considering how best he could shepherd this new genius to certain fame and millions of dollars. Martin was, fortunately, in a uniquely influential position. Not only was he the editor of the electrical field's top American journal, he was also the current president of the prestigious American Institute of Electrical Engineers (an organization all of four years old). He was therefore well versed in the feuds and cutthroat rivalries of the electrical universe and knew how best to introduce such a large and brilliant star into its heavens at such a delicate and stormy moment. Martin departed Tesla's Liberty Street lab in a great thrill of excitement, planning his campaign to launch Nikola Tesla. The English editor's first task was to get others equally enthralled with Tesla's system.

First, the AC polyphase machines needed to be tested and their revolutionary nature acknowledged by an outside expert of high standing. Martin arranged for Professor William Anthony, an eminent academic of electrical engineering at Cornell University, to come to Liberty Street and meet Tesla and his machines. Then machines were sent to him and several others for further testing. In March of 1888, Professor Anthony excitedly wrote a friend that he "was shown the machines under pledge of secrecy as applications were still in the Patent Office. . . . I have seen an armature weighing 12 pounds running at 3,000, when *one* of the (ac) circuits was suddenly reversed, reverse its rotation so suddenly I could hardly see what did it. In all this you understand there is *no* commutator. The armatures have no connection with anything outside. . . . It was a wonderful result to me . . . in the form of motor I first described, there is absolutely nothing like a commutator, the two (ac) chasing each other round the field do it all. There is nothing to wear except the two bearings."[26] So, just as Martin hoped, Tesla's name and riveting invention were filtering out among the people who mattered. Professor Anthony, a completely disinterested party, had judged Tesla's motors the equal in efficiency of existing direct current models.

Now Martin recommended to Tesla that he prepare a lecture to establish himself before the electrical world. Tesla demurred. When the first seven of Tesla's fourteen foundation patents were granted May 1, 1888, Martin again urged the Serbian inventor to make formally known to his electrical peers his magnificent breakthrough. Again Tesla excused himself politely, pleading exhaustion from the tremendous exertion of designing and constructing his whole complex system so swiftly. Professor Anthony then joined T. C. Martin in pressing Tesla to speak as soon as possible. Franklin Pope, editor of the *Electrical Engineer* and a Westinghouse engineer and patent attorney, had also been invited to Liberty Street, and he added his voice. Some years later, Martin would write that he "had great difficulty in inducing Mr. Tesla to give the Institute any paper at all. Mr. Tesla was overworked and ill, and manifested the greatest reluctance to an exhibition of his motors, but his objections were at last overcome. The paper was written the night previous to the meeting, in pencil, very hastily, and under the pressure just mentioned."[27] Martin brushed aside Tesla's worries about discussing aspects of his system for which he had not yet even filed patents. Nikola Tesla needed to establish his preeminence in this field. So on the cool Tuesday evening of May 15, Tesla traveled up to Columbia College on Madison and 47th Street, where the American Institute of Electrical Engineers was convening.

The meeting that evening commenced with various laudatory remarks celebrating Martin's energetic term as president. Then Nikola Tesla, tall, slender, his hair parted in the middle above a wide forehead, stood before the assembled electricians, a sea of men attired in high hats and dark frock coats, the crowd interspersed with the occasional interested lady. Tesla, with his high cheekbones, looked like a foreign, somewhat eccentric aristocrat in his preferred swallowtail coat. Speaking in his excellent but accented English, he first thanked his benefactors, Professor Anthony and Mr. Martin and Mr. Pope, men any ambitious electrician would wish as patrons. Then he excused his wan and weary appearance and the inadequacy of his presentation. "The notice," he said in his high voice, "was rather short, and I have not been able to treat the subject so extensively as I could have desired, my health not being in the best condition at present. I ask your kind indulgence, and I shall be very much gratified if the little I have done meets your approval."

Standing behind his shiny AC induction motors, Tesla began his talk, making his point by starting and stopping the machines and showing through drawings and diagrams also how they worked. "The subject which I have the pleasure of bringing to your notice is a novel system of electrical distribution and transmission of power by means of alternate currents, affording peculiar advantages, particularly in the way of motors, which I am confident will at once establish the superior adaptability of these currents to the transmission of power and will show that many results heretofore unattainable can be reached by their use; results which are very much desired in the practical operation of such systems, and which cannot be accomplished by means of continuous currents."[28] Tesla went on to tell all present that what they took quite for granted—the presence of commutators and brushes on existing motors to redirect the naturally produced alternating current into direct current when it entered the machine—would from here on in be unnecessary. He had invented a motor that was like no other, one that operated in a system expressly designed for it by him. Consequently, from this time forward, "alternate currents would commend themselves as a more direct application of electrical energy."[29]

Scientist and Tesla biographer Robert Lomas notes that others who came before Tesla found that the magnetic fields produced by alternating current entered motors and "just churned about, not turning the motor. What Tesla did, was to use two alternating currents that were out of step with each other [polyphase]. Like the propelling waves of legs that move a millipede forward, the magnetic fields worked together to push the rotating shaft of the motor around. By using more than one set of currents, he could ensure that there was always a strong current available to power the motor. As one of the currents died away, the other would continue to move the motor round. The magnetic field rotated and carried the motor round with it, and it did so without using any electrical connections to the rotating shaft."[30]

After Tesla delivered his lecture, Martin stepped forward to propose that the distinguished Professor William Anthony say a few words. The professor, in turn, bestowed his prestigious blessing on this brilliant Serb's astonishing new motor: "I confess that on first seeing the motors the action seemed to me an exceedingly remarkable one." He briefly discussed the technical advantages—few wearing parts—and the motor's efficiency. Then the well-known electrical inventor Elihu Thomson, whose fast-growing firm Thomson-Houston had also entered the AC central station business six months earlier in the fall of 1887, stood up. Thomson, a tall man with deep-set eyes and a thick brush mustache, was not—like Edison, Westinghouse, or Tesla—an inventor who had created original and pathbreaking technologies. But he was immensely skillful at improving and making commercially viable the work of the field's pioneers. Thomson-Houston, seeing the demand for Westinghouse AC systems, had begun offering their own line of central stations, prompting an outraged George Westinghouse to swiftly sue them for infringing on his Gaulard-Gibbs patents. Within a couple of months, the two companies reached terms, with Thomson-Houston agreeing to pay a $2-per-horsepower royalty on each transformer it produced. (About the same time, Hiram Maxim's old firm, United States Electric, also began selling AC central systems. The Westinghouse response to this infringement was so bellicose, U.S. Electric opted instead to let itself be bought by the flourishing Westinghouse electrical empire.)

When Professor Thomson stood up after the Columbia College AC talk, he complimented Mr. Tesla on "his new and admirable little motor." But what Thomson really wished to establish was that he, too, had been working on an AC motor. "I have, as probably you may be aware, worked in somewhat similar directions and towards the attainment of similar ends. The trials which I have made have been by the use of a single alternating current circuit—not a double alternating current—a single current supplying a motor constructed to utilize the alternation and produce rotation."[31] However, Thomson's AC motor depended—as had all others attempted up to this time—upon that troublesome object the commutator. Tesla understood exactly what Thomson was trying to do—establish precedence. He gracefully parried the challenge, declaring himself flattered to be noticed by someone as eminent as Professor Thomson, "being foremost in his profession." Tesla was highly deferential, acknowledging, "I had a motor identically the same as that of Professor Thomson, but I was anticipated by him."

But he also honestly suggested that Thomson would be hard put to claim any kind of equality or anticipation here. Tesla pointed out that Thomson's "peculiar form of motor represents the disadvantage that a pair of brushes must be employed to short circuit the armature coil."[32] Tesla had thought briefly along Thomson's lines but had soared forward, eliminating the commutators with his completely original solution—magnetic fields continuously pushing round the motor's core. Martin deftly cut off discussion while Tesla held the advantage. But this bristly public exchange marked the start of a lifelong antipathy. As Nikola Tesla returned to his seat, the assembled electricians comprehended uneasily and somewhat resentfully that a new titan had risen unbidden among them, eclipsing much of what they had done, making irrelevant many of their dearest labors. His name was Nikola Tesla, and the ambitious and influential Thomas Commerford Martin was his prophet.

Tesla's first lecture, "A New System of Alternate Current Motors and Transformers," was all Martin could have hoped, catapulting Tesla to instant fame in the engineering world. This paper, printed in all the foremost engineering journals, quickly became a landmark for its lucid description of an entirely new kind of very simple "induction" motor. Engineers and the press were astonished at the originality, simplicity, and promise of his AC design. Edison viewed it as but a variant on a technology that was unsafe and unfit for use in human habitations.

It is likely that George Westinghouse had learned about Tesla's revolutionary rotating motors and AC system before the Serb inventor made his dazzling public debut in mid-May before the engineers. After all, Franklin Pope, editor of the *Electrical Engineer* and a Westinghouse employee, had visited the Liberty Street lab at Martin's behest. But it was not until Westinghouse read Tesla's landmark lecture that he took action. He quickly dispatched H. M. Byllesby, a onetime Edison engineer lured away to become a Westinghouse vice president, to visit Nikola Tesla in Manhattan and see if the now famous motors merited such huzzahs. On May 21, 1888, Byllesby wrote his boss that he had met up with Tesla's backers, engineer Alfred S. Brown and lawyer Charles Peck, and then proceeded to the Liberty Street lab with them. There he had met Tesla and witnessed several demonstrations, all of which, he admitted, were somewhat over his head. "His [Tesla's] description was not of a nature which I was enabled, entirely, to comprehend. However, I saw several points which I think are of interest. In the first place, as near as I can get it, the underlying principle of this motor is the principle which Mr. Shallenberger is at work at this present moment. The motors, as far as I can judge from the examination which I was enabled to make, are a success. They start from rest and the reversion of the direction of rotation can be suddenly accomplished without any short-circuiting."[33]

After the demonstration, Byllesby and his escorts returned to Alfred Brown's office to talk business. The Westinghouse executive inquired about the possibility of purchasing the patents. He learned they were held by the Tesla Electric Company and that already Peck and Brown had an offer from a San Francisco capitalist of $200,000 plus $2.50 per horsepower on all apparatus. Reportedly, Cornell professor William Anthony was joining this syndicate. If Westinghouse intended to match or better this offer, they needed to know by Friday at the latest. Byllesby was aghast. He wrote the home office, "The terms, of course, are monstrous; and I so told them. . . . I told them there was no possibility of our considering the matter seriously but that I would let them know before Friday. . . . In order to avoid giving the impression that the matter was one which excited my curiosity I made my visit short."[34]

It's not clear if Peck and Brown really did have such a lucrative offer, for a week later, they were willing to grant Byllesby a $5,000 option for six weeks. George Westinghouse began seriously consulting with his in-house engineers and patent experts. What Tesla and his partners did not know was that a Westinghouse representative—the peripatetic Guido Pantaleoni—was once again in Europe on an AC mission, this time seeking to buy an AC motor patent from his Italian engineering professor, Galileo Ferraris. A month before Tesla's talk before the American Institute of Electrical Engineers, even as Tesla's patents were being considered at the U.S. Patent Office, Professor Ferraris of Turin had given a lecture laying out his own version of an alternating current motor. There was, however, a monumental difference between Tesla and Ferraris, and that was that the Italian electrician viewed his effort purely as a tantalizing and amusing toy, while Tesla had designed a machine and system intended for heavy-duty commercial work. Martin had been properly impatient that Tesla declare his great discovery in a high-profile way, for others were indeed working away on the dilemma. With AC systems spreading across Europe and America, the pressing need for a working motor had become more than well-known. Among engineers, AC motors were in the air. Mere weeks before Tesla's talk, Westinghouse engineer Oliver Shallenberger had solved one of the outstanding gaps in the company's AC lighting system, the lack of a meter to measure electrical use. That meter was also based on the rotating effects of out-of-phase currents, and Shallenberger had begun to experiment with a possible motor. While Westinghouse continued to survey the general status of AC motors, he instructed Pantaleoni to buy the Ferraris patent for the small sum of $1,000.

The wiry William Stanley would complain later in life (when George Westinghouse was dead and gone) that Westinghouse never really appreciated the possibilities of alternating current in the pioneering days and never compensated him fairly. Now, in the wake of Tesla's triumph, Stanley claimed to his boss that he, Stanley, had already invented an AC motor. "I have built an AC system on basically the same principle," he said. However, like Elihu Thomson, Stanley was overlooking the inconvenient fact that his AC motor was still using commutators and brushes. Only Nikola Tesla had designed an AC induction motor free of those troublesome, sparking objects. On July 5, 1888, Westinghouse, his option running out, wrote to one of his lawyers and partners, "I have been thinking over this motor question very considerably, and am of the opinion that if Tesla has a number of applications pending in the patent office, he will be able to cover broadly the apparatus that Shallenberger was experimenting with and that Stanley thought he had invented. It is more than likely that he will be able to carry his date of invention back sufficient time to seriously interfere with Ferraris and that our investment there will probably prove a bad one.

"If the Tesla patents are broad enough to control the alternating motor business, then the Westinghouse Electric Company cannot afford to have others own the patents."[35] Early hopes of using the Ferraris patents as leverage now evaporated, and the Westinghouse people simply had to go along with what Brown and Peck were asking, which was, in fact, far less than the original $200,000. They sought and received $20,000 in cash and $50,000 in notes (payable in three installments), plus the $2.50 royalty per horsepower on every AC Tesla motor, with $5,000 minimum paid in royalties the first year, $10,000 the second, and $15,000 the third. Westinghouse was his usual phlegmatic and pragmatic self: "With reference to the Tesla motor patents, the price to be paid seems rather high when coupled with all of the other terms and conditions, but if it is the only practicable method for operating a motor by the alternating current, and if it is applicable to street car work, we can unquestionably easily get from the users of the apparatus whatever tax is put upon it by the inventors."[36]

Even if the California offer was real, the Tesla Electric Company partners may have preferred to sell to George Westinghouse. In an era of robber barons, Westinghouse had developed a reputation as a fair but indomitable, no-nonsense businessman who defended his patents ferociously. He had already sued Thomson-Houston over his transformer and forced them to make a royalty deal. And he had simply bought up United States Electric when they dared to trespass. In the sharklike atmosphere of Gilded Age capitalism, Nikola Tesla and his partners well knew that they needed such a fearless fighter if they were ever to see more than three years of royalties. Tesla very much admired Westinghouse's qualities as a businessman. He said once, "No fiercer adversary than Westinghouse could have been found when he was aroused. An athlete in ordinary life, he was transformed into a giant when confronted with difficulties which seemed insurmountable. When others would give up in despair he triumphed. Had he been transferred to another planet with everything against him he would have worked out his salvation."[37]

Nor could Tesla and his investors overlook, as they considered their options, the overt hostility of Thomas Edison toward AC. The War of the Electric Currents was only likely to escalate as the stakes rose, and Tesla's much needed AC motor put him squarely in the enemy camp. That being so, Tesla would later say that George Westinghouse was "in my opinion, the only man on this globe who could take my alternating system under the circumstances then existing and win the battle against prejudice and money power. He was a pioneer of imposing stature, one of the world's noblemen."[38] Tesla, who had little trouble envisioning that the whole world would soon be operating on millions of AC-generated horsepower, viewed the deal for his patents as quite fair, even if he had to give five-ninths to his partners. Like Edison, Tesla wanted great wealth not for itself, but so he would be completely free to think, invent, and develop his ideas. The ardent idealist, he saw himself as finally bestowing his great gift on the world. Years earlier, when he had first conceived of the whirling magnetic field, he declared to the doubting Szigety, "No more will men be slaves to hard tasks. My motor will set them free, it will do the work of the world."

Thus, in late July 1888, Nikola Tesla quit the heat of Manhattan, ferried across the breezy Hudson, and boarded the comfortable cars of the Pennsylvania Railroad for the ten-hour journey to Pittsburgh, where he had agreed to serve as a Westinghouse consultant. By bestowing his new all-important AC induction motor upon George Westinghouse's rapidly expanding electrical empire, Tesla was eliminating the one great remaining advantage of Edison's DC system. The War of the Electric Currents was about to be joined in earnest.

CHAPTER 7

"Constant Danger from Sudden Death"

On the warm late afternoon of Tuesday, June 5, 1888, weary New York City commuters ascended to the downtown elevated station as newsboys bellowed out their best headlines of murder, mayhem, and politics. Yet only those riders who purchased the venerable *New York Evening Post* to read for their trek back home would have seen a long, fevered, and highly bellicose letter to the editor titled "Death in the Wires." The letter began, "The death of the poor boy Streiffer, who touched a straggling telegraph wire on East Broadway on April 15, and was instantly killed, is closely followed by the death of Mr. Witte in front of 200 Bowery and of William Murray at 616 Broadway on May 11, and any day may add new victims to the list."

The letter writer, one Harold P. Brown, denounced the wretched deadly spiderwebs of thousands of electric wires strung haphazardly above the city's busiest streets. But Brown wished not simply to castigate the officials for the well-known dangers of unsafe wires, but to thunder of new perils. "Several companies who have more regard for the almighty dollar than for the safety of the public, have adopted the 'alternating' current for incandescent service. If the pulsating [arc] current is 'dangerous,' then the 'alternating' current can be

Thomas Edison at his desk dictating into the Edison Business Phonograph described by no adjective less forcible than *damnable.*" Declared Brown, "The only excuse for the use of the fatal 'alternating' current is that it saves the company operating it from spending a larger sum of money for the heavier copper wires, which are required by the safe incandescent systems. That is the public must submit to *constant danger from sudden death* in order that a corporation may pay a *little larger dividend.*" He called for the outlawing of all AC above 300 volts to "prevent the wholesale risk of human life." (It was perhaps not immaterial that the *Evening Post* was owned by longtime Edison investor Henry Villard, soon to be Edison president.) Suddenly, the War of the Electric Currents had been moved onto far more dangerous, high-stakes terrain. Brown was seeking not to dissuade potential customers, but to *outlaw* outright AC.

Who was this Harold Brown? Until that Tuesday, he was a thoroughly obscure New York engineer and electrical consultant, a complete nobody. At the time he composed his denunciatory letter to the *Post,* Brown described himself on his professional stationery as an electrical engineer, designer of apparatus for special purposes, contractor for arc and incandescent electric lights and steam power, and creator of life-protecting apparatus for arc light dynamos. Now, on June 5, 1888, he had suddenly charged forth against AC. What prompted Brown's wrathful attack remains something of a mystery, despite extensive probings by numerous historians. He had no apparent ties at this stage to the Edison camp. Nor did he bear any discernible personal or professional vendetta or grudge against members of the AC group, including Westinghouse. Perhaps it was simply that this little-known member of the New York electrical fraternity saw a golden opportunity for fame and glory as the aggrieved voice of the gathering anti-AC crusaders. Until now, the Edison forces had contented themselves with written vituperation and proxies arguing against patronizing AC in genteel forums like the Chicago Electric Club. Suddenly, with Harold Brown, they had a man of action, an enraged fighter who would lead the public charge in the anti-AC holy war.

Nor was Brown a mere blowhard. His strident urging of a swift legislative end to high potentials—"No alternating current with a higher electromotive force than 300 volts shall be used"—would indeed put the AC men out of business. He swiftly seized new and strategic ground: The famously dilatory New York City Board of Electrical Control convened days later on Friday, June 8, another warm and lovely late spring day, and Brown himself appeared, a young and rather harmless-looking fellow, his dark hair parted neatly to one side and a tame handlebar mustache highlighting a biggish nose. Yet when Brown spoke he was very much the firebrand, insisting his now notorious letter be read verbatim into the record. His proposed safety rules were to be published in the minutes and marked copies sent for comment to an array of eminent electric companies and electricians, including George Westinghouse. Events had now escalated beyond angry phrasemaking to bellicose and economically threatening efforts to legislate AC out of existence.

Many hours west in Pittsburgh, where the summer heat and humidity concentrated the city's famous grime and soot into an unhealthful black celestial umbrella, Brown's main target, George Westinghouse, was biding his time. He put off answering the New York City Board of Electrical Control's request for comment. Instead, on June 7 he sat in his large office with its lovely Persian rug and wrote a personal note to Thomas Edison in West Orange, in part about rumors of mergers between their two companies. But his main purpose was to propose peace: "I believe there has been a systematic attempt on the part of some people to do a great deal of mischief and create as great a difference as possible between the Edison Company and The Westinghouse Electric Co., when there ought to be an entirely different condition of affairs.

"I have a lively recollection of the pains that you took to show me through your works at Menlo Park when I was in pursuit of a plant for my house, and before you were ready for business, and also of my meeting you once afterwards at Bergman's factory; and it would be a pleasure to me if you should find it convenient to make me a visit here in Pittsburgh when I will be glad to reciprocate the attention shown me by you."[1] Edison responded in a noncommittal fashion on June 12: "My laboratory work consumes the whole of my time. . . . Thanking you for your kind invitation to visit you in Pittsburg."[2] And that was all. Not long after, Edison's sales force began accusing their Westinghouse rivals of lying about AC advantages, a tactic that so infuriated Westinghouse, he briefly considered suing.

His olive branch rebuffed, George Westinghouse was heard from publicly for the first time when the New York City Board of Electrical Control reconvened at Wallack's Theater on Monday, July 16. Unlike Thomas Edison, who at this point operated quietly behind the scenes, George Westinghouse hid behind no one. Certainly, by mid-July of 1888, he had no further illusions. Thomas Edison and his company were serious about their holy war against AC. The time had come to strike back. So Westinghouse's letter to the New York City Board of Electrical Control was a skillful, hard-hitting piece of PR. First, he excused his tardy response by pleading the tremendous press of business. In less than two years, his company and its licensee Thomson-Houston had installed 127 AC stations, 98 of which were up and operating already. Of those 98 AC plants, a third had already expanded. The plant in Pittsburgh was the "largest incandescent lighting station in the world." With so much business, explained Westinghouse, "it has been considered inexpedient, heretofore, to take any notice of, or make any reply to, the criticisms and attacks of some of the opposition electric lighting companies."

Westinghouse, a corporate titan well seasoned in the most cut-throat of Gilded Age business warfare, declared himself amazed at a "method of attack which has been more unmanly, discreditable and untruthful than any competition which has ever come to my knowledge." If they wanted to fight a dirty war, Westinghouse could lob his own incendiary devices. Was the issue safety? Well, not one Westinghouse central station had sustained "a single case of fire of any description from the use of our system. Of the 125 central stations of the leading direct current company [Edison] there are numerous cases of fire, in three of which cases the central station itself was entirely destroyed, the most recent being the destruction of the Boston station; while among the almost innumerable fires caused by this system, among the users, may be mentioned the total destruction of a large theater at Philadelphia."

An octet of pro-AC affidavits had also fluttered into the quiet offices of the electrical control board in the preceding weeks, and these were now duly read aloud at Wallack's Theater. All followed along the lines of one from Philadelphia, wherein an electrical worker named W. L. Wright described his inconsequent encounter with the supposedly fatal AC current. Working on some wires in a damp basement, read his testimonial, and forgetting that a 1,000-volt current was active in it, "I took hold of the socket while standing on the wet ground, when I received a shock which threw me on my face with my hand underneath me, and still handling the socket.... When I came to my senses I was sitting in the cellar held up by two of the men. In the meantime, an ambulance had been called. I went down to the electric light station and waited there for fifteen or twenty minutes to receive my money; it being pay-day, and then went home." The company insisted he visit a doctor, who dressed his burned hand. "These burns healed very slowly; but I have not felt in any way any of the after effects from this shock, such as are usually felt from high tension direct current machines.... I feel sure that had I received this kind of shock from a direct current machine of any of the ordinary types ... it would have been fatal."[3]

Harold Brown was not present at the Wallack's Theater meeting, being conveniently away in Virginia on business. This was probably just as well, for the opposition forces were in a highly hostile mood, questioning everything from Harold Brown's fitness and training as an electrician to his motives. Where, his critics reasonably demanded, was Brown's proof? On what basis was he urging that their electrical enterprises be banned?

Brown had catapulted from utter obscurity in two brief weeks to become the self-appointed, impassioned champion of one of America's most revered icons—Thomas Alva Edison. What must Edison have thought when he read Brown's passionate diatribe against Edison's own most hated enemies? We do not know, but the beleaguered and indignant Brown tells us he was determined to prove his case and so called upon Edison, "whom I had never met before, and asked the loan of instruments.... To my surprise, Mr. Edison at once invited me to make experiments at his private laboratory, and placed all necessary apparatus at my disposal."[4] (Brown was probably the only electrician in Gotham surprised by Edison's hearty embrace of his mission.) From this moment forth, Edison eagerly aided and abetted this self-appointed crusader against the "damnable current," a man whose stated goal was the legislative end to Edison's biggest rivals, the AC companies. Brown needed to respond to his critics, he said. He needed to prove his case. Edison offered Brown not just space in his extraordinary new laboratory, but the help of his most trusted lieutenants, Charles Batchelor and newly hired British scientist Arthur Kennelly, son of the Bombay harbormaster and future eminent professor at MIT and Harvard.

Edison's new laboratory had been built over in New Jersey on ten acres just half a mile down the hill from his new estate, Glenmont, in the quiet valley of West Orange. Edison had been a widower briefly before becoming engaged to the young and lovely Mina Miller. He and Mina had first gone to see the sprawling Queen Anne–style château Glenmont in December of 1885. On a cold wintry day they had crossed the Hudson River and then driven out to the snow-covered garden suburb of Llewelyn Park. There they inspected the brick-and-wood mansion's many gables, admired the vast entry hall and glittering chandeliers, the broad curving staircases, gleaming wood floors, rich stained-glass windows, and palatial living rooms and marveled that it was all lavishly and conveniently furnished, including walls bedecked with oil paintings and niches filled with statuary. Outside, the rolling grounds white with the recent snowfall boasted numerous greenhouses and dormant flower beds certain to be beautiful come spring and summer. Built for $235,000, Glenmont was on the market for a bargain $50,000. (The original New York millionaire owner had been caught embezzling from his firm, the Arnold Constable department store, and had fled overseas.) The lovely Mina thought it a fittingly opulent new home for the nation's greatest inventor, and Edison was anxious to please her.

Charles Batchelor then supervised both the purchase of nearby land for Edison's new laboratory and its construction. Edison's up-to-the-minute invention factory, ten times larger than ramshackle

Menlo Park, included a vast and graceful central building with sixty thousand square feet, housing a machine shop, glassblowing, chemical, and photographic departments, electrical testing rooms, and stockrooms. Here Edison put his stately wood-paneled office and library, his huge rolltop desk arranged just below a two-story gallery that held ten thousand scientific volumes. No longer so given to playing the hick, Edison appreciated the importance of impressing his multitudes of visitors. A quartet of other buildings provided further space. The laboratory was stocked with "eight thousand kinds of chemicals, every kind of screw made, every size of needle, every kind of cord or wire, hair of humans, horses, hogs, cows, rabbits, goat, minx, camel." Edison, ever the kidder, told one reporter he "ordered everything from an elephant's hide to the eyeballs of a United States Senator."[5]

Edison's early ambitions of being far more than an inventor were undimmed. He yearned to succeed as an industrial titan. And he had come to see from developing his electrical networks the commercial advantages of a well-stocked, well-financed laboratory that could quickly parlay ideas into products. It was here, in quiet West Orange, an hour by ferry and train from noisy, dirty, moneyed Manhattan, that the energetic and eager Brown set himself to proving during numerous late night experiments in Thomas Edison's state-of-the-art facility that AC was truly the "damnable" current, one that should be made illegal.

By late July, Harold Brown felt prepared to confront his critics with the kind of definitive scientific evidence they had been clamoring for. Engraved invitations went forth to the members of the New York City Board of Electrical Control, representatives of all the electric light companies, and numerous others in the electrical fraternity, inviting one and all up to Madison Avenue and 50th Street for a demonstration in "Prof. Chandler's Lecture Room at the School of Mines, Columbia College," on Monday, July 30.[6] The day was warm, but seventy-five electrical gentlemen and numerous reporters gathered in the large, airy college lecture room with a sense of delightful anticipation, for Brown had already shown that he was a lively sort not likely to turn the other cheek. "No intimation of the character of the exhibition had been given," noted one journal. How would he make his case?

Brown, his hair neat and glossy, his mustache trim, began by saying that "he had been drawn into the controversy by his sense of right. He represented no company and no financial or commercial interest." He then discussed the differences between the alternating and the continuous current and stated he had proved "by repeated experiments that a living creature could stand shocks from a continuous current much better."[7] This would explain the large wooden cage with copper wires interlaced between the bars. At this point, Brown vanished for a moment into a side room and then reappeared, leading a large black retriever dog. After muzzling the animal, he put him in the large cage, strapped him into place, and locked the cage door. The crowd of men holding their straw hats and light-colored derbies stirred and murmured. Brown said that the dog, which looked quite a brute, was in perfect health but of vicious disposition. He weighed seventy-six pounds. The dog woofed through his muzzle.

Arthur Kennelly, the Edison chief electrician, served as Brown's assistant, as did Dr. Frederick Peterson, a doctor who specialized in treating patients with electricity, and a few others. They were needed to hold the struggling black dog as wires were attached to the furry right foreleg and left back leg, each being already wrapped with some waterlogged material. The dog's "resistance" was found to be 15,300 ohms. Harold Brown began by applying 300 volts of DC. The dog seemed startled and unhappy. When the power was upped to 400 volts of DC, the large black creature struggled and yelped pitiably inside his cage. The crowd of men shifted in their seats, and there were audible murmurs of disapproval in the lecture hall growing warm with summer heat and bodies. At 700 volts DC, the dog's violent thrashings broke the restraints and he had to be restrapped.

Brown ignored the rising feeling against his cruel display and increased the voltage to 1,000. "Many of the spectators left the room," reported one journal, "unable to endure the revolting exhibition."[8] The poor beast contorted in pain, and some in the audience began loudly telling Brown to cease. At this juncture, Brown turned off the direct current. He told his restive audience, "He will have less trouble when we try the alternating current. As these gentlemen say, we shall make him feel better." A Siemens Brothers alternator was hooked up, and 330 volts of AC was administered to the quivering and terrified retriever, which quickly collapsed in a horrible heap, dead. At that moment, a reporter for the *New York World* stood up. He fiercely protested any further such torturing of dogs, which emboldened an agent for the American Society for the Prevention of Cruelty to Animals (ASPCA) to step forward and forbid Brown from executing another dog. Seeing how the audience had grown hostile rather than enlightened, Brown stopped.

The shaken electricians stood up, put on their hats, and began to disperse unhappily from the lecture hall. Despite Brown's assertion that he had now provided the demanded proof, few in the audience concurred, for the black retriever had obviously been much weakened by his initial electrical torture with DC before he received the killing zap of AC. Brown bitterly resented being thwarted from quickly killing a dog with straight AC. He blamed the "treachery" of the AC forces. He reassured the departing audience that he had many other dogs and had experimented on enough in the past month to be quite certain of AC's superior fatalness. Harold Brown's final remark as the demonstration ended was that "the only places where an alternating current ought to be used were the dog pound, the slaughter house, and the state prison."[9]

This electrical dog show was what Brown had been preparing for all those unusually cool July nights across the river at West Orange. He had gathered his dogs by paying a twenty-five-cent bounty to local children. Kennelly and Charles Batchelor had frequently helped out. The latter had been severely shocked himself when holding down a puppy. He felt "body and soul being wrenched asunder . . . the sensations of an immense rough file thrust through the quivering fibers of the body."[10] Brown learned that he could kill a dog with only 300 volts of AC, but with DC he needed 1,000 volts.

Four days after Brown's first truncated demonstration, he returned again uptown, arriving at the Columbia School of Mines on Friday with three big caged dogs in the suddenly soggy heat of early August. This time, few but his helpers, public health officials, and newspaper reporters were present in the stifling lecture hall. In short order, Brown and his minions dispatched a sixty-one-pound mongrel with AC of less than 300 volts, a ninety-one-pound Newfoundland after eight seconds, and a fifty-three-pound setter-and-Newfoundland mix who survived four excruciating minutes before finally collapsing, his tongue lolling out. In the sticky heat, among the smells of one of the dead dogs being dissected, Brown felt most pleased and vindicated. "All of the physicians present," said Brown, "expressed the opinion that a dog had a higher vitality than a man, and that, therefore, a current which killed a dog would be fatal to a man."[11]

First thing the next day, from his warm office on Wall Street, where the noises of the street drifted in the windows open for any breeze, Harold Brown wrote triumphantly to Arthur Kennelly out in West Orange, reporting, "We made a fine exhibit yesterday, as you will see from all the papers, and I had the report of the proceedings signed by all present and sent to the associated press throughout the country. I missed you, but as no representative of the alternating current concerns favored us, it was just as well that there should be no Edison man there. . . . Whatever action the Board of Electrical Control may take, it is certain that yesterday's work will get a law passed by the legislature in the fall, limiting the voltage of alternating currents to 300 volts."[12] At the bottom of this typewritten letter, Brown wrote in a nice even hand, "I have lost 12 lbs over this struggle and am all worn out, but am going to the mountains today to rest." The Edison camp must have been delighted with Harold Brown's tremendous get-up-and-go and his natural flair for publicity, though his confident prediction of a state law limiting volts to 300 that fall was overly optimistic.

Once rested up in the cool, pine-scented air of the mountains, Harold Brown returned to launch the most macabre of all battles in the fast escalating War of the Electric Currents. The New York State Legislature, having designated electrocution as its official state mode of capital punishment, was now seeking technical advice from the Medico-Legal Society of New York. How best did one electrically kill condemned prisoners? The chairman of this new committee happened to be Dr. Frederick Peterson, Harold Brown's able assistant in both of his ghoulish dog-killing demonstrations at Columbia College. Brown was now hell-bent on one thing: getting AC designated as the ideal form of electrocution.

Most Americans that fall had turned their attention to the drama of presidential politics, with Democrats cheering on President Grover Cleveland, the Big One who backed Civil Service reform, curbed dubious Civil War pensions, and infuriated big business by opposing higher import duties. This bland, fat chief executive had also charmed the nation by marrying late in life the beautiful young Frances Folsom in a White House wedding. The Republicans rallied around Ohio senator Benjamin "Little Ben" Harrison, a short, gray-bearded grandson of the nation's ninth president, who held the right ideas about high tariffs and wielding American clout. Harrison would win by a hair and thanked providence. But Pennsylvania's all-powerful Republican boss, Matthew Quay, wondered if Harrison knew "how close a number of men were compelled to approach . . . the penitentiary to make him President."[13]

While the tight election contest stirred electoral passions, Harold Brown and Dr. Peterson retreated to Edison's prestigious West Orange laboratory and began further dreadful experiments on how to most efficiently kill living things with electricity. With these findings in hand, both attended the November 15 meeting of the Medico-Legal Society, where Dr. Peterson said either direct or alternating current would do the job, "but preferably the latter."[14] The society would announce their electrocution decision at their December 12 meeting. Not content to wait for the society's decision, Harold Brown swung right back into action, orchestrating a new demonstration that would silence once and for all critics who scoffed that killing dogs was not at all comparable to killing humans. Brown needed to execute creatures more akin in size to grown men.

Once again, the Edison people happily made available the famous West Orange lab for what they termed a "matter of very great importance."[15] On the overcast, chilly afternoon of December 5, Brown and Dr. Peterson were admitted to the Edison complex in West Orange. Assembled in the back in a brightly lit room were numerous reporters and Edison men, several important physicians from the Medico-Legal Society, and two members of the New York State Death Commission: Buffalo dentist Dr. Southwick, who first drew Edison into the issue, and the chairman, Elbridge T. Gerry, long active in the New York Society for the Prevention of Cruelty to Children and its sister society for animals, the ASPCA. Gerry was author of the state's electrocution bill. Brown's greatest coup that day was, without question, the august presence of Thomas A. Edison, who heretofore had served only as a silent and reclusive sage (rather than general) for the DC forces. World renowned, a national icon, Edison instantly imbued the lethal proceedings with great legitimacy and an inevitable glamour. Now, here in the celebrated Edison laboratory, these distinguished guests would see with their own eyes just how deadly the alternating current really was.

First came a soft-eyed calf bought from the local butcher. It walked docilely onto a sheet of tin laid on the laboratory floor, its hooves making a loud crackling noise on the metal. Tied to a nearby post, the 124-pound calf was cut on the forehead and upper spine, and sponge-covered plates were fastened to those places. The tin "rug" was attached to wires fed by an alternator, all this being first-rate Edison apparatus. The AC was then zapped up to 700 volts, and after thirty seconds the calf collapsed heavily and died. A second calf weighing 145 pounds was electrocuted after only five seconds. The pièce de résistance was now brought forth: a large, healthy horse weighing 1,230 pounds. Here was a beast far bigger and stronger than any criminal. It whinnied lightly as copper wires were wound around its forelegs. The men stood back, for no one wanted to be struck by the horse's flailing hooves if things went wrong. When the voltage hit 700, the horse slumped to its knees, dead. This impressive display of lethal AC electricity and the presence of Edison were all that Brown could have hoped for.

The next morning, *The New York Times* solemnly reported, "The experiments proved the alternating current to be the most deadly force known to science, and that less than half the pressure used in this city [1,500 to 2,000 volts] for electric [arc] lighting by this system is sufficient to cause instant death. After Jan. 1 the alternating current will undoubtedly drive the hangman out of business in this State."[16] Indeed, at the December 12 meeting of the Medico-Legal Society at the Palette Club on West 24th Street, the group unanimously adopted the electrocution committee's proposal of "death by alternating current" and its recommendation that the criminal be executed in "a recumbent position, on a table covered with rubber, or the sitting position, in a chair especially constructed for the purpose."[17] New York State was on its way to being the first government in history to execute condemned criminals with electricity.

Out in Pittsburgh, where the roaring iron and steel furnaces spewed forth the usual fiery layers of filthy smoke, George Westinghouse read the newspaper stories about Brown's latest electrical slaughter and muffled his fury so he could write a careful, reasoned reply. This long letter ran in various New York papers the day after the Medico-Legal Society had voted to endorse alternating current for electrocutions. George Westinghouse pointed out that even as Mr. Brown categorically claimed to have proved that anything over 300 volts of AC was deadly, "a large number of persons can be produced who have received a one-thousand volt shock from alternating currents without injury." Once again, Westinghouse emphasized his company's huge success. The 1888 Edison annual report showed central station orders totaling forty-four thousand lights for the whole year. Contrast that, he suggested, with his firm's orders *just* for October of forty-eight thousand lights.

As was his way, Westinghouse minced no words, scoffing that "the business would not have had this enormous and rapidly increasing growth if there had been connected with it the dangerous features which Mr. Harold P. Brown and his associates of the Edison company so loudly proclaim.... We have no hesitation in charging that the objects of these experiments is not in the interest of science or safety."[18] In one of the most bizarre rejoinders of all time, the egregious Harold Brown wrote to the newspapers and challenged his Pittsburgh nemesis to an electric duel! Like a gentleman of old throwing down a gauntlet, Brown insulted Westinghouse by charging that the great industrial leader cared only about "the pecuniary interests . . . of the death-dealing alternating current" that had "crippled, paralyzed or otherwise injured for life a number of men." Then Brown laid out the nature of this unique duel: "I challenge Mr. Westinghouse to meet me in the presence of competent electrical experts and take through his body the alternating current while I take through mine a continuous current. . . . We will commence with 100 volts, and will gradually increase the pressure 50 volts at a time, I leading with each increase, until either one or the other has cried enough, and publicly admits his error."[19] Westinghouse did not deign to respond.

As the year 1888 drew to an end, the Edison forces had scored some notable and important victories in the War of the Electric Currents. First and foremost was Harold Brown's triumph in having alternating current designated as the official New York State "executioner's current." Equally important, from the DC standpoint, most public discussions of electricity now revolved largely around questions of safety, with the Edison systems always emerging the shining example of truly safe electricity—low voltage and buried wires. The fact that Edison's central plant electricity was expensive and unsuitable for anyplace but high-density cities was rarely a part of the discussion.

The spring of 1889 delivered further delightful victories for the Edison forces. Brown was hired by the New York State prisons as their electrical expert. He would design its electrocution apparatus, guaranteeing that Westinghouse machines would soon be used and indelibly linked to an odious death. The month of March brought the dramatic implosion of the French copper corner after eighteen expensive months. Monsieur Secretan, historian Kenneth Ross Toole explains, "had, first of all, forgotten the junkman. His scheme was predicated on control of the world's supply of copper. But with cop-

FROM A COPPER-PLATE ENGRAVING IN J. G. DOPPELMAYR'S 1744 TEXT

Stephen Gray's demonstration of the Electrified Dangling Boy.

COURTESY OF THE LIBRARY OF CONGRESS

Franklin's experiment, which actually took place in September 1752.

COURTESY OF THE NEW-YORK HISTORICAL SOCIETY

The urban spider webs spun by telegraph lines as lampooned by *Harper's Weekly*, May 14, 1881. Soon, electric-light wires would join the haphazardly strung webs.

COURTESY OF THE U.S. DEPARTMENT OF THE INTERIOR, NATIONAL PARK SERVICE, EDISON NATIONAL HISTORIC SITE

Thomas Edison (*left front in dark skullcap*) and his Menlo Park crew in the second story of the laboratory.

COURTESY OF THE NEW-YORK HISTORICAL SOCIETY

An 1880 portrait of J. Pierpont Morgan, the Wall Street banker who helped finance Thomas Edison's electric company.

COURTESY OF THE U.S. DEPARTMENT OF THE INTERIOR, NATIONAL PARK SERVICE, EDISON NATIONAL HISTORIC SITE

Thomas Edison prided himself on safely burying his company's electric wires under the streets of Manhattan, an arduous enterprise shown in this *Harper's Weekly* print.

COURTESY OF THE U.S. DEPARTMENT OF THE INTERIOR, NATIONAL PARK SERVICE, EDISON NATIONAL HISTORIC SITE

A cutaway print shows the three floors of the Edison Electric Light Company's Pearl Street Station, where coal-fed steam engines powered the direct current generators, visible above.

REPRINTED FROM *ELECTRICAL WORLD*, JULY 11, 1891

Nikola Tesla lecturing before the AIEE at Columbia University, May 20, 1891.

Nikola Tesla demonstrates one of his wireless electric lights.

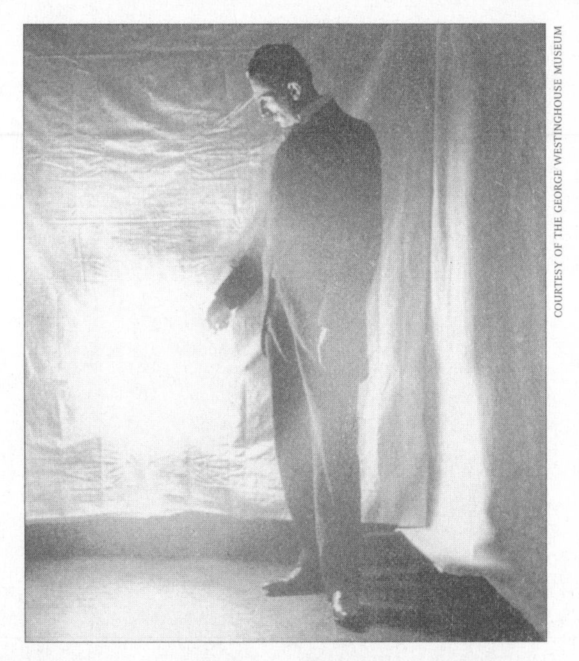

COURTESY OF THE GEORGE WESTINGHOUSE MUSEUM

Nikola Tesla posing with his wireless bulb.

COURTESY OF THE NIKOLA TESLA MUSEUM

COURTESY OF THE GEORGE WESTINGHOUSE MUSEUM

George Westinghouse and his young wife, Marguerite, in a formal portrait.

COURTESY OF THE GEORGE WESTINGHOUSE MUSEUM

The Westinghouses visit Niagara Falls.

COURTESY OF THE GEORGE WESTINGHOUSE MUSEUM

George Westinghouse working. He did not like being photographed; this was taken without his knowledge.

NEW YORK WORLD

Illustration for July 22, 1894, *Sunday World* story reading, "Nikola Tesla, showing the Inventor in the Effulgent Glory of Myriad Tongues of Electric Flame After He Has Saturated Himself with Electricity."

per at 17 cents, in very short time scrap dealers had thrown at least 70,000 long tons on the market [about a quarter of the world's annual consumption]. . . . Secretan's second miscalculation was with regard to consumption. When the price of copper soared, many consumers simply stopped buying it."[20] The electrical world turned out to be far less voracious a captive customer than assumed. The copper syndicate found itself sitting atop one hundred thousand tons of gleaming red metal it could not sell for what it had paid. The French banks and the major world producers quickly reached an agreement that would leave copper at twelve cents for the foreseeable future.

The fall served up yet more unalloyed victories for the DC forces. For four years, Edison's electric light bulb patent cases had been dragging along. In some courts, Edison had received severe setbacks, in others some encouragement. In 1886, Westinghouse had felt so certain Edison's patent was doomed that he indulged himself and struck back by filing a retaliatory suit. The Westinghouse case, in the guise of Consolidated Electric Light, which held the original Sawyer-Man light bulb patents, had been filed against McKeesport Light Company, an Edison entity. Then, on October 4, 1889, Justice Bradley of the United States Circuit Court in Pittsburgh delivered a serious setback to the light bulb infringers. The good judge upheld Edison's longtime assertion that what differentiated his light bulb from all its many unworkable predecessors was "high resistance in the conductor with a small illuminating surface and a corresponding diminution in the strength of current. This was accomplished by Edison . . . and was really the grand discovery in the art of electric lighting, without which it could not have come into general use in houses and cities. . . . But for this discovery electric lighting would never have become a fact."[21] Westinghouse sought to buy time by appealing the case.

When Nikola Tesla journeyed to Pittsburgh to help develop his AC induction motor and polyphase system, he had finally met George Westinghouse, now forty-one. He admired him right away. Wrote Tesla, "Even to a superficial observer, [Westinghouse's] latent force was manifest. A powerful frame, well-proportioned, with every joint in working order, an eye as clear as crystal, a quick and springy step—he presented a rare example of health and strength. Like a lion in the forest, he breathed deep and with delight the smoky air of his factories."[22] Tesla toured the impressive Westinghouse electrical shops, met the engineers, and then returned briefly to New York to wrap up his affairs before returning to Pittsburgh to work as a consultant with the Westinghouse Electric Company.

Almost certainly, Tesla was a guest for a time under the Westinghouse roof in leafy Homewood at Solitude. The handsome white brick villa with its white window awnings lowered against the summer heat was surrounded by an attractive array of flower and vegetable gardens. To one side, a long grape arbor provided a filtered shady retreat. The leaves of large ginkgo trees waved and shimmered in the sun. Across the railroad tracks, truck farmers were tending tomatoes and vegetables. The otherwise gracious Westinghouse home had one jarring aspect—the festoons of electric wires drooping along the ceilings and up the stairway. Naturally Westinghouse had installed electricity in Solitude, but he had insisted all the wires remain freely accessible so he could test out new improvements as they came along. The dynamo and generator were far back out in the stable, and the subway that connected the generator to the villa was big enough for a man to walk in.

After the sudden acclaim and overnight wealth of 1888, Tesla's ensuing year in Pittsburgh was a sharp reminder of the perilous journey between invention and commercial success. First off, it turned out that Nikola Tesla's much ballyhooed AC induction motor did not, as George Westinghouse had hoped, have any value for traction work, which eliminated its use for the fast-growing and lucrative electric streetcar business. Then, it was worryingly obvious that the polyphase induction motors as shown at Columbia College did not mesh easily with the Westinghouse single-phase AC central lighting stations. Tesla had early on concluded that the ideal frequency was 60 cycles, and all his AC induction motors were so designed. The Westinghouse engineers, however, had designed all *their* AC central lighting stations to operate at more than twice that frequency, or 133 cycles. Tesla did not endear himself to his new colleagues when he insisted that the central stations would have to be retrofitted, for otherwise they would never get a workable AC motor.

The Westinghouse engineers were loath to concede Tesla's point and undertake such a major revamping. Like many pioneers of the electrical fraternity, the Westinghouse electricians resented this flowery fellow declaring that everything they knew about making electricity and running motors was now passé and irrelevant. They resented his sudden wealth and fame and the widespread belief by knowledgeable men that he was going to be as big as, maybe bigger than, Edison. Even a world-celebrated writer like Mark Twain had been quietly confiding to his diary in November of 1888, "I have just seen the drawings and descriptions of an electrical machine lately patented by a Mr. Tesla, and sold to the Westinghouse Company, which will revolutionize the whole electric business of the world. It is the most valuable patent since the telephone." Mark Twain, a great writer who was a notable flop in his own various forays into nonliterary business, never wrote truer words. And that assertion was what was so immensely galling to the Westinghouse engineers. As the months passed, they and some of the electrical press had the satisfaction of concluding that Tesla was full of hot air. It did not help that Shallenberger's new AC electric meter was quickly and easily adapted to existing Westinghouse AC systems, with impressive results. As soon as customers saw bills based on usage, they began turning off unneeded lights. Westinghouse central stations equipped with meters now had to generate only a half to two-thirds the amount of electricity as those central stations operating still without meters. The savings to the company were dramatic.

Charles Scott was too young and too much of a newcomer at Westinghouse to harbor any animosity. Assigned to work with Tesla as his assistant, helping to build AC induction motors and test them, he was overjoyed. "It was a splendid opportunity for a beginner, this coming in contact with a man of such eminence, rich in ideas, kindly and friendly in disposition. Tesla's fertile imagination often constructed air castles which seemed prodigious. But, I doubt whether even his extravagant expectations of the toy motor of those days measured up to actual realization."[23] Young Scott also admired how in these trying months, Westinghouse himself was always "suggesting, inspiring, directing, urging. Each step was a progress toward a universal system of power distribution. That was his great vision and ambition. . . . [It is hard to] realize how little was then known to imagine the magnitude of what might develop when means were found for making larger generators and larger transformers which would not short-circuit or overheat."[24]

And so, thwarted and frustrated in Pittsburgh, Tesla prepared to retreat to Manhattan. "Having worked one year in the shops of George Westinghouse, Pittsburgh," he would later say, "I experienced so great a longing for resuming my interrupted investigations that, notwithstanding a very tempting proposition by him, I left for New York to take up my laboratory work."[25] That was in the fall of 1889, and Tesla was soon ensconced in a new laboratory on Grand Street. Tesla, reported his friend and biographer, John J. O'Neill, was "thoroughly disgusted. . . . He felt his advice concerning his own invention was not being accepted." He also told O'Neill, "I was not free at Pittsburgh. I was dependent and could not work. To do creative work I must be completely free. When I became free of that situation ideas and inventions rushed through my brain like Niagara."[26] One suspects that many among the entrenched electrical brotherhood experienced a highly delicious frisson of schadenfreude, for Nikola Tesla had returned from Westinghouse without managing to produce a commercial AC induction motor that could operate with the firm's two hundred central stations.

Thus, as October brought pleasant fall weather to the clamorous streets of Manhattan and society prepared to whirl into high gear with the Patriarchs' Ball at Delmonico's, Edison could feel happy about the drop in copper prices, his clear-cut light bulb patent win in the courts, and Tesla's humbling in the real world of making things work. There was also the unexpected appearance of a new and powerful player. All through the War of the Electric Currents, both Edison and Westinghouse were continually in the public eye, whether through their proxies or firsthand. Notably absent from this very public fray was the other major AC company, the Thomson-Houston Company of Lynn, Massachusetts, which had expanded from arc lighting into AC incandescent central stations. But Charles Coffin of the Thomson-Houston Company, ever the mover and the shaker, was yearning to enter the fray.

CHAPTER 8

"The Horrible Experiment"

As the streaked gray sky lightened on Friday morning, March 29, 1889, one John Hort drunkenly sauntered forth from the noisy warmth of a bar on Buffalo's tough waterfront into the morning cold. A bantam of a fellow with a dark bushy beard and heavy-lidded eyes, Hort, twenty-eight, was a successful huckster, a seller of fruits and vegetables. A habitué of the port's many lowlife saloons, Hort often passed whole afternoons bellied up to one polished bar or another, drinking morosely. Sometimes he would stumble out, perch himself on the barrels in front of a nearby mission house, and spend hours staring vacantly and twiddling his thumbs. Thursday night had been just another hard night of drinking. For hours John Hort had swilled buckets of stale beer and slung back cheap whiskey with John "Yellow" DeBella, his boarder and employee. Through the rowdy noise of the saloon, Hort had been unusually loquacious, haranguing all those around that his wife, Tillie, was a wanton whore.

Hort, thoroughly drunk and in vile humor, walked home through the cold, misty dawn. A little before 8:00, he reached his ground-floor apartment at 526 Division Street, a big run-down cottage, and flung open the door to the dingy rooms. There in the small kitchen stood Tillie, thirty-one, an attractive enough woman still in her morning

---

An early electric chair

wrap, and her four-year-old daughter, Ella. The warm kitchen smelled of baking potatoes and frying eggs. Hort fixed Tillie, who was standing at the stove holding the frying pan, with a drunken stare, yelled at her that she was a harlot, and then staggered out to the barn, where his six workers were loading his fruit and vegetable wagons and hitching up the restless horses. Hort's young peddlers knew to steer clear when their boss was soused and simmering mad. One noticed him walk across the hay-covered stable floor, grab an ax off its holder on the barn wall, and head back to his apartment. Soon the cool morning was rent with curdling shrieks, shattering dishes, a rhythmic whacking, then silence and low moans.

The landlady, Mrs. Mary Reid, rushed to the Horts' front door and called, "Mrs. Hort," several times, but she heard only the gurgled groans. Then she ran around to the side, there to see Hort staggering forth from his back door, his hands, his arms, his very beard, smeared with blood. "Mr. Hort," she asked wildly, "what have you done?"

"I have killed Mrs. Hort."

She looked terrified. "No, you have not."

Coolly he answered, "Yes, I have, and I'll take the rope for it."

As Mrs. Reid ran to get help, Hort returned inside. A man visiting his father in a nearby house put on his hat and came running in response to Mrs. Reid's tearful shouts for help. Later, he would recount, "I opened the door of the kitchen and looked in. A woman was upon the floor on her hands and knees, blood all over her and her hair hanging down, her body swaying backwards and forwards. I shut the door." He steadied himself and then opened the door again. "A man stood before me. He was wiping his hands on something. They were bloody. He stepped over the body." The horrified visitor said to Hort as he walked out, "This is brutal, man. Let us go for a doctor."

But Hort was continuing back toward the barn. One of his young peddlers was standing there petrified as he saw his bloodied boss. "Go for a policeman," Hort instructed him dully, "I have killed my wife." When the peddler boy hesitated, Hort said again, "I've done it and I expect to take the rope." Then Hort wobbled slowly off to Thomas Martin's saloon, where he was demanding whiskey when a police officer arrested him. "I want the rope," said the dazed Hort, "the sooner the better."

Back by the barn, a horse-drawn ambulance had clattered into the yard and the unconscious, bloodied Tillie Hort, her head etched with ax wounds, was carried out and taken to Fitch Hospital, where the efforts of surgeons accomplished little. She expired half an hour after midnight. By April 2, *The Buffalo Evening News* was able to report that Hort "laid it all to jealousy and bad temper. He said he went into the house from the barn with the hatchet in his hand, and without any words hit her on the head."[1] But in fact Hort was not John Hort, but William Kemmler, and Tillie was not really his wife, but his paramour, and her last name was Ziegler. The two had run away from their despised mates in Philadelphia with little four-year-old Ella and sought a new life in booming Buffalo. Instead of a new life, the man the local press quickly labeled the "South Division Street hatchet fiend" was about to achieve invidious notoriety with a new death. Within a week of Tillie's vicious murder, *The Buffalo Evening News* blared on page one the strange and terrible fate awaiting this fiend: ELECTRIC DEATH. THE HORRIBLE EXPERIMENT WHICH MURDERER KEMMLER WILL HAVE TO SUFFER IF CONVICTED.

The Buffalo police electrician waxed most philosophical and skeptical: "These so-called experts say they know that they can kill a man instantly. I would like to know how they can tell that. . . . Sometimes you will notice that a man or woman is killed by electricity by accident. Then again they will only be shocked. How do you account for that? . . . They may put a murderer in a metallic chair and when the circuit is turned on it may only paralyze him. What a horrible death that would be. So what will they do if the shock does not kill?"[2]

Through William Kemmler's four-day trial in early May, the "hatchet fiend" sat hunched over and silent. A jury convicted him on May 10, and in the midst of a thunder-and-lightning storm on May 13, 1889, the judge sentenced him to die by electricity, a coincidence duly observed by the local press. The case attracted much attention nationally purely because of the unique form of death awaiting Kemmler. Back in New York City, Harold Brown, now the official New York State expert on electrical execution, was alerting his allies that "a conviction under the new law was reached at Buffalo on Friday—a brute who chopped a woman into bits with an axe."[3] Brown had long since taken to calling AC "the executioner's current." It was an odd coincidence that the first victim of the newfangled electric chair was to be from Buffalo, for that was also the home of Dr. Alfred P. Southwick, the dour dentist and thanatologist so instrumental in replacing the hangman's rope with the electric current. It was also the first city where Westinghouse had installed an AC central station to light up the four floors of fancy goods offered by the luxurious Adam, Meldrum & Anderson department store.

Now, at last, in the spring of 1889, Edison and Brown had their official human victim. With that, the War of the Electric Currents entered its most ghoulish and macabre phase. Arrayed on the DC side were, of course, Harold Brown and Thomas Edison, both delighted that a human being was to be officially and specifically electrocuted with the "man-killing" Westinghouse generators. This would be a stupendous public relations coup, indelibly linking alternating current to death and criminality. On the other side, naturally, stood George Westinghouse, equally determined to thwart the ignoble defilement of his machines, whose whole purpose was to bring lovely light and clarity into the daily lives of men and women.

At the time of William Kemmler's trial in Buffalo, all the money the accused ax murderer had in the world was about $500 from the selling of all his vegetable wagons and horses. But Kemmler said very definitely he would not spend any of that on lawyers, because he intended to turn over this sum for Tillie's burial—in a silver-handled casket, no less—and for the future care of Tillie's little girl, Ella. Yet soon after Kemmler's death sentence, one of the era's rising legal stars, former New York congressman W. Bourke Cockran, materialized as Kemmler's champion. Cockran would forcibly argue that death in the electric chair was a violation of federal and state constitutional prohibitions against cruel and unusual punishment. Were his handsome fees as William Kemmler's champion paid by Westinghouse? It was widely assumed so, but never definitively proven.

For this most cold-blooded of all the DC versus AC battles, it would be Cockran, not Westinghouse, who led the public attack. Cockran, thirty-five, was already well seasoned in battles political and legal, a celebrated orator of the Gilded Age who would soon return to Congress to serve another decade. A top-level sachem in New York's

Tammany wigwam, Cockran had landed up in Manhattan at age seventeen, an Irish immigrant with the unlikely advantage of an elegant French education. Tall, commanding, with a leonine head, deep-set eyes, and a clean-shaven, expressive face, Cockran was always a firm friend of the down-and-out. But he was also a highly paid corporate lawyer, acting, for instance, as counsel to Joseph Pulitzer, owner of the fabulously successful and influential *New York World* newspaper. Cockran's later legal clients included many of the great corporations of the day—railroads, utilities, steamship lines. Yet the ardent opposition of this silver-voiced attorney (who spoke still with an Irish lilt) to capital punishment led him to represent a number of murderers during his illustrious legal career.

Even as Cockran was preparing his appeal of Kemmler's electrocution, Harold Brown was pushing forward with the grisly practical details of the electrocution apparatus. Every step of the way, he required, and received, Thomas Edison's all-important support. First in early March, when Brown had had to perform a few demonstrations for the New York State prison authorities, an Edison official aiding Brown wrote desperately to his boss, Thomas Edison, "I have been trying for the past week to buy, borrow or steal a Westinghouse dynamo but have been unsuccessful. I am afraid therefore that we shall have to trespass again upon your good nature.... Would it be possible to rewind your Siemens alternating dynamo so that we can get at least 1000 volts?"[4] It all went smoothly, and the state said it would order three dynamos for Auburn, Sing Sing, and Dannemora Prisons, and it would purchase them from Brown, but the terms were quite strict: The state would pay the estimated $7,000 cost of the dynamos only when "the first execution proves that the plant is suitable for the purpose."[5] From his new prestigious Wall Street address, Brown wrote a wheedling letter to Edison on March 27, stating he would need $5,000 to be able to afford to proceed and "the people at 16 Broad Street [Edison corporate headquarters] do not feel like undertaking the matter unless you approve of it. Do you not think it worth doing, as it will enable me, through the Board of Health, to shut off the alternating current circuits in the State?"[6]

At this critical moment, the wheeling-dealing Charles Coffin of the

Thomson-Houston Company quietly stepped forward. At first glance, the firm's surreptitious entry into the fray was decidedly odd, since Thomson-Houston had been an AC proponent and for two years the firm had paid a licensing fee to Westinghouse. But that all had ended the previous year when Westinghouse lost a patent suit over the Gaulard-Gibbs transformers. Freed from the licensing alliance with Westinghouse, Thomson-Houston was engaged in active talks about a possible merger with Edison Electric. So Coffin began secretly helping Brown to buy used Westinghouse alternators for the three prisons. This was a critical logistical victory. The DC forces now had their "man-killing" AC machines—manufactured by Westinghouse. On the Fourth of July, while others were enjoying patriotic speeches and parades of Civil War veterans, Brown was devoting himself to electrical death, informing Coffin defensively, "I have withstood tremendous pressure and tempting offers to use other machines for that purpose but have nevertheless kept my agreement to the letter."[7]

When on May 23 a fully sober Kemmler was led off the Buffalo train at Auburn, he was escorted straight across the tree-lined street. Before him loomed Auburn Prison, a gray stone fortress surrounded by a thick twenty-foot-high wall with manned guard towers. The prison's grim and ominous appearance had been softened somewhat over the years by a luxuriant covering of green ivy alive with gaily twittering brown sparrows. Located in the small town of the same name on the heavily traveled Erie Canal halfway between Buffalo and Albany, Auburn Prison was the second built in the New York State system and had opened in 1817. Aside from the busy chirping and twittering of the sparrows, the prison was unnaturally quiet, even though it housed 1,200 inmates. All were garbed in its famous black-and-white-striped uniforms, and all were subject to the prison's much admired system of extreme regimentation. "Absolutely no communication was allowed among the men," writes Thom Metzger. "Their eyes were to be downcast at all times. They did the 'Auburn' shuffle marching in lockstep with one hand on the next man's shoulder wherever they went."[8] Amid the habitat of more than one thousand men, one heard only the muffled sound of lockstep movement. Oddly, William Kemmler, as a condemned man, was exempt from this regime. In his tiny cell, he was allowed to wear regular clothes and hear human voices. The guards who kept him under continual watch read Bible stories and popular novels aloud, perhaps comforted by the sound of a human voice in that grimly silent place.

As June turned to July in that summer of 1889, attorney Bourke Cockran's constitutional appeal of Kemmler's death in the electric chair slowly moved forward. Cayuga County judge Edwin Day authorized Buffalo attorney Tracy C. Becker to serve as a referee, taking testimony to determine whether death by electricity would be less painful than the traditional hangman's noose or if electrocution was more cruel and unusual and therefore unconstitutional. Harold Brown was scheduled to testify, of course, which created some uneasiness in the Edison camp. A fortnight before the hearings were to begin, the Englishman Arthur Kennelly, Edison's chief electrician, contacted Brown. "At Mr. Edison's instance," Kennelly advised, they wanted Brown to be aware that "the only argument of any weight which can be urged against electrocide on the score of a cruel punishment is that its application may burn the flesh of the criminal at the points of contact, and that the amount of current which can be given without such mutilation is not yet known."[9]

The first day of hearings was Monday, July 9, in Bourke Cockran's well-appointed offices at the prestigious and splendid Equitable Building. Through the open windows, for it was summer, came the muted cacophony of Broadway below, newsboys yelling the day's headlines, hawkers tempting passersby with corn on the cob, teamsters lashing their horses forward. For the next two months, the silver-tongued Cockran would hammer pleasantly away at two fundamental questions: Were the various electrical experts genuinely knowledgeable in the aspects of electricity at issue here? and How could anyone guarantee instant and painless death, death that was not cruel and unusual, to William Kemmler when there were so many instances of people subjected to huge amounts of electricity who had survived? Like any great orator, Bourke Cockran understood the power of humor and public fun. So while a major part of his legal strategy was to produce a stream of people who questioned the certainty of quick, painless death from electricity, they themselves often having personally survived great shocks from lightning and wires, he also introduced as evidence (and comic relief) the amazing dog Dash.

The first witness was, fittingly enough, Harold Brown. Once on the stand, Brown received a thorough grilling, albeit delivered with Cockran's silvery Irish lilt. Cockran began by establishing that Brown, who described his professional status as an electrical engineer, was not particularly well trained in that field or well regarded. He was not, Brown conceded, a member of the American Institute of Electrical Engineers, nor did he have any schooling beyond high school. Brown asserted that his thirteen years working for such companies as Western Electric and Brush Electric were equal to any degree. He had been on his own for five years. "My particular business at present is designing apparatus for people who require it, or standing as an expert between the purchaser of electric apparatus and the company supplying them . . . or as an expert in advising in matters in which electricity is used." He insisted that "I at present am entirely independent of any company."

Brown seemed quite pleased when Cockran began inquiring about the dynamo machine for Auburn Prison. "That is actually there?" Oh, indeed, Brown said smugly, and it was a Westinghouse alternating current dynamo. Through two days of questioning, Brown conceded little. He did have to admit that there was one dog, Ajax, who refused to die during Brown's dog-killing experiments despite numerous electrical jolts. But generally Brown remained adamant that he was well enough versed in practical electricity to guarantee Kemmler a quick, painless death. All in all, reported *The New York Times,* Brown "underwent the ordeal" of testifying "without being disconcerted."[10] Certainly Brown's own fortunes had risen considerably since he'd appointed himself to lead the anti-AC battle. His offices had moved from the unfashionable boondocks of West 54th Street to the golden precincts of 45 Wall Street. And he was now a recognized ally of America's most prominent inventor and electrician.

On Monday, the hearings recommenced in Cockran's offices. That day's witnesses served Cockran's purposes well. Daniel Gibbens, one of the feckless commissioners from the New York City Board of Electrical Control who had watched Brown's dog killings, turned out to be skeptical of quick and certain electrical death. He grimaced just remembering the dog trials. "It was one of the most frightful scenes I have ever witnessed. The dogs writhed and squirmed and gave vent to their agony in howls and piteous wails." As for electricity itself, Gibbens said, it was so unpredictable in its effects on different animals and people, "just as the effect of whisky varies when used by different men."[11]

Then Cockran brought on an odd character named Alexander McAdie, a Harvard graduate who had worked at the U.S. Signal Service Laboratory in Washington, D.C., there devoting himself to the study of lightning. Cockran asked the young man if he thought the electric chair would work. He responded haltingly, "Its deadly effect would depend upon the subject's resistance and upon the route through the body.... It might only paralyze one half of his body and leave the other half unharmed.... It might kill him, and if it didn't kill him instantly, it might carbonize him—burn him up.... Yes, I think it would char his flesh."[12] The strange Mr. McAdie, who had left the weather service to study lightning on his own, described standing atop the pinnacle of the Washington Monument during a thunderstorm and deliberately taking lightning through his own body—and here he was to tell the tale. One can imagine Cockran's glee at this damaging testimony. "Carbonize" Kemmler—if that wasn't cruel and unusual, what was?

To keep up the lively doubts generated in the morning session, after lunch Cockran brought on one of his most entertaining witnesses: Dash the dog, a "splendid looking... big fellow, a cross between a Scotch collie and a St. Bernard." Dash, it was related, had been thrown four feet in the air by a dangling Western Union wire supercharged with errant AC, knocked unconscious, feared dead, and then revived many hours later. Dash was living canine proof that a big mammal could be knocked out cold, taken for dead from electric shock, and gradually come back to life. Might Dash the dog foretell Kemmler's dreadful fate of rising from the seeming dead? This was disquieting information to the state, which wished Kemmler's life permanently and electrically extinguished. Surely putting a man to death twice would be considered cruel and unusual? The day ended with the cheering news that Elbridge T. Gerry, head of the legislative New York State Death Commission (which promoted electricity over the rope), would now be able to detach himself from the summery delights of yachting at Newport to testify at the next day's session.

Cockran's puckish approach to the deadly question of painless finality had stirred up enough naysayers and doubters to get the state and Harold Brown worried. On July 17, the day after Dash the dog's winning appearance, Harold Brown was anxiously requesting the ultimate reinforcement for this battle of judicial opinion: the appearance of the world-famous electrical wizard Thomas A. Edison. Who would give a second thought to a McAdie or a Gibbens if Edison said the electric chair was a sure thing? In the year and a half since Thomas Edison had written that fateful letter to Dr. Southwick endorsing electricity in general—and Westinghouse machines in particular—for electrocution, the great inventor had never once personally appeared in a public forum or uttered a public word on the whole matter, allowing others always to wage his battles. Now for the first time he was flushed out from the shadows. Without his influence, Harold Brown and the state might lose to the silver-tongued Cockran.

On July 23, a cool and rainy day in an unusually cool summer, Edison sallied forth for the first time to assume personal leadership of the DC forces. He ascended to Bourke Cockran's law offices, which were filled with eager spectators hoping to see and hear America's most celebrated inventor in the guise of star witness. Harold Brown attended as his aide-de-camp. Such was the great man's deafness that the yelled questions and Edison's equally loud answers "might have been heard on the street," reported *The New York Times.* At one point, a smiling Edison arose and dragged his chair closer to Cockran so he could hear through his better ear. There was a great deal of (loud) discussion once again about what the average man's resistance was in ohms, since presumably resistance was an integral reason some men survived big electric shocks and others keeled over dead. How could anyone know for sure? But Edison did. He said he had made numerous experiments of resistance on the 250 men in his lab before coming to testify. This revered scientist and inventor, the man who had brought the electric light to the world, was unequivocal in his answers to Bourke Cockran's questions. Asked Cockran loudly in his Irish brogue: "In your judgment, can artificial electric current be generated and applied in such a way to produce death in human beings in every case?"

"Yes," said Edison, fingering his unlit, half-smoked cigar.

"Instantly?" Cockran asked.

"Yes." Edison's only caveat was that "the culprit's hands [should be placed] in a jar of water diluted with caustic potash and connecting the electrodes therewith."

"How much of a current do you think it would take to burn a man?" Cockran asked very loudly. *The New York Times* reported the following exchange.

Edison thought a moment and answered probably " 'several thousand horse power . . . you'd probably burn him up.'

" 'Have a nice little bonfire with him, would you?'

" 'Oh, no,' said Edison, 'Just carbonize him.'

" 'Well, Mr. Edison, with this tremendously wicked Westinghouse dynamo,' Mr. Cockran threw all the sarcastic power at his command into the question—'when it has been worked up to its most thoroughly wicked point, how long do you think it would take to burn a man?'

" 'His temperature would rise 3 or 4 degrees above the normal and after a while he'd be mummified.'

" 'Mummified,' cried Mr. Cockran gleefully. 'Now we are getting the true inwardness of electrical science. How?'

" 'The heat would evaporate all the fluids in his body and leave him mummyized.' "[13]

Naturally, Bourke Cockran inquired about the nature of Edison's relationship with Harold Brown, the official state electrical execution expert. Had Edison, for instance, ever given him a letter of recommendation? Not at all, said Edison, obviously forgetting or ignoring a March 22, 1889, testimonial he had provided at Brown's behest so Brown could prove his bona fides to the mayor of Scranton. His dealings with Mr. Brown, said Edison, were strictly limited to Brown's use of Edison's West Orange laboratory, a privilege granted to numerous other engineers and scientists. After a few more questions, Bourke Cockran dismissed Edison, first lighting his half-chewed cigar stub for him. The newspaper headlines the next day were all that Harold Brown and the Edison Company could have hoped for: EDISON SAYS IT

WILL KILL, THE WIZARD TESTIFIES AS AN EXPERT IN THE KEMMLER CASE, HE THINKS AN ARTIFICIAL CURRENT CAN BE GENERATED WHICH WILL PRODUCE DEATH INSTANTLY AND PAINLESSLY IN EVERY CASE—ONE THOUSAND VOLTS OF AN ALTERNATING CURRENT WOULD BE SUFFICIENT.

Cockran did a yeoman's job of trying to show that the great and beloved Edison was woefully ignorant on this particular aspect of electricity—its effect on the human body and its ability to kill swiftly and painlessly. Yet Edison's usual cocky manner and absolute assertions carried the day. Historians Terry S. Reynolds and Theodore Bernstein argue, "Edison's reputation probably overrode Cockran's exposure of [Edison's] ignorance of the effects of electricity on living organisms." Certainly some newspapers regarded his testimony as critical. The *Albany Journal*, for example, noted: "The Kemmler case at last has an expert that knows something concerning electricity. Mr. Edison is probably the best informed man in America, if not the world, regarding electric currents and their destructive powers."[14]

Edison and his lovely young wife, Mina, sailed off to Paris ten days later for a two-month visit. On the Continent, Edison the international celebrity was feted at elaborate and adulatory banquets, presented official state honors from France and Italy, and received standing ovations at the Paris Opera, with the entire bejeweled crowd chanting, "*Vive Edison! Vive Edison!*" Hailed by the local press and officials as a genius, Edison reaped phenomenal publicity for his company's products at the important Paris Exposition. His exhibition enthralled huge crowds daily with its twenty-five "perfected" phonographs speaking in dozens of languages, along with all manner of electrical lights and devices. Edison adored the Eiffel Tower, erected for the exposition by Alexandre-Gustave Eiffel as a fantastic specimen of modern engineering and lit up at night from top to bottom. "The tower is a great idea," Edison told the dozens of reporters hanging on his every lively phrase. "The glory of Eiffel is in the magnitude of the conception and the nerve of the execution. I like the French, they have big conceptions. The English ought to take a leaf out of their books. What Englishman would have had this idea? What Englishman could have conceived the Statue of Liberty?"[15] However, the stylish Parisian way of life annoyed Edison. "What has struck me so far chiefly is the absolute laziness of everybody over here," he commented bluntly. "When do these people work? What do they work at? People here seem to have established an elaborate system of loafing. I don't understand it at all."[16]

Back across the Atlantic in suddenly hot and sticky Manhattan and then in cooler Buffalo, Bourke Cockran plugged away for a few more sessions, taking the testimony of physicians who had attended men killed by electricity or hearing from men who had survived lightning strikes. Cockran elaborated his basic themes with verve, but there was no one he could produce for his side with Edison's fame or stature. When Edison the world-famous celebrity weighed in so definitively for the electric chair, the battle was almost certainly lost.

While New Yorkers sweltered in a blanket of oppressive August heat, the electricians had gathered farther north in the cool and watery precincts of Niagara Falls to vent their spleen against Edison, Brown, and the planned electrocution. One man angrily told his fellow members of the National Electric Light Association, "We are here for the purpose of advancing the uses of electricity, to make it rejuvenate the world, to carry it forward as a civilizing agent, not as an instrument of torture. . . . I say, let us here condemn that action. . . . Let it not be trumpeted over this country that the dying groans of that criminal cursed electricity with its last sound."[17] The assembled agreed to dispatch emissaries to the governor of New York to push for repeal of death by electricity.

This was rather encouraging to the Westinghouse forces, and late August of 1889 brought yet another small but delicious triumph. In its Sunday edition, the *New York Sun* ran an exposé on Harold Brown, with the headline FOR SHAME, BROWN! and a subhead that told the story: "Paid by One Electric Company to Injure Another." Someone had broken into Harold Brown's Wall Street office and stolen forty-five letters from his locked rolltop desk. These missives showed that for some time he had indeed been advised, aided, abetted, and paid by the Edison Company and Thomson-Houston, both explicit rivals of Westinghouse. As the *Sun* wrote, "Brown is known not to be a wealthy man, and that he could afford to devote all his time thus purely for the benefit of the human race at large with little thought of self, has been a mystery to those acquainted with him."[18] Yet little changed. Brown complained to the district attorney, requested an investigation, and offered a $500 reward for information about the thief. To the rest of the press, Brown still blustered, "I am exposing the Westinghouse system as any right-minded man would expose a bunco starter or the grocer who sells poison where he pretends he sells sugar."[19]

On September 11, referee Tracy C. Becker submitted to Judge Edwin Day the full record of 1,025 pages of transcribed testimony, describing the widely varying electrical near death and actual death experiences, with all the contradictory opinions about how quick and painless Kemmler's electrocution would be. Just over a month later, on October 12, a few days after Thomas Edison returned in triumph from Europe, Judge Day ruled against Cockran, who immediately appealed to the Supreme Court of New York.

Even as the Kemmler case dragged on through the courts, there came a great and unexpected coup for Edison: the most spectacular high-voltage death yet. It was a death so hideous, so public, so highly visible—occurring just blocks from City Hall at lunch hour—that it galvanized the high-voltage safety debate as nothing before had managed to do. It was, unbelievably, the second public roasting of a lineman in three days in lower Manhattan. The first death in the wires on October 9 killed a fellow so dissolute and brutish that his wife had long since fled with their six children to a refuge in the country. The strong implication was that the first lineman was working while drunk, became careless, and managed to electrocute himself. He toppled dead onto the pavement. An ugly tale. But two days later, on Friday, October 11, came yet worse. Far, far worse.

This time, on a lovely fall noon a handful of Western Union linemen were working forty feet up a towering wooden pole at Chambers and Center Streets, a block from the magnificent Tweed Courthouse, cutting out dead wires from the great spiderweb of lines looping hither and yon from poles to buildings and back. Far below, the thick lunchtime crowds surged along the sidewalks and across the streets, weaving in between the horsecars and teamster carts. One of the linemen, John E. H. Feeks, high above the milling throngs, was standing astride the fourth crossbar from the bottom of the fourteen bars laddering up the top of the light pole. He reached through a tangle of wires to cut out a dead wire "when he was suddenly seen to shiver and tremble as though he had received a violent shock. He put out his right hand and seized a wire as though to steady himself, and immediately there was a flash of flame under his hand. Then bright sparks and tongues of blue flames played all about his hand and a small cloud of smoke curled up into the air. His right hand next slipped from the wire and he fell forward across a network of wires which caught him across the throat and face and held him suspended some forty feet above the ground. The man appeared to be all on fire. Blue flames issued from his mouth and nostrils and sparks flew about his feet. Then blood began to drop down from the body on the pole and a great pool formed on the sidewalk below. . . . There was no movement to the body as it hung in the fatal burning embrace of the wires. A great crowd of people collected and stood awestricken and fascinated by the fearful sight."[20] As some men yelled for help, mesmerized spectators jammed the sidewalks and roadway, blocking all the passing streetcars and teams. From every nearby window and roof, people craned through the haze of electrical wires to see this grisly sight of the smoking, sparking man. Feeks's body was so interwoven with the numerous wires that he swayed but did not fall. His corpse was held horribly aloft for forty-five minutes, like a poor fly stuck in the spider's sticky web, until at last the current was turned off and his blackened body extricated and lowered to the silent street.

Public outrage at this latest accidental electrocution reached fever pitch. Feeks was a solid citizen and husband, long employed. A tin cracker box donated by nearby Coogan's Saloon was nailed to the deadly light pole for donations to the deceased man's pregnant wife, already mother to one child. Reported the *Times,* "Men and women whose dress was of the poorest came in a constant stream and dropped in money. Newsboys, bootblacks and Italian fruit vendors brought pennies and nickels. Drivers on the Madison avenue cars stopped their teams and ran to the box to drop in a dime." In ten hours the amazing sum of $822.23 was gathered. Three days later, that figure had risen to $1,873.50. (It's unlikely that Feeks earned more than $12 a week himself.)

Mayor Hugh Grant rose from his sickbed and came to City Hall to order the shutting down of all the high-voltage electric arc lights in Manhattan while the companies were forced to remove, repair, and upgrade the jungle of overhead wires. The citizenry, who had not experienced such pitch black cloaking the city in decades, since before there was gaslight, were most distraught, and a great hue and cry arose at the loss of the man-made light. "Again last night did the city seem to have gone into mourning for its lost brilliancy," lamented *The New York Times.* "Darkness and gloom were everywhere."[21] Several unhappy facts quickly emerged: The New York City Board of Electrical Control had made little headway in getting built the electrical subways mandated by law for all the city's electrical wires. The three Tammany-appointed commissioners each took home the breathtaking salary of $5,000 a year and rarely did a day's work. The press had a field day of high dudgeon: OF NO USE TO THE PUBLIC, THE DILATORY BOARD OF ELECTRICAL CONTROL or NOT ON THE PUBLIC'S SIDE, THEY BLOCK THE MAYOR'S EFFORTS TO HASTEN THE SUBWAY WORK—THE CITY IN DARKNESS.

The heinous Feeks affair propelled Thomas Edison at long last fully into the open. Emerging from the shadows whence he had directed Harold Brown, Edison stood forth in all his radiance as the true general and leader of the DC forces. For the first time, Edison personally issued his own clarion battle cry in his holy war against the "executioner's current": Death to AC! In the November issue of the prestigious *North American Review,* Edison thundered that the "the martyrdom" of Feeks offered dramatic and horrible witness of what the future held—*unless* electrical pressures (voltage) were legally limited. Edison had heard the "popular cry" to bury the wires and restore safety to the city streets. But this was no solution, he asserted. Burying AC wires "will result only in the transfer of deaths to man-holes, houses, stores, and offices, through the agency of the telephone, the low-pressure systems, and the apparatus of the high-pressure current itself." DC should never be higher than 700 volts. As for the safe voltage for AC, "I myself have seen a large healthy dog killed instantly by the alternating current at a pressure of one hundred and sixty-eight volts . . . it is difficult for me to name a safe pressure." He then told how his own company had purchased the ZBD patents. "Up to the present time I have succeeded in inducing them not to offer this system to the public, nor will they ever do so with my consent. My personal desire would be to prohibit entirely the use of alternating currents. They are as unnecessary as they are dangerous."[22]

As the War of the Electric Currents grew uglier and fiercer, George Westinghouse decided in the fall of 1889 to hire a Pittsburgh newspaper reporter named Ernest H. Heinrichs to promote his companies and their achievements. On Heinrichs's first day, Westinghouse came by to wish him success and explain his purpose. "All I want to see is that the papers print [things] accurately. The truth hurts nobody."[23] One November morning soon thereafter, Heinrichs was installed at his desk in the brown, turreted nine-story Westinghouse Building scanning a New York newspaper with an article and an editorial attacking AC and Westinghouse. The young man became so incensed, he leaped up and rushed into his boss's office without knocking. Westinghouse was sitting in his big upholstered chair at the gigantic wooden dining table that served as his desk. He was calmly reading that selfsame newspaper. He saw that Heinrichs was agitated and what he was clutching. The Pittsburgh industrialist cocked his great head and asked Heinrichs, "Well, what's the hurry?"

"Don't you think we ought to say something against these slanders and false statements?"

Heinrichs would always remember how Westinghouse eyed him for a few seconds, the wooden wall clock above the mantel ticking through the silence. Then Westinghouse smiled.

"Heinrichs, they tell me you are quite a Whist player. Is that so?"

He admitted a fondness.

"Well, then, you know the meaning of the expression, 'Don't play the other fellow's game.' "

Heinrichs found this thoroughly puzzling. What did whist (which was akin to bridge) have to do with Edison and his calumnies? Westinghouse explained, "Now seriously speaking, all this opposition to the Alternating Current is doing our business a great deal of good. We are getting an invaluable amount of free advertising. . . . As a practical, commercial proposition the Alternating Current system is so far superior to the direct current that there is really no comparison. . . . By keeping up this agitation about the deadly Alternating Current, they are playing our game and we are taking the tricks. . . . They hope that by their power, their influence, they can accomplish the arrest of the march of progress. This, by the very laws of nature cannot be done. . . . As to the attacks made against me personally, of course they hurt, but my self respect and conscience do not allow me to fight with such weapons. Besides, I feel that my moral reputation and my business reputation are too well established to be hurt by such attacks. However, I am preparing an article for the *North American Review* in answer to Mr. Edison's charges against the Alternating Current system, but beyond that I shall have nothing to give you for publication. . . . By letting the others do all the talking, we shall make more friends in the end than if we lower ourselves to the level of our assailants."[24]

The December issue of the *North American Review* could have done nothing to improve Edison's venomous feelings toward Westinghouse, for his adversary had penned a blunt and steely "Reply to Mr. Edison." The AC-DC battle was just the latest in a long "struggle for the control of the electric light business [that] has never been exceeded in bitterness by any of the historical commercial controversies of a former day. Thousands of persons have large pecuniary interests at stake, and, as might be expected, many of them view this great subject solely from the stand-point of self interest." Westinghouse tried to put it into perspective with the following: In the year 1888, sixty-four people in New York City were killed in streetcar accidents, fifty-five by omnibuses and wagons, twenty-three by illuminating gas, and all of five by electric current. This was not exactly an orgy of wanton and careless killing. The bold Westinghouse described Edison's cherished DC central plant as "regarded by the majority of competent electrical engineers as in many respects radically defective; so defective, in fact, that, unless the use of alternating currents can be prohibited, it seems destined to be wholly supplanted by the more scientific and in all respects (so far as concerns the users or occupants of buildings) far safer inductive system."

The by now familiar arguments were lobbed back and forth about copper costs, transformers, who had or had not survived shocks of what amounts. But Westinghouse ended his counterattack with two new and terrible strikes to the Edison forces. The first was quite self-inflicted by the Edison soldiers and therefore all the more painful. At the annual August meeting of the Edison Illuminating Companies, held in the water-cooled environs of Niagara Falls, New York, Westinghouse reported, the manager of the Detroit Edison Station had introduced a resolution, which passed. It asked the parent company to provide "a flexible method of enlarging the territory which can be profitably covered from their stations for domestic lighting *by higher pressures* and consequently *less outlay of copper* than that involved by the three-wire system." Edison's own troops were breaking ranks and asking for AC! Westinghouse's final and rather devastating salvo was simply that "for three years past the purchasers of apparatus for electric lighting, who are at perfect liberty to buy from any company, have, for the most part, preferred to use the alternating system, so that today the extension of that system for central station incandescent lighting is at least five times as great as that of direct current."[25]

Having emerged from the shadows, Thomas Edison did not retreat again. He now proceeded to use the full force of his monumental fame and prestige to persuade the public and politicians that there was safe electricity, which was his—low-voltage DC, whose transmission lines were safely buried—and dangerous electricity—high-voltage AC, which was carried on open wires. His goal: Such public fear of AC that it would be legally banned from use in the United States. He would thereby eliminate Westinghouse from the field and recover his own company's primacy, which was faltering. Predictably, the next battlefields for the escalating War of the Electric Currents were the state legislatures, where Edison and Brown hoped to ruin Westinghouse by governmental bans against high-voltage electricity.

The first clash was at Richmond, Virginia, capital of the Old South. Westinghouse hired powerful lawyers and one of Edison's longtime enemies, Professor Henry Morton of the Stevens Institute of Technology, to serve as an expert. On February 12, 1890, Edison himself appeared as the first witness to testify before the Virginia State Senate. The hearing room was packed with many men and women craning for a glimpse of America's most beloved inventor. Edison's worsening deafness made it hard for him to hear and, thus, to answer the committee's questions. The famously witty raconteur was not as eloquent as hoped. Edison was followed by Professor Morton, who long ago had very publicly pooh-poohed Edison's invention of the light bulb. Now, Morton denigrated his old enemy on new grounds, asserting that AC was, contrary to Edison's alarmist views, a perfectly benign force when handled responsibly.

But the most compelling witnesses turned out to be the local arc-lighting men, who rushed to defend their flourishing businesses from these battling Yankees. Some were even sons of the old Confederacy and thus gained instant sympathy. "The first of these gentlemen who was called upon had but one leg and used a crutch. . . . He expressed himself fluently and with great force. . . . In closing, he derided the suggestion that 3,000 volts was dangerous and exclaimed, 'Why gentlemen, the pennyroyal bulls of Fairfax County are far more dangerous than that current.' "[26] The Westinghouse men quickly saw that here were their best allies, far more persuasive to state representatives than eminent northerners such as Thomas Edison or Professor Morton. The Edison people had failed to take into account the powerful arc light lobby, for almost every American city of any size now had some sections with arc lights. Those local companies would be destroyed by a high-voltage ban. The Edison DC forces would lose in Virginia, but this did not deter Edison and Brown from pressing on, presenting their case—sometimes illustrated by Brown's ghoulish dog shows—in other states and Canada, determined to shut down AC via state legislatures.

As 1889 turned to 1890, Judge Dwight of the Supreme Court of New York had rejected Kemmler and Cockran's appeal, seeing nothing cruel or unusual in death by electricity. Cruel punishments, he wrote in his opinion, included such deaths as "burning at the stake, breaking on the wheel, being fired out of a cannon, hanging in chains to die of starvation, or disemboweling and crucifixion."[27] When Cockran heard the disappointing (but not unexpected) verdict, he immediately announced, "It will be taken to the Court of Appeals."[28] By spring of 1890, Cockran had once again lost. This time the court pointed out that the question of cruelty had been thoroughly addressed by the New York State Death Commission when it chose electricity over the hangman's rope. By late April, the newspapers were having a field day, churning out stories preparatory to the nearing electrocution, describing Kemmler's efforts to make a will, his sagging spirits, and the construction by "stripeds" of his plain pine coffin. Kemmler passed the time in his small cell reading simple children's Bible stories, playing a pigs-in-clover puzzle (marbles tilted into holes on a small board), and laboriously scrawling his signature on small cardboard cards that he gave to the warden's wife and favored guards. These were much sought after by autograph collectors.

After Cockran's appeals had come to naught, the prison issued a statement said to be from Kemmler: "I am ready to die by electricity. I am guilty and I must be punished. I am ready to die. I am glad I am not going to be hung. I think it is much better to die by electricity than it is to be hung. It will not give me any pain. I am glad Mr. Durston is going to turn the switch. He is firm and strong. If a weak man did it, I might be afraid. My faith is too firm for me to weaken. They say I am not converted. I don't care what they think. I know what I've got. I am happy to die. I have never been so happy in my life as I have been here." Rampant rumors that Kemmler had gone stir-crazy under the strain of waiting for his electrocution may have prompted this report to show Kemmler's state of mind: resigned but sane.

And what of "The New Instrument of Execution"? In the same November 1889 issue of the prestigious *North American Review* that featured Thomas Edison's high-minded crusade urging that AC be outlawed, Harold P. Brown was describing (in an article with this very title) what the electric chair would do and how it would work. He imagined the coming event: "The condemned criminal's cell is visited by the prison authorities and his hands and feet are saturated with the weak potash solution which so rapidly overcomes the skin's resistance; during this space of thirty seconds or less the electrical resistance may be measured.... Shod in wet felt slippers, the convict walks to the chair and is instantly strapped into position, his feet and hands are again immersed in the potash solution contained in a foot-tub connected to one pole and in hand-basins connected to the other. With this perfect contact there is no possibility of burning of the flesh and thus reducing the effect of the current upon the body.

"Dials of electrical instruments indicate that all the apparatus is in perfect order and record the pressure at every moment. The deputy-sheriff closes the switch. Respiration and heart-action instantly cease, and electricity, with a velocity equaling that of light, destroys life.... There is a stiffening of the muscles, which gradually relax after five seconds have passed; but there is no struggle and no sound. The majesty of the law has been vindicated, but no physical pain has been caused."[29]

In late April, the twenty-five witnesses to the state's first official electrocution began to gather in Auburn. Harold Brown was conspicuous by his absence. Almost certainly, his unmasking a year earlier by the *Sun* as an Edison/Thomson-Houston corporate lackey had diminished his usefulness to those parties. When his original contract as the state's expert on electrical execution had expired on May 1, Brown did nothing to extend it. Perhaps the warden banished him. In any case, after all his eager, relentless pursuit of electrical death by Westinghouse AC, Harold Brown now made a great about-face, claiming to reporters, "You may rest assured that I am glad to be relieved of the unpleasant responsibility."[30] Dr. Alfred P. Southwick, the Buffalo dentist and chief instigator of the electrical death penalty, was naturally present at Auburn as an enthusiastic witness. A big man with a fringe of white beard, he, like many others, was taken aback by the malevolent appearance of the chair itself, a brutish-looking oversize oaken armchair with wide, flat arms, a crude footrest, and a perforated wooden seat. There were numerous thick leather restraint straps—and most disturbing, a heavy leather mask that enveloped and covered the criminal's face, pressing it back into a neck brace that would have a saturated sponge. Southwick explained, "I am opposed to so much paraphernalia, but the present arrangement will have to do, because we cannot afford to suffer failure. The whole world is watching the result of this experiment, and if we neglect any precautions there might be a slip and the system would therefore be condemned. I am fully convinced that Kemmler's death will be instantaneous. . . . I anticipate no disfigurement at all."[31]

Just as all the preparations were being completed, a law clerk showed up in Auburn. Kemmler's case had been further appealed! A new lawyer, Roger Sherman, who specialized in appeals, was taking up the cudgels. He came to the prison to see Kemmler but was not allowed a visit. He left more legal papers and swiftly departed back down to Manhattan. Sherman denied he was in the pay of George Westinghouse, but he would not say who was paying his handsome fees. When Dr. Southwick and the other witnesses, including referee Tracy C. Becker, also from Buffalo, heard the electrocution was off for the moment, they were thoroughly chagrined. To assuage them and provide some sort of event preparatory to Kemmler's death, Warden Charles Durston authorized the killing of a calf with the electric chair apparatus. Within the next month, the Kemmler case had reached the U.S. Supreme Court, with Sherman once again arguing that death by high voltage was cruel and unusual. But once again, in early August 1890, the Kemmler-AC forces were turned down. Chief Justice Melville Fuller said the death planned would have to be "something inhuman and barbarous—something more than mere extinguishment of life."[32] Meanwhile, at the Edison headquarters, gloating executives proposed that from here on in, "as Westinghouse's dynamo is going to be used for the purpose of executing criminals, why not give him the benefit of this fact in the minds of the public, and speak hereafter of a criminal as being 'westinghoused,' or (to use it as a noun) as having been *condemned to the westinghouse* in the same way that Dr. Guillotine's name was forever immortalized in France?"[33] The Edison officers were savoring this most monstrous and momentous of victories in the ongoing War of the Electric Currents.

In the stifling August heat, Auburn's best hotel, the four-story Osborn House, began to fill up with out-of-town reporters, mainly from New York City. The grimy railroad freight depot across from the high-walled prison proved convenient for a special Western Union office, equipped with fourteen lines just to New York City. With each arriving train, more people debarked and surged toward the grim, walled fortress of Auburn Prison. Tellingly, the warden's wife was seen to leave the prison and board a train out of town. She had a certain fondness for this condemned man she had taught to read and preferred not to be present at the prison when he died. She had also departed in late April when execution appeared imminent. On August 5, there was mounting excitement amid the oppressive hot and humid weather. All day, subdued crowds gathered outside the prison walls, pushing forward and clinging to the heavy iron bars of the entry gate. Young men shimmied up telephone poles and the tall leafy trees and peered over the twenty-foot prison walls toward the vine-covered prison building. Nearby rooftops and windows were lined with solemn spectators and reporters. That evening at 7:00 P.M., as the air cooled, many of the official state witnesses, serious men of medicine and law, bearded and clad in lighter summer suits, walked importantly from the Osborn House through the silent assembly of curious townfolk and in through those heavy barred gates. Some arrived by train and hurried across to the prison. And so, on that evening of August 5, some believed that the arriving witnesses were gathering for the long delayed execution. It was but a test, however, with one of the physicians volunteering to sit in the chair and experience the low-voltage run-through. The official witnesses then emerged and strolled back to the hotel in good spirits, asking to be awakened very early.

August 6, 1890, dawned the palest of blues in Auburn, and a cooling breeze riffled pleasantly through the city's many trees. It looked to be a glorious sunny summer day. Shortly after 6:00 A.M. the official state witnesses could be seen walking in straggling groups through the quiet village streets toward the fortress-style prison. They threaded through the hundreds of curious waiting townspeople and pacing newspaper reporters gathered outside the heavy iron gates. Today, finally, William Kemmler, the "South Division Street hatchet fiend," would become the first man in history to die in an official state electrocution. Each witness presented his pass and walked into the prison yard. Inside the prison itself, Warden Charles Durston looked distinctly distracted and unsettled. He had already been in earlier to see Kemmler, whose thick bushy beard and mustache had been neatly clipped. The murderer had been sitting on his cell bunk, attired for electrical death in new dark gray sack trousers, vest, and jacket, suspenders, a white shirt, and jaunty black-and-white-checked bow tie. He had spent a great deal of time combing his hair, carefully arranging a Hyperion curl on his forehead. Durston, who had often expressed distaste at his assigned role of electrocution overseer, now entered Kemmler's cagelike cell for the second time that early morning and read him the state death warrant. "All right," said Kemmler, "I'm ready."

Then his jailer from Buffalo appeared to say good-bye and was invited to eat breakfast with the condemned man. Just then, two ministers walked in who had visited frequently in the fourteen months that Kemmler's case had dragged through the labyrinthine appeals process. They all knelt now on the hard stone floor and prayed quietly. Breakfast was then served to the subdued group. Before they could leave Kemmler's tiny cell, the Buffalo jailer had the awkward task of cutting a slit down the seat of Kemmler's pants so the electric chair's electrodes would make ideal contact. For the same reason, he also nervously shaved a big patch atop the condemned man's head. As the jailer manipulated the razor, Kemmler spoke to him: "They say I'm afraid to die, but they will find that I ain't. I want you to stay right by me, Joe, and see me through this thing and I will promise you that I won't make any trouble."[34] With Kemmler's hair razed away on top, Warden Durston descended with him to the electrocution chamber at 6:32 A.M. For reasons unknown, the previous day Durston had moved the heavy wooden electric chair from its original site in an upstairs room to an isolated basement room. Now, the dynamo was a thousand feet away in the prison marble shop, and all communication with its operators would be based on bells.

Warden Durston ushered Kemmler into a death chamber incongruously bright with morning sunlight. Cheerful shafts were streaming through two windows seven feet above the rough-planked wood floor. This small basement room had once been used to process new prisoners, who had washed at a sink and bathtub in the corner before donning their "stripes." Several months earlier, the room had been painted a soothing pale gray. Two double-armed gaslight fixtures hung from the ceiling. The twenty-five official witnesses, eminent physicians and lawyers, as well as two select reporters, sat in uneasy silence.

"Gentlemen," said Durston, his voice trembling, "this is William Kemmler. I have warned him that he has got to die and if he has anything to say he will say it."

The condemned man bowed to the witnesses, all arrayed in a horseshoe of wooden chairs facing the malevolent-looking oversize electric chair. Then Kemmler said, "Gentlemen, I wish you all good luck. I believe I am going to a good place, and I am ready to go. I want only to say that a great deal has been said about me that is untrue. I am bad enough. It is cruel to make me out worse."[35] The bearded bantam, his eyes close set, bowed once again to the witnesses, turned, and removed his gray suit jacket. He began to sit in one of the regular chairs, then realized his mistake and sat instead in the electric chair. The sun illuminated his sallow bearded face. Warden Durston remembered he had to check that Kemmler's clothes had holes clear through at the bottom of his back, again for proper attachment of electrodes. The shirttail did not. So Durston fished out a scissors and awkwardly cut through the tucked-in shirttail. "Are your suspenders all right?" asked the warden as he put down the scissors.

"Yes, all right," said Kemmler.

"Well, then, Bill, you'd better sit down here."

Kemmler matter-of-factly settled back into the electric chair, while the warden began to attach the rear electrode with its suction cup and wet sponges. "Now take your time and do it all right, Warden," instructed Kemmler, cool as could be. The witnesses looked at one another in astonishment. "There is no rush," he continued. "I don't want to take any chances on this thing, you know."

"All right, William," Durston said grimly. Hands shaking, he was now attaching the leather masklike headpiece, also designed with a suction piece and wet sponge.

As Durston stepped back, Kemmler shook the mask and said in a muffled voice, "Warden, just make that a little tighter. We want everything all right, you know."

Durston complied and then moved on to strapping in the arms and legs, his hands trembling as he buckled one after the other. "All right," he said, and stepped back to the threshold of the death chamber. "Is it ready?" Reassured, he turned toward Kemmler, now just an ominous leather mask and maze of thick leather straps and wires warmed by the streaming sunshine. The freshening breeze outside rustled the thick ivy and the green grass. The sparrows twittered furiously.

Two physicians had come up to inspect the straps. One said, "God bless you, Kemmler, you have done well." Many of the witnesses had tears in their eyes at this dreadful scene and now murmured, "You have, Kemmler." At that moment the district attorney of Buffalo rose from his chair, looking quite green. Excusing himself, he walked unsteadily toward the closed door to the corridor, opened it, and went out. Later, the other men learned he had fainted in the hallway.

Back in the death chamber, the warden conferred briefly with the two physicians, who then sat back down. Durston surveyed his twenty-four tense witnesses and two reporters and said, "Very well, gentlemen," then walked toward the room where he would signal the electricians at the dynamo. The electric lights in there were on, meaning the dynamo was running. Kemmler's hands could be seen clutching the wide arms of the electric chair. The bright sunlight lit up the dust motes floating slowly about. A heavy shade at one window was lifted by the straying breezes and fell against the bars. The sparrows in the ivy sang noisily. In the room, all was tense silence.

"Good-bye, William," said Durston, and there was heard a low click.[36] The switch was thrown in the far-off dynamo room. Then, reported *The New York Times,* Kemmler's body jolted upright and strained terrifyingly against the straps. He became "as rigid as though cast in bronze, save for the index finger of the right hand, which closed up so tightly that the nail penetrated the flesh on the first joint and the blood trickled out on the arm of the chair." When Kemmler's face turned ashen, the prison physician, who had crept forward from his chair to watch, said, "He is dead," leading the warden to signal for the switch to be turned off. Seventeen seconds had passed, and it was now 6:43 A.M. The assembled witnesses collectively exhaled in relief that the awful deed was done and turned their heads away from Kemmler as Durston detached the electrode from his scalp. Two physicians leaned forward to examine Kemmler. The other doctors gathered around and dented Kemmler's flesh to judge his state. Dr. Southwick smiled broadly as he came away from the fresh corpse. "There," he exclaimed to a knot of witnesses who had quietly withdrawn to the far end of the chamber, "there is the culmination of ten years' work and study. We live in a higher civilization from this day."[37]

But the blood was continuing to ooze from Kemmler's small finger wound. His heart still had to be beating. The physicians around the limp figure recoiled as one yelled in horror, "Great God! He is alive!" Another ordered, "Turn on the current." "See, he breathes," gasped a third. When Dr. Southwick and the others whirled around at these cries, they saw that Kemmler's body was still limp, but his chest was heaving up and down. He seemed to be struggling for breath, and foam was seeping horribly from his masked mouth hole. "For God's sake, kill him and have it over!" screamed one witness. The Associated Press reporter fainted on the wood floor, and several men carried him to a bench, where they fanned him. Durston had turned chalk white. He fumbled and reattached the scalp electrode. As the current flowed anew and Kemmler again went horribly rigid, "an awful odor began to permeate the death chamber." Kemmler's hair and skin were being visibly singed. A blue flame played briefly behind his neck. His clothes caught fire, but one of the doctors quickly extinguished them. "The stench," reported the *Times,* "was unbearable." After several minutes, the current was turned off, and as purple spots mottled Kemmler's hands, arms, and neck, the doctors again declared him dead. The room reeked of burned meat and feces. The nauseated witnesses signed the death warrant for Warden Durston and then trailed out into the stone corridors, silent, shaken, several sick, the Erie County sheriff so distraught that tears trickled down his face.

Three hours later, when the doctors had sufficiently recovered to perform an autopsy, they found that rigor mortis had stiffened Kemmler into a permanent sitting position. Examination of the body showed scorch marks wherever the electrodes and buckles touched the body. Kemmler had been "roasted" as well as a piece of overdone meat. Once the autopsy was complete and numerous organs removed, Kemmler's baked corpse was taken and buried at night in the prison courtyard with great quantities of quicklime to dissolve all ultimate traces. Southwick excused the bungled electrocution as inevitable on a first try. Otherwise, he was rhapsodic. "I tell you, this is a grand thing, and is destined to become the system of legal death throughout the world."[38] The next day, New York's many newspapers devoted page after page to the world's first official electrical execution, including every ghastly detail. *The New York Times*'s front-page headline charged, FAR WORSE THAN HANGING; KEMMLER'S DEATH PROVES AN AWFUL SPECTACLE. The *New-York Daily Tribune* subhead read, "Errors and Misunderstandings Made the Execution Painful to the Witnesses." Thomas Edison was quoted as saying, "I have merely glanced over an account of Kemmler's death and it was not pleasant reading."[39] He blamed the doctors present for misplacing electrodes—they should have been attached to the palms of Kemmler's hands—and general "bungling." Bourke Cockran said, "It is a sort of ghastly triumph for me. The experts against me on the trial figured it all out such that such a shocking thing was impossible and yet it has just happened. . . . After Kemmler's awful punishment no other state will adopt the electrical execution law."[40]

As for George Westinghouse, he said, "I do not care to talk about it. It has been a brutal affair. They could have done it better with an axe. My predictions have been verified. The public will lay the blame where it belongs and it will not be on us. I regard the manner of the killing as a complete vindicator of all our claims."[41] Almost forty years later, Nikola Tesla still abhorred the electric chair, calling it "an apparatus monstrously unsuitable, for the poor wretches are not dispatched in a merciful manner but literally roasted alive. . . . An individual under such conditions, while wholly bereft of the consciousness of the lapse of time, retains a keen sense of pain, and a minute of agony is equivalent to that through all eternity."[42] Suspicion was rife that Westinghouse had somehow managed to mastermind this botched execution. As for Harold Brown, one of the most ardent and combative officers in the War of the Electric Currents, he was strangely absent from all the bitter aftermath and hullabaloo. He disappeared from public sight, never to be heard from again.

CHAPTER 9

1891:
"Fear Everywhere of Worse to Come"

November 15, 1890, dawned crisp and azure in Manhattan, one of those delicious fall Saturdays where the very air shimmers sweetly, full of life's promise and yet tempered by autumnal tristesse. The money men in their mansions on upper Fifth and Madison, settling into their breakfasts, knew better than anyone the nation's greatness, its commercial promise, its phenomenal wealth. The steady waves of immigrants flowing in through Castle Garden and then washing out across the land had pushed America's population up to a mighty sixty-three million, according to the recent census. Politicians had taken to boasting that the United States was now a "billion dollar" country. It was, in fact, far richer than that, thanks to its relentless can-do commercial spirit and the reckless ambition of men like Thomas Edison and George Westinghouse. In the postbellum United States, the national wealth had soared to $65 billion, more than the accumulated holdings of all the aristocrats and commercial classes of Great Britain, Germany, and Russia combined. By 1890, in America, nearly $40 billion was invested in land and buildings, $9 billion in the sprawling network of railroads, and $4 billion in manufacturing and mining.

But as the New York money men unfolded their daily newspa-

A formal portrait of George Westinghouse, 1906

pers that glistening November morning, there was only visceral dread. For weeks the rumors of a big London investment firm on the brink of collapse had been creeping quietly about the chaotic floor of the New York Stock Exchange, multiplying, growing more sinister, more fantastic, infesting the always volatile atmosphere with sensations of doom. Now here, today, the worst of the whispered disasters was openly aired. On the *World*'s front page, for all its vast readership to see, was the story in public print: DEPRESSED BY RUMORS, WALL STREET AGITATED BY REPORTS FROM LONDON, THE BARING BROTHERS, IT IS SAID, INVOLVED IN THE RECENT TROUBLES. As the *World* so rightly observed, "When London trembles, the tremor is felt the world over. Therefore New York quaked." English investors leery of American silver policy had been flocking to the bonds known as "Argentines," eager to earn the easy riches of that far-off South American nation, but a revolution had now rendered worthless all those considerable investments.

When the financiers converged on their elegant Wall Street offices that fine fall morning, the lurking skeletal rumors solidified into monstrous and hideous facts. As *The New York Times* reported in the lead story of its fat Sunday edition, Saturday morning had brought the bleak and mind-boggling news that Baring Brothers & Company, "the greatest banking house of all the world, a firm whose business connections have extended for a century or more to the uttermost limits of the habitable globe [whose] . . . signature has stood always and everywhere for an absolute guarantee," was indeed tottering on the brink of bankruptcy. "This confession was sensational beyond anything that Wall Street had even suspected or dreamed of." One *New York Times* financial writer elaborated on the nightmarish news: "If the solvency of the Bank of England had been questioned, it could not carry a more severe shock." The ensuing days featured placating stories about the brilliant rescue of Barings and its reorganization and soothing assertions about the general good health of American business.

But the stunning and ominous collapse of Barings was not so easily dismissed. The American banks of the Gilded Age were not unlike Potemkin villages: august and solid on the face of it, hushed and impressive marble-columned temples to mammon. But behind that stolid, expensive architecture flapped flimsy financial structures, liable to collapse under the slightest economic ill wind. And the cataclysm of Barings was a gale-force reminder of how dependent American finance was on foreign investors. Despite the good face the newspapers put on the matter, every businessman knew in his gut that the Barings debacle would take a toll. Several smaller banking houses failed, "spreading everywhere fear of worse to come."[1] In this laissez-faire era before meaningful bank legislation, many decades before stringent federal bank regulation and individual deposit insurance, no authority or institution backed the banks, and the merest whiff of financial bad weather understandably sent jittery depositors to besiege their own branch and retrieve whatever cash they could. These panicked runs set off a chain of inevitable destabilization. Even before depositors began jostling and lining up, some banks began calling in their loans, knowing that only large stacks of greenbacks would reassure their clientele. And so, the economic disasters rippled out from each affected bank, sending fear seeping inexorably deeper into the American financial psyche.

George Westinghouse had just gone up to his new Lenox, Massachusetts, country retreat to settle a real estate matter when his Pittsburgh office flashed an urgent telegraphic alert of the Barings catastrophe. Just ten days earlier, Westinghouse and his wife, Marguerite, whose health depended on regular doses of the pine-scented Berkshires, had finally moved into their turreted and gabled Gilded Age "cottage," Erskine Manor. Surrounded by acres of elegant greensward, old shade trees, and brooding evergreens, the Westinghouses, their young son, and many relatives and guests could sit on the wide covered porches of the rustic new mansion and pleasantly contemplate the ever changing moods of the rolling Berkshire Hills. Two years earlier, Walter Uptegraff had become part of the household, serving as a full-time secretary, family accountant, and general confidant. Old Mrs. Westinghouse, eighty-one, whose late-life passion was the card game whist, had also come to live with her son, while brothers and sisters, young nieces and nephews, often stayed for months. Formal dinner parties were the routine in the Berkshires, as they were in Pittsburgh.

"After the guests had gone and the family had retired," recalled Uptegraff, George Westinghouse "would go to his study where he would work on his inventions until one or two o'clock in the morning. During these quiet hours he always liked to have someone with him (myself) to whom he could open his inmost self without restraint; to explain what he was trying to accomplish by showing the drawings he was working on; to compliment the listener by inviting suggestions. He required but four or five hours of sleep and would come down to an early breakfast bursting with energy, eager to get at the day's work as quickly as possible."[2] That fall of 1890, George Westinghouse had all reason to be enjoying his sylvan new estate and reveling in the phenomenal success of his substantial industrial companies, especially his newest and most cherished: the fast-growing electric company, which was finally beginning to fulfill his Olympian dreams.

Just the previous month, the *Electrical Engineer* wrote, "Reports from all over the country show that this year is the greatest the electrical industry has ever known.... [The Westinghouse] works are now being operated at their utmost capacity day and night."[3] Despite Edison's war against alternating current, the month of September 1890, soon after William Kemmler's botched electrocution, had been a banner sales month for the Westinghouse Electric Company's alternating current central stations (operating incandescent lights). In October, the city of Baltimore, Maryland, had ordered a 6,000-light AC system; Elmira, New York, put in for 1,500 lights, as did Nebraska's state capital, Lincoln. And these were just three of many orders. Westinghouse had also expanded into arc lights and electric streetcars, and those departments were equally booming. In four short years, since it was established in early 1886 as George Westinghouse's fifth industrial enterprise, total annual sales at the Westinghouse Electric Company had soared from $150,000 to more than $4 million.

The telegraphed news of the Barings collapse ended any sense of well-earned satisfaction. Westinghouse quickly packed, and he and Uptegraff got in a buggy to the train station to board his private railcar back to Pittsburgh. All the big electric companies were financially shaky, for they were fast-expanding, capital-intensive businesses in an era when capital was hard to come by. Before 1890, the New York Stock Exchange, like other exchanges around the country, listed only railroad stocks. So raising adequate long-term capital was vexing. Many local electric firms and towns made initial payment partly in cash and also by giving the Edison or Westinghouse companies stock in their new local lighting companies. The previous March, Edison president Henry Villard complained bitterly to Drexel, Morgan that Edison Electric was growing so fast, its working capital was "entirely inadequate. Instead of one million, several millions are imperatively wanted to meet the demands of the several manufacturing departments."[4] Westinghouse faced exactly the same problem, and he knew better than anyone the parlous state of his indebtedness as his train steamed through Pennsylvania's strike-and-violence-plagued coalfields and past the still recovering hamlets savaged more than a year earlier by the fearful Johnstown flood.

The next morning, Westinghouse strode into the imposing nine-story Westinghouse Building on Pittsburgh's Penn Avenue, a street notable for the presence of trees, wearing his usual somber, vested suit and high collar and carrying his ever-present umbrella. He knew all too well that he needed at least $500,000 in cash to pay his immediate creditors. The company's current assets totaled $2.5 million, its short-term liabilities $3 million. Before the Barings bankruptcy, closing this gap would have been accomplished quietly. Unlike Edison, who had from the first been linked to the powerful Morgan financial forces on Wall Street (though they were generally far too cautious and tightfisted), George Westinghouse had amassed capital as needed by personal appeals to friends and existing stockholders. Biographer Henry Prout explained, "He had such an impelling personality and such a remarkable power for effectively presenting his case that he frequently secured large sums of money from men of wealth."[5] The majority of Westinghouse stockholders were leading citizens of Pittsburgh who had invested and prospered as the indomitable industrialist and entrepreneur created one flourishing company after another.

In those grim November days of 1890, the economy was souring swiftly, even as Westinghouse grappled with his money woes head-on. He and his board of directors appealed immediately to their own stockholders to raise the desperately needed $500,000, by doubling the electric company's capital stock and offering shares at a 20 percent discount. But fear was in the air even as Pittsburgh was becoming the nation's greatest producer of steel, recounted Francis Leupp, and "general commercial conditions were so depressing that the response fell far short of what he had hoped."[6] A delegation of employees from the original Westinghouse Electric Company appeared, offering to work for half pay until the crisis was solved. Their boss was deeply touched, but he declined. Next, Westinghouse convened an informal meeting of Pittsburgh's leading bankers, men "who had profited by the business brought them through the industries he had built up in the city or had attracted thither from outside."[7] Forceful and charming as always, he stood before them and confidently reviewed for these potential rescuers the excellent health of his seven other companies and, most relevant, the large future prospects of the at-risk electrical company. All told, his eight companies were worth $23 million and earned $16.5 million, of which an impressive $4.2 million was profit. The bonded debt was just over $1 million. Accounts receivable, however, stood at more than $6 million, and materials on hand were an unproductive $2.5 million.

Ironically, the Westinghouse Electric Company had just posted its best year ever. Westinghouse fervently believed that "in spite of its temporary embarrassment, [it] was destined for a career of unparalleled prosperity." Westinghouse had, after all, just begun to light up and power the world. Every day, cities all over America were sending in orders, not just for light, but for electric streetcars. Once Westinghouse Electric got its elusive AC induction motor working, the sales potential was stupendous. And then there was the rest of the globe. Had not Westinghouse already installed 750 incandescent lights in Havana, Cuba? And was the company not active in Canton, China?

Now Westinghouse, his large walrus mustache bristling, asked these men, who certainly had grown rich (or merely richer) through his eight companies, to see him through in his hour of need. So certain was he of the solidity and future of the electric company, he declared himself prepared to put up his own beloved Homewood mansion, Solitude, along with several other adjacent suburban parcels, as collateral for the half-million loan he so urgently needed. The assembled bankers murmured in a friendly manner and agreed to appoint a committee to review the books.

On December 10, the Westinghouse Electric Company publicly stated that due to the "difficulty in raising new money . . . owing to the stringency in the money market," the board of directors would raise the required $500,000 by creating and selling preferred stock. Over the next two weeks, thirty Pittsburgh businessmen and seventeen banks, many having been part of the friendly group present a few weeks earlier, subscribed from $2,000 to $35,000 each to the needed $500,000, which then entitled them to stock, even as Westinghouse Electric stock, originally valued at $50, plummeted to $13 a share. The day before Christmas, the *Electrical Engineer* reported the heartening news that "only minor details are to be looked after before a deal is closed. This will put the electric company—certainly a money-making concern when properly managed—on a sound footing and it would seem that the improvement noted in the stock today reflects those changes."[8]

The phrase *properly managed* reflected an ominous turn of events. Believing Westinghouse cornered, one of the bankers saw an easy chance to gain partial control of this highly valuable industrial property. He said to his colleagues, "Mr. Westinghouse wastes so much on experimentation, and pays so liberally for whatever he wishes in the way of service and patent rights, that we are taking a pretty large risk if we give him a free hand with the fund he has asked us to raise. We ought at least to know what he is doing with our money."[9] When the bankers demanded a voice in management, Westinghouse explained in a genial and pleasant way that this was impossible. He had always run his own companies, they were flourishing—but for this immediate need to pay off electric creditors—and he had no intention of being second-guessed or told what to do. The two sides went back and forth at some length, until Westinghouse said he must have an answer: Either they were providing the loan no strings attached or they were not. The bankers looked at one another and said, no, they must insist on naming a general manager.

"Realizing what this meant to him, they waited almost breathlessly to note its effect," wrote Francis Leupp. "To their astonishment, instead of being staggered, he rose with a smile, remarking, 'Well, thank God I know the worst at last!' " It was not Westinghouse, imperturbable as ever, who was shaken, but the overreaching bankers. Westinghouse told several jokes, bade the silent bankers good day, and left the room. They had just witnessed the deservedly famous Westinghouse courage. As his old friend and biographer Henry Prout explained, Westinghouse was a great one for consulting others, "then he made up his own mind, and nothing milder than an earthquake could budge him. We have seen him sitting like a rock, serene, gentle, and unmoved when every member of the board of directors was against him. Whether he was determined or just obstinate depends upon your point of view."[10] Westinghouse had been right far too often when those he judged pusillanimous had been wrong. Why should he stop trusting his own formidable instincts?

That very Friday night, as a snowstorm engulfed the eastern seaboard, George Westinghouse again boarded the *Glen Eyre*, his private railway car complete with sleeping quarters, dining room, kitchen, and office, and as the snow rapidly carpeted the countryside, he steamed toward Gotham, the heart of America's brittle financial system. By the time they came across on the Pennsylvania Railroad ferry from Jersey City the next morning, the snow had stopped falling and the Hudson was bobbing with chunks of floating ice. The city looked festive swathed in holiday greens, and its usual noise and bustle were muffled by the thick mantle of clean, sparkling snow. But the pretty winter scenes were merely irksome to the cursing horsecar operators, who found draymen and their big teams also driving on the cleared horsecar rails, the only usable portion of the snowed-in city streets. Few avenues or side streets were passable, even as shopkeepers and "useful men" were gamely shoveling aside the snow. Far uptown, spirits were more frolicsome, as hundreds of sleighs glided colorfully along behind handsome horses, bells jangling, all thronging the uptown avenues and Central Park drives. One "dude" had three big bays pulling his elegant getup, each horse trimmed smartly in fur and prancing plumes. But whether New Yorkers cursed the snow or reveled in it, Westinghouse was embarked on a far more serious mission: the seeking of New York money. This would be just the first of numerous sallies into the magnificent marbled edifices of Wall Street, the nation's gilded financial heart. Westinghouse had never before met many of the nation's big money men, but they certainly knew of him, and he had his young and brilliant New York lawyer, Paul D. Cravath, to smooth his way. The survival of Westinghouse's best-loved and most promising company, the fulfillment of his very electrical dreams, was now at stake.

Back in Pittsburgh, where the fresh snow was promptly glazed squalid gray with soot from a hundred belching chimneys, the bankers smugly assumed that it would just be a matter of time before Westinghouse returned ready to proceed on their terms. When this had not transpired by mid-January, and Westinghouse was telegraphing back from Manhattan and describing his prospects as "rosy," the Pittsburgh bankers and businessmen began returning the $520,000 in uncashed rescue checks. Nor did they share Westinghouse's perennial rosy optimism. *The New York Times* reported that in Westinghouse's headquarters city, "financiers and stockholders now favor the appointment of a receiver for the Westinghouse Electric Company." Meanwhile, the first of the creditors filed lawsuits—one a small bank seeking $2,000, another a local steel company after its $800. The human jackals were beginning to circle.

In Manhattan, Westinghouse had found a possible savior in August Belmont, son and partner of the fabled founder of one of Wall Street's most respected investment houses. August Belmont & Company was a major power, for it served as the American branch of the legendary Rothschild banks. The Belmonts were strong Democrats and famous horsemen. On January 4, 1891, *Electrical World* carried a statement from the Westinghouse Electric Company describing the new rescue plan (it made no mention as yet of New York bankers), which entailed selling a new block of preferred stock, having existing shareholders turn in 40 percent of their old stock and accept less valuable new stock, and paying as many creditors as possible with new preferred stock. It listed various other steps taken to restore fiscal health, including "elimination of doubtful values and the book value of patents." George Westinghouse was described as "now devoting himself to the placing of this stock and is very hopeful of success."

This story only obliquely hinted at one of the many sticking points for the Westinghouse rescue: Tesla's lucrative AC patent royalty deal. George Westinghouse was well-known for having a soft spot for inventors, always remembering his many early rebuffs when trying to sell his air brake. The general feeling was that he was too generous. The New York bankers felt that Westinghouse's championing of Tesla's as-yet-unworkable alternating current motor was one more contributing cause in the company's woes. The industrialist had paid $20,000 in cash to Nikola Tesla, then $50,000 in notes. He had laid out yet more big money trying to develop Tesla's AC induction motor and the rest of his system, paying the Serbian inventor handsome consulting fees. And all this had come to very little. The top Westinghouse engineers had not cared for Tesla or his system. They had balked when he was in Pittsburgh, and were still balking, at designing a system that operated on the 60-cycle frequency Tesla insisted was necessary. The industry was in its infancy, but already there were various established cycles for electrical equipment, and 60 cycles was not one of the common frequencies. Tesla had run into the problem that even if one system (his) was better, sheer inertia, habit, and the cost of the new could combine to thwart technological improvement. The hard truth was that Westinghouse had spent a great deal of money and did not seem even close to making Tesla's system or his AC induction motor work on a commercial basis. Nor was that all. Westinghouse also had paid sizable legal fees to defend his Tesla patents against all sorts of interloping engineers and inventors claiming priority. Westinghouse had sunk large sums into Tesla, and now his money-short company was imperiled.

Nikola Tesla had returned disillusioned from Pittsburgh in the fall of 1889, just over a year earlier, lingered briefly in Manhattan, and then sailed off to France to visit the Paris Exposition, with its sprawling and marvelous electrical exhibits. From the still wondrous City of Light, he would travel east to Austria-Hungary to reunite briefly with his many sisters and their Eastern Orthodox priest husbands and to comfort his dying mother. Tesla was an immigrant success story now, one of the self-made Gilded Age rich, thanks to his AC patents. The inventor's ocean voyage back to Europe was a luxurious contrast to his penniless seaboard arrival five years earlier. In Paris, *le tout monde* could talk of nothing but the soaring Eiffel Tower and the amazing American wizard Thomas Edison, whose multilingual phonographs dazzled all visitors to the exposition's huge Edison display. Edison, also in Paris, was in excellent spirits, relishing the great acclaim that greeted his every public appearance and his own ascension into the ranks of the American millionaires.

That summer, Edison had happily allowed Edison Electric president Henry Villard—onetime crack Civil War correspondent, the Union Pacific builder who had driven in the golden spike, U.S. representative for big German banks, and longtime Edison investor—to reorganize the Edison Electric Light Company and its various manufacturing entities into Edison General Electric, capitalized at $12 million. The Drexel, Morgan investors were very handsomely rewarded—their original $1 million investments were now deemed worth $2.7 million in Edison GE stock. Edison emerged flush with $1.75 million in stock and cash for the manufacturing shops he had created. Wrote Edison gratefully to Villard, "I have been under a desperate strain for money for 22 years, and when I sold out, one of the greatest inducements was the sum of cash received, so as to free my mind from financial stress, and thus enable me to go ahead in the technical field."[11] Edison GE president E. H. Johnson enthused to Edison, "We shall speedily have the biggest Edison organization in the world with abundant capital when goodbye Westinghouse et al."[12] By 1889, Edison General Electric had indeed become a major American corporation, employing three thousand men in its three main shops and bringing in revenues of $7 million a year, with profits of almost $700,000. During this reorganization, Villard had also sold $4 million in new Edison GE stock, mostly to his German-backed North American company and to the Morgan group. So Edison had every reason to feel buoyant during his triumphant 1889 Parisian visit.

Tesla's own European sojourn involved meetings with numerous eminent electricians, and as he sailed back into New York Harbor in the late fall of 1889, he was afire with new electrical projects and plans. He established a commodious and well-equipped laboratory on the fourth floor of 33-35 South Fifth Avenue (today West Broadway), a run-down commercial block just west of City Hall. Edison had decamped to the countryside and his magnificent West Orange lab, while the Edison General Electric corporate offices were moved to 16 Broad Street, next to the Sub-Treasury Building, in the more logical Wall Street district, leaving 65 Fifth Avenue to serve as an elaborate and fashionable showroom of electrical lighting wonders, ablaze with chandeliers and such.

When Tesla returned to Manhattan, he was the picture of European elegance, a tall, slender man in the prime of life, impeccably dressed in the finest hand-tailored Parisian styles right down to his cane, spats, and soft-leather shoes. He was well able to afford his indulgence of tossing out his fawn-colored kid gloves and silk handkerchiefs after a week of use. Many who met him noticed the intense blue of his eyes and the size of his hands and their unusually long thumbs. Tesla had also acquired a taste for the convenience of luxury hotel living while residing in Pittsburgh. After inspecting numerous Manhattan hostelries, he elected to live in the Astor House, Gotham's first luxury hotel. Situated on Broadway and Vesey, it was conveniently near his lab. He began to dine nightly at Delmonico's, America's most famous restaurant, a gustatory temple that for six decades had served exquisite French cuisine and wines. Tesla preferred the mutedly elegant uptown Delmonico's at Fifth Avenue and 26th Street, located across from the civilized greenery of the handsome and leafy Madison Square Garden Park, a pleasant refuge from Manhattan's tumult, an oasis of spreading elms, graveled paths, statuary, and fountains.

As a regular solo customer of Delmonico's, the polite and gracious inventor had let the highly trained staff know that he liked to have eighteen pristine napkins in a stack at his table. Very discreetly,

Tesla used them to wipe the germs off each piece of heavy silverware, sparkling china, and crystal stemware before he partook of the chef's many delights. Tesla's germ phobia had developed after a fellow scientist allowed him to observe through a microscope the many normally invisible creatures inhabiting unboiled water. Tesla would later explain, "If you would watch only for a few minutes the horrible creatures, hairy and ugly beyond anything you can conceive, tearing each other up with the juices diffusing throughout the water—you would never again drink a drop of unboiled or unsterilized water."[13] So Tesla was ever vigilant in limiting his exposure to these vile microscopic bugs.

Tesla's new wealth and the luxurious lifestyle he now adopted with such relish in no way affected his complete obsession and devotion to his electrical dreams. All through 1890, even as he prided himself on his exquisitely tailored clothes and daily savored his excellent Delmonico's dinners, he still labored his usual seven days a week in his laboratory, working through the night and retreating to the Astor House in the wee hours for five hours' rest—reportedly only two of these being actual sleep. During the day at his lab, Tesla could hear the loud rumbling of the passing elevated, spewing sparks and cinders on the unwary below, while down on the grungy street lined with ash barrels and piles of refuse arose the angry shouts of the teamsters and their massive teams of horses negotiating their way to and from the nearby wharves. Tesla had one or two helpers and sometimes his partner, Alfred Brown, to work with him and assist. Sadly, his old friend and colleague Szigety had died while still a young man. Despite the disillusionment with his own Westinghouse experience, Tesla maintained cordial relations with the firm and remained hopeful that the problems with his AC system and induction motor would finally be overcome. He offered whatever advice he could to the Westinghouse Electric engineers, visited from time to time when he was picking up special equipment for his lab, and sent along possible clients. He was most distressed when he heard in late 1890 that all development work on his motor had been halted.

Tesla would have been well aware in early 1891, as was the whole electrical fraternity, that Westinghouse Electric was in severe financial straits. So perhaps he was not surprised when Westinghouse appeared at his fourth-floor laboratory, where Tesla was making amazing advances with high-frequency research. The month that Barings practically went belly-up, setting off the whole Westinghouse crisis, Tesla had managed to light up protofluorescent lamps of his own invention using only ambient electrostatic waves. These unique light bulbs had no filaments, only gases that responded to the highly charged electrical atmosphere, and were connected to no electrical wires. When Westinghouse came to Tesla's lab, wearing as always his dark formal vested suit, he got right down to his unfortunate task. Ever a mixture of bluntness and charm, he elaborated upon the crisis and asked Tesla to repudiate his contract and forgo his patent royalties. Tesla described this critical episode in his own life to his first biographer thus:

" 'Your decision,' said the Pittsburgh magnate, 'determines the fate of the Westinghouse Company.'

" 'Suppose I should refuse to give up my contract; what would you do then?'

" 'In that event you would have to deal with the bankers, for I would no longer have any power in the situation,' Westinghouse replied.

" 'And if I give up the contract you will save your company and retain control so you can proceed with your plans to give my polyphase system to the world?'

" 'I believe your polyphase system is the greatest discovery in the field of electricity,' Westinghouse explained. 'It was my efforts to give it to the world that brought on the present difficulty, but I intend to continue, no matter what happens, to proceed with my original plans to put the country on an alternating current basis.'

" 'Mr. Westinghouse,' said Tesla, drawing himself up to his full height of six feet two inches and beaming down on the Pittsburgh magnate who was himself a big man, 'you have been my friend, you believed in me when others had no faith; you were brave enough to go ahead and pay me . . . when others lacked courage; you supported me when even your own engineers lacked vision to see the big things ahead that you and I saw; you have stood by me as a friend. The benefits that will come to civilization from my polyphase system mean more to me than the money involved. Mr. Westinghouse, you will save your company so that you can develop my inventions. Here is your contract and here is my contract—I will tear both of them to pieces and you will no longer have any troubles from my royalties. Is that sufficient?' "[14] And so, ever the romantic, the idealist, the brilliant dreamer of elysian electrical dreams, Tesla grandly sacrificed those possible future profits to Westinghouse's own electric survival.

Was Tesla the major cause of Westinghouse's financial woes? Probably not. Westinghouse was well-known for being generous to young inventors of all sorts, but he had also purchased numerous businesses that ate up capital and were not yet generating much income. He was always an enthusiastic litigator on many fronts, especially patent infringement, and his long-running and still ongoing battle with Edison over the incandescent bulb had racked up huge legal fees. But Tesla was certainly a factor and just one of many savings the financiers demanded. Moreover, at this stage of his burgeoning career, Tesla probably felt he could well afford to be magnanimous with Westinghouse, who was struggling to preserve his electrical empire. For the Serbian inventor's own deep delvings into the mysteries of high-frequency electricity were just beginning and already were yielding all manner of fascinating secrets. Tesla, along with his fervent and loyal admirers, including editor T. Commerford Martin and Professor William Anthony, was firmly convinced that his AC system and motor were just the rich beginning, the mere debut of many magnificent and hugely lucrative inventions: Electricity was in its infancy, and Tesla was destined to be its greatest explorer, unveiling for the world the inner workings of a universe pulsing with invisible energy.

As Tesla advanced into more and more remote electrical terrain—the unknown peaks and valleys of very high frequency alternating current—T. C. Martin again urged him to publish his findings and then present his new work in another lecture. The rumors were rampant among the electrical fraternity about Tesla's doings: hard-to-credit tales of wireless lights weirdly glowing in Tesla's laboratory all through the night or talk that the strange young inventor was taking tens of thousands of volts through his body, standing there smiling and unharmed as he allowed crackling electrical flames to engulf him, like some medieval Serbian saint surrounded by a pulsing silvery blue aura. Tesla was harking back to the long-ago electrostatic shows of Dangling Boys. Martin was as important and influential as ever in the electrical field, having jumped ship from the *Electrical World* in early 1890, after squabbling with the publisher, to run the *Electrical Engineer*, which thereby immediately became the leading journal. Martin, always debonair with his huge handlebar mustachio and dashing British charm, cajoled the ever reluctant Tesla to emerge from the long nights at his lab and share with the world his rare and strange discoveries. Finally, Nikola Tesla agreed to speak on May 20, 1891, at the Wednesday night meeting of the American Institute of Electrical Engineers, the same prestigious forum where his first talk had catapulted him to fame among his peers.

Again the meeting was held at Columbia College up on Madison Avenue in the school's electrical workshop. Tesla stood up on a stage, his apparatus arrayed on a wooden table before him. Many of the nation's leading electricians crowded in, anxious to see what Tesla had got up to now. Was there anything to all the odd stories? The place was abuzz with excitement. T. C. Martin's coeditor, Joseph Wetzler, was in that eager crowd and wrote for readers of *Harper's Weekly:* "Mr. Tesla held his audience in complete captivity of attention and admiration for over three hours."[15] Tesla explained to his rapt audience that what he had been investigating in a broad sense was how to unleash the infinite energy "nature has stored up in the universe." He had constructed in his downtown laboratory "alternating current machines capable of giving more than two million reversals of current per minute" and then had been experimenting with them to investigate a range of questions. But for his Columbia talk, he would confine himself to just one topic, "the production of a practical and efficient source of light." Tesla had developed a series of powerful AC machines that generated great electrostatic waves—or thrusts—that produced varying kinds of "discharges," eerie and luminous electrical emanations, ranging from the most delicate and threadlike fans to whirling pinwheels to great spurts of streaming, glowing flames that spread about a whole machine, like an unearthly corona.

Speaking in his high, nervous, but perfect (accented) English, Tesla demonstrated his numerous different electrostatic machines and explained the specific kinds of light each produced. Tesla had also been experimenting with new kinds of light bulbs that operated on completely new principles. "It cannot be denied that the present methods [of illumination], though they were brilliant advances, are very wasteful," he explained. "Some better methods must be invented, some more perfect apparatus devised." He then amazed his audience by showing a light bulb with a single looped filament. He then set that filament to *spinning* even as it was glowing brightly. But Nikola Tesla's showstoppers, his most amazing wonders, were the long-rumored bulbs *without any filament* that indeed glowed as brightly as any incandescent bulb and yet were unattached to any wire or machine. The tall, pale Mr. Tesla in his slender swallowtail coat held them aloft in his hand, somewhat like the Statue of Liberty, and they shone. He explained how it was that his whole laboratory operated using just such magical lights, protofluorescents. "I suspend a sheet of metal a distance from the ceiling on insulating coils and connect it to one of the terminals of the induction coil, the other terminal being preferably connected to the ground. An exhausted tube [one with a vacuum] may then be carried in the hand anywhere between the sheets or placed anywhere, even a certain distance beyond them; it remains always luminous." Thomas Edison's proudest invention was his incandescent bulb. And indeed, it had lit up the world as he had promised. Now Nikola Tesla was declaring Edison's technology passé, obsolete, a primitive and inefficient solution that deserved to be replaced.

Outside in the late spring evening, it had long since grown dark. For three hours, Nikola Tesla had mesmerized his well-educated audience of electricians with his demonstrations of completely new electrical effects, machines, and light bulbs. As he completed his lecture, he said, "Among many observations, I have selected only those which I thought most likely to interest you. The field is wide and completely unexplored, and at every step a new truth is gleaned, a novel fact observed. How far the results here borne out are capable of practical applications will be decided in the future. As regards the production of light, some results already reached are encouraging and make me confident in asserting that the practical solution of the problem lies in that direction. . . . The possibilities for research are so vast that even the most reserved must feel sanguine of the future."

Tesla understood that many branded him a "visionary" for his deep belief that in time energy would be easily extracted from the universe around us. But he pointed out, "We are whirling through endless space with an inconceivable speed, all around us everything is spinning, everything is moving, everywhere is energy. There *must* be some way of availing ourselves of this energy more directly. Then, with the light obtained from the medium, with the power derived from it, with every form of energy obtained without effort, from the store ever inexhaustible, humanity will advance with giant strides. The mere contemplation of those magnificent possibilities expands our minds, strengthens our hopes and fills our hearts with supreme delight."[16]

With that Tesla bowed modestly and stood beaming as the audience rose to its feet and thunderously clapped its amazement at what they had heard and seen. *Electrical World* declared it "one of the most brilliant and fascinating lectures that it has ever been our fortune to attend."[17] Joseph Wetzler, awed and astounded, proclaimed in *Harper's Weekly* that with this second lecture of his career, Nikola Tesla had "at one bound placed himself abreast of such men as Edison, Brush, Elihu Thomson, and Alexander Graham Bell." Was it any wonder that Tesla so generously gave up all rights to AC royalties when he and his admirers believed he was well on his way to unleashing in completely novel ways the energy of the whole universe for man's needs? A mere decade hence, AC might well be as outmoded as DC.

As May turned to June, Westinghouse's determined quest for money continued. Out in West Orange, Thomas Edison might well have gained great pleasure from Westinghouse's financial woes in that grim spring of 1891, were he not in much the same predicament. Several months after his agreeable autumn of continental adulation in 1889, Edison was writing the ever charming, ever busy Edison General Electric president, Henry Villard, that the previous summer's reorganization (which had made him a millionaire) was not, after all, such a great arrangement. Before, Edison wrote, he had "an income of $250,000 per year, from which I paid easily my Laboratory expenses. This income by the consolidation was reduced to $85,000, which is insufficient to run the Laboratory. . . . I am placed in such a position that my active connection with the Lighting business costs half my time . . . [and] has produced absolute discouragement."[18]

Villard was well aware from his previous efforts to merge Edison's company with rival firms that Thomas Edison bristled at such talk. The inventor had angrily dismissed the notion that mergers (and ensuing sharing of contested patents) would solve any problems by asserting, "The Company with the best and cheapest machinery will do the business, patents or no patents. Fact is, Mr. Villard, that all electric machinery is entirely too high now. These high prices hurt the business." As for merging with Thomson-Houston, that was absolute anathema. Edison had already denounced its men for "having boldly appropriated and infringed every patent we use."[19] Edison's AC war against Westinghouse made the Pittsburgh magnate and his firm an unlikely merger mate. Moreover, Edison fervently believed that were he to combine with any of his hated rivals, his inventive fertility would dry up. "If you make the coalition my usefulness as an inventor is gone. My services wouldn't be worth a penny. I can only invent under powerful incentive. No competition means no invention."[20]

In the wake of the Barings debacle, however, Villard had seen his German-funded North American Bank, source of much Edison GE electrical capital, collapse. Moreover, Edison now held only 10 percent of his own company's stock. So about the same time that Westinghouse began desperately seeking Wall Street money, Villard once again began secretly talking to Charles Coffin, the former shoe salesman whose business brilliance had steadily forged Thomson-Houston into a major electrical power. Villard was still intent upon a merger of the two firms, with an eye to ending the draining and costly patent wars (sixty lawsuits were wending their expensive way through various courts) and to gaining access to Thomson-Houston's well-developed AC lighting system at a time when Edison salesmen were clamoring for AC. And while the incandescent light bulb patent was still on appeal in the courts, the general feeling was that Edison would win. Thomson-Houston would very much need the rights to that all-important bulb. Then, of course, there were the obvious advantages of size and scale.

Yet nothing could proceed without the blessing of J. P. Morgan, who had in the past decade become one of the most powerful financial figures on Wall Street and, hence, in America. As Morgan sat in his office smoking monstrous Havana cigars, busily and efficiently combining what he viewed as wastefully competitive railroads into "trusts," he provoked the growing ire of those disgusted by an arrogant plutocracy of unchecked big money. But in February of 1891, J. P. Morgan could see no purpose in an Edison merger with Thomson-Houston as proposed by Coffin's Boston financiers. "The Edison system," Morgan wrote them, "affords us all the use of time and capital that I think desirable to use in one channel. If, as would seem to be the case, you have the control of the Thomson-Houston, we will see which makes the best result. I do not see myself how the two things can be brought together, certainly not on any such basis as was talked about a year or more ago."[21] Villard gamely kept in touch with Coffin but held out little hope for the time being.

Charles Coffin tried on numerous occasions to woo Westinghouse into a merger of their two firms. Clarence W. Barron, a talented financial journalist, happened to stop by Westinghouse's factory when the magnate was in an irritable mood, and with one mention of Coffin, out poured a blunt Westinghouse diatribe: "Mr. Coffin has a very swelled head. He talks about making [his] . . . Company bigger than the Standard Oil Company. . . . Coffin will make a man about ten different propositions in ten minutes. I have had many interviews with him. I suppose you know how he takes people into his inside room and then locks the door on them, so as not to be interrupted.

"At one of our interviews, I think it was in New York, he asked me if I would be willing to go into any electrical combination of which I was not the head. I said most emphatically that I would not go into any electrical combination of which I was to be head . . . [nor] of which he was to be head. He told me how he ran his stock down and deprived both Thomson and Houston of the benefits of an increased stock issue. . . .

"I said to Coffin, 'You tell me how you treated Thomson and Houston, why should I trust you after what you tell me?'

"At another time . . . we were at a hotel in New York and Coffin proposed, I forget the exact details of the arrangement, that we should form a combination, so that whatever occurred, whether the companies were successful or unsuccessful, we should make money under any circumstances. I said to Coffin . . . that I was not in the habit of robbing my stockholders. There was no reply."[22] As Barron heard, Westinghouse had nothing but contempt for his very successful rival.

In the early months of 1891, George Westinghouse was spending week after week in New York trying to save his company. Westinghouse, who hated newspaper puffery, even cooperated with *The New York Times,* which ran a flattering profile lauding his "inventive genius . . . [and] the boldness and scope of his business undertakings" and his abilities to "offhand, tell approximately the condition of every one of his companies, even down to the items which figure in the assets and liabilities. He is a hard worker and not infrequently is at his office until late at night."[23] Nonetheless, January turned to February and the Westinghouse Electric news was only bad: "Boston men are gathering enough stock together to control the company."[24] The ever upbeat Westinghouse dismissed these rumors as sprung from some patent-sharing agreements with Coffin's Thomson-Houston, a Massachusetts firm. But even George Westinghouse could not deny that the deadline for selling preferred stock had to be extended to March 1, 1891. Then that date had to be pushed back again, this time to March 20. Yet even as Westinghouse exerted his huge energies on rounding up reluctant investment capital, he found himself and his board suddenly ousted from one of his other companies, Union Switch and Signal. The industrial jackals, scenting weakness and distraction, were closing in to snatch whatever prizes they could. In the end, the jackals were beaten back and Westinghouse was reenthroned, but it was a rude reminder of vulnerability.

Finally, on May 4, the annual stockholders meeting was convened in Pittsburgh at the imposing Westinghouse Building. Fifty anxious investors assembled to hear what George Westinghouse would have to say. Westinghouse was present and looked as hale and hearty as ever, not at all like a man who had spent the past four months desperately seeking to save his company. Prior to the meeting, each stockholder had received a circular that provided welcome confirmation of what had otherwise been only broadly rumored: A syndicate consisting of August Belmont & Co. of New York, Lee, Higginson & Co. of Boston, and Brayton Ives, president of Western National Bank, was proposing to reorganize the Westinghouse Electric Company and its associated interests. However, as soon as Westinghouse opened the meeting, he politely made clear that negotiations were still under way, and thus the board proposed to adjourn the annual meeting to two weeks hence.

The Belmont plan was straightforward: The electric company would continue to be capitalized at $10 million. The stockholders of $7 million worth of shares were being asked to sacrifice 40 percent of their stock, to be sent to the Mercantile Trust Company of New York. Another $3 million of stock issued but not sold would be combined with this surrendered stock to create a $6 million pool of capital. Of this, $4 million would be converted into 7 percent stock, of which $3 million would be directed to paying off the debt and the rest given over for expansion. Another $500,000 would be held in reserve. The remainder would be used to purchase two companies already controlled by Westinghouse—U.S. Electric Lighting and Consolidated Electric Light Company. Westinghouse had acquired these companies largely for their incandescent light bulb patents, crucial weapons in the ongoing patent wars. They would become part of the bigger, properly financed Westinghouse Electric & Manufacturing Company. Now it was up to the shareholders. Would they relinquish enough stock to save the company? Two weeks hence, the postponed annual stockholders meeting was put off yet again, a most unpromising sign.

Even as George Westinghouse struggled to maintain control of his electric company, the entire electrical community was riveted, in that unsettling summer of 1891, on an altogether different—but equally momentous—matter. What had come to be known as the Seven Years' Incandescent Light Bulb War was heading into its most decisive phase. Ever since 1889, when Thomas Edison had triumphed in Pittsburgh's circuit court, the electrical companies and their bevies of lawyers had hoped to overturn Judge Bradley's clear-eyed decision that Edison alone had created a practical, functioning light bulb based on the novel and original long-burning high-resistance filament set in an exhausted glass globe. Westinghouse and the U.S. Electric Lighting Company immediately appealed the decision, determined to spin out the whole process for as long as possible. For it was well understood that the minute the light bulb case was conclusively won in favor of Edison General Electric, the company intended to cut off all access to its light bulbs, save to its own beleaguered licensees. The electrical community had long since been formally warned that the Edison people would seek the right to extract in damages from infringing firms the terrifying sums of "$25 for each lamp in an original installation and $2.50 for each renewable lamp, independent of the fact that such decision would necessarily render your plant inoperative."[25]

True, the Edison light bulb patent would run out in 1894, but that was little solace to the thousands of electrical men whose businesses faced extinction without an Edison-style bulb. So when federal judge William Wallace of the Southern District of New York sustained the earlier decision on Tuesday, July 14, 1891, he unleashed low-level panic in the electrical ranks. Of course, Westinghouse would appeal, but Judge Wallace was reputed to be "appeal proof," so it was not a future victory that was hoped for, only time. Nor were the triumphant remarks of the Edison lawyers reassuring. *The New York Times* reported the next day, "They say that hitherto half of the incandescent lamps in the country have been manufactured by other companies." The Edison legal team envisioned $2 million a year in new royalties pouring into the firm's coffers. The black day of reckoning loomed.

All this was sweet vindication for Thomas Edison, who gloried in his long-awaited judicial triumph, the refutation of all those who dismissed his historic light bulb as mere tinkering with others' breakthroughs. Above all, Edison hoped this excellent decision would squelch for good Villard's tiresome talk about merging with the inventor's patent-stealing rivals. The postdecision statements of Edison officer and attorney Major Sherbourne Eaton were music to Edison's deaf ears: Now that the Edison Company's rightful light bulb monopoly had been reasserted, said Major Eaton, "the prospect of consolidation is far more remote than ever, as the company would have nothing to gain and everything to lose by such an operation."[26]

And who would not savor Judge Wallace's much-commented-upon rebuke of Edison's longtime nemesis Professor Morton Smith for daring to "belittle his [Edison's] achievement" when testifying as an expert witness? In truth, Edison himself was immensely jaded on the subject of patents, having seen his company launch hundreds of lawsuits to little visible result thus far (aside from enriching attorneys). The disgusted inventor could cite endless absurd claims and decisions against him, like "the foreign patent lost because the patent office in that country discovered that something similar had been used in Egypt in 2000 B.C.—not exactly the same device, but something nearly enough like it to defeat my patent."

The moment Judge Wallace released his opinion, *Electrical Engineer* issued an extra that reprinted the decision in full. All ten thousand copies were quickly snapped up. Editor T. Commerford Martin obviously hoped to calm the panicked electricians and set a beneficent tone when he subsequently wrote in an Edison lamp editorial: "As to the attitude of the Edison General [Electric] Co., we can only hope and believe that the corporation will exercise its victory with the moderation which is always the best proof of the right to power." Martin also correctly pointed out that aside from as-yet-unknown judicial vicissitudes, the future would also be shaped by ever changing technology. Commerford, still very much Nikola Tesla's great champion, proposed that the brilliant Serb's "work is the most striking illustration of the possibilities that lie before us. Mr. Tesla gave us a motor without a commutator; and it would be strangely in keeping if he gave us now a lamp without a filament."[27]

Even as the electricians were digesting the big Edison patent win, the bold George Westinghouse once again proved all his naysayers wrong, pulling off what most deemed impossible: He had saved his cash-strapped electric company, emerging stronger than ever *and* still fully in charge. On July 15, 1891, in Pittsburgh, at long last, the stockholders agreed to yield up hunks of their holdings to pay off the company's immediate debt. A new high-powered board was installed, including such financiers as Bostonian Charles Francis Adams and August Belmont. The very issue of *Electrical Engineer* that opined about Edison's lamp triumph also congratulated Westinghouse on his own considerable coup, especially as it came in the immediate wake of his resounding light bulb loss to Edison. "The only inference possible," noted the electrical journal, "is that the men of affairs . . . are well satisfied as to the outlook. We are ourselves inclined to look upon the Westinghouse Company as now one of the most formidable in the field, and as being far more likely today to get business and do it profitably than it was in the time of its inflation and extravagance." Westinghouse's lawyer, Paul D. Cravath, still marveled years later at this triumph of reorganization. Westinghouse, he said, "found it difficult to work with so-called financiers. What seemed to him to be their lack of vision and faith was always annoying to him. . . . In at least two great financial crises, when the financiers had given up the task as hopeless, Mr. Westinghouse, by his faith, by his untiring energy, and by the exercise of a power to influence men that I have never seen equaled, was able to weather the financial storm, raise enormous sums of money, and restore his enterprises to a sound financial position when his critics and most of his friends were certain that he was facing a crushing defeat."[28] Once again, Westinghouse had emerged victorious.

The light bulb win did not prove to be the great windfall Edison

General Electric had hoped. The final appeal dragged on, and the big Wall Street investors once again became restive as they watched their Edison stock drop from $120 a share to $90. By mid-December of 1891, as famine, typhus, and smallpox ravaged parts of far-off Russia, *The New York Times* was reporting that Henry Villard's days as president of Edison General Electric were numbered. "He is a strong talker and has wonderful personal magnetism," noted the paper. "J. Pierpont Morgan, however, is a hard man to dazzle."[29] So the rumors multiplied, were quashed, and then came to life again. The jovial Alfred O. Tate had been serving as Edison's personal secretary since May of 1883, and he had heard the rumors ebb and flow so often, he paid them little mind. But on February 5, 1892, Tate was sitting at his desk at the Edison Building at 16 Broad Street when one of his old friends, journalist Herbert Sinclair, whose beat was Wall Street, walked in. Tate relates what happened in his memoir:

" 'Alf,' he said as he seated himself, 'do you know that the Edison General [Electric] and Thomson-Houston are going to amalgamate?'

" 'Herby,' I replied, 'that's an old yarn that was buried long ago. Where did you resurrect it?'

" 'Now listen to me,' he answered, 'I know what I am talking about. Charlie Coffin and Villard are in Morgan's office right now and we are all waiting at Henry Clews' office for the story. We have been told that we can break it after three o'clock.' "

Tate jumped up and said he had to leave immediately to catch the ferry to Orange. "I'll have to go right out to see Edison. He knows nothing about this."

It was a bone-chilling bright winter day as Tate raced through the well-dressed Wall Street crowds, threaded through the jams of horse-cars and trucks full of huge barrels, past the oyster sellers and hot-coffee carts, and down to the Hoboken ferry. The wharves were full of ships, their masts jaunty with many flags. The weather was so frigid that upriver in Poughkeepsie, ice yachts were out racing across the great frozen riverine expanse. Tate would say in his memoir, "I have always regretted the abruptness with which I broke the news to Edison but I am not sure that a milder manner and less precipitate delivery would have cushioned the shock. I never before had seen him change color. His complexion naturally was pale, a clear healthy paleness, but following my announcement it turned as white as his collar.

" 'Send for [Samuel] Insull,' was all he said as he left me standing in his library. Insull [his treasurer] was sent for. What passed between them I do not know. Edison never again made any reference to the subject except on one significant occasion."[30]

Back on the other side of the Hudson, J. P. Morgan in his smoke-filled office had come to support the consolidation of Edison General Electric and Thomson-Houston for the most obvious and compelling of business reasons: the bottom line. In 1891, Edison General Electric had sales of $11 million and profits of $1.4 million, or an 11 percent return. Thomson-Houston had sales of $10 million and profits of $2.7 million, or a 26 percent return. Charles Coffin boasted to the investment bankers trying to push the merger that he was "beating the stuffing out of them [Edison] all along the line."[31] Business was booming for both companies, but once again Edison General Electric was sitting on a $4 million pile of local electric stocks, unable to translate this paper into cash. In notable contrast, the Boston bankers for the Thomson-Houston people had prudently marketed and steadily sold these sorts of securities so they could recoup that capital. And, of course, there was the lure of the Edison light bulb patent.

Nonetheless, Coffin had begun to wonder why they should sell their company to the Edison forces when Thomson-Houston was doing so well. In meetings at the Boston mansion of Morgan associate and Vanderbilt family member Hamilton McK. Twombly, one Thomson-Houston executive surprised the Morgan forces by saying, "We don't think much of the way the Edison company has been managed."[32] When their lack of enthusiasm was conveyed to Morgan in Manhattan, he ordered, "Well, send them down here to talk to me."[33] Coffin, with his thick brush mustache and hard-charging manner, went as bade to hold his meeting with Morgan. The balance sheets of the competing companies deeply impressed Morgan with Coffin's business savvy, for here was a man who produced twice the profit of the Edison people. With that, Morgan could quite see that it made more sense for the better managers—the men of Thomson-Houston—to buy out Edison General Electric.

Coffin then demurred, proposing a consolidation but one that left him in charge. Capitalized at $50 million, this financial reworking ranked in 1892 as the nation's second-largest industrial merger. Edison stock was converted at a rate of one to one, Thomson-Houston stock at the rate of three shares for five new shares. While this sounded generous, the reality was that the smaller, lesser-known company was taking over the great name of the electric field. Edison shareholders now controlled only $15 million in the new company, compared with the $18 million controlled by the Thomson-Houston shareholders. The remaining $17 million went into the treasury as future capital. The Edison forces were routed in more than mere dollars. Most hurtful of all, the new company's generic name—General Electric—tossed aside the names of the founders of both firms. He, Edison, the father of electricity, his famous name, was being dropped. And so it was that J. Pierpont Morgan, whose house had been the first in New York to be wired for electricity by Edison but a decade earlier, now erased Edison's name out of corporate existence without even the courtesy of a telegram or a phone call to the great inventor. Edison biographer Matthew Josephson wrote, "To Morgan it made little difference so long as it all resulted in a big trustification for which he would be the banker."[34] Edison had been, in the vocabulary of the times, Morganized.

The *New-York Daily Tribune* reported two weeks later that "Edison was so disgusted with the turn affairs had taken he proposed to withdraw entirely. . . . He feels much aggrieved over what he considers the mismanagement of the company and the sacrifice of his interests."[35] This was, of course, completely true, but Edison hated to look like a sap, so the very day that story appeared he quickly adopted his usual cocky public face. He did not want the world or his enemies to know his company had been sold out from under him without so much as a word from the money men. For the many reporters, Edison now assumed his best cavalier manner, explaining that he was already on to bigger, better things. "I cannot waste my time over electric-lighting matters, for they are old. I ceased to worry over those things ten years ago, and I have a lot more new material on which to work. Electric lights are too old for me. I simply wish to get as large dividends as possible from such stock as I hold. I am not businessman enough to spend my time at that end of the concern. I think I was the first to urge the consolidation."[36] He insisted that the new company, which would dominate three-quarters of the electric market, was no trust or monopoly.

The influential editor of the *Electrical Engineer,* T. Commerford Martin, who once briefly worked for Edison at Menlo Park, wrote an editorial in the wake of the announcement titled "Mr. Edison's Mistake." He pondered what had brought about Edison General Electric's demise. Without question one could blame "the discontinuous, frequently changing, though often able, organization and management of the Edison interests," he said. But was not the far greater cause, he proposed, "the attitude taken, and persistently held, by Mr. Edison towards alternating current distribution? He could see no merit in that system. But upon its advent, its possibilities were promptly perceived by others. . . . Since its introduction for long-distance service, six years ago, it has practically driven the direct system from the field of much central station business. Mr. Edison set his face against it as a flint from the first, and has sought on every possible occasion to discredit it through the weight in the community of his justly great name. But the tide would not turn back at his frown." Here in this first and most furious of battles pitting one new modern technology against another, the superior technology had thus far prevailed. But despite Edison's obstinate opposition to AC, which T. C. Martin believed had cost Edison his company, Martin also believed Edison's greatness would endure. "His name and fame are too deeply impressed upon the world to incur any risk of obscuration through any changes of business or methods."[37]

Understandably, Edison took little pleasure in the fall of 1892 over his final victory and vindication in the Seven Years' Incandescent Light Bulb War, when yet another court sustained his lamp patent. This was slender solace to the wounded wizard in his New Jersey redoubt. Edison's secretary, Alfred O. Tate, went out to see Edison several months after the patent decision was handed down, wanting some technical information about a battery project. He found Edison standing alone at his big rolltop desk in the magnificent two-story book-lined library. Atop the desk was sculptor Aurelio Bordigo's spritelike statue holding aloft a lamp. Edison himself had purchased *Genius of Electricity* during those happy months in late 1889 at the Paris Exposition. When the young Tate put his electrical question to the great electrician, Edison answered vehemently, "Tate, if you want to know anything about electricity go out to the galvanometer room and ask Kennelly. He knows far more about it than I do. In fact, I've come to the conclusion that I never did know anything about it. I'm going to do something now so different and so much bigger than anything I've ever done before that people will forget that my name ever was connected with anything electrical."

Tate was utterly stunned at Edison's bitterness and vehemence. "I knew that something had died in Edison's heart and that it had not been replaced by the different and bigger thing to which he had referred. His pride had been wounded. There was no trace of vanity in his character, but he had a deep-seated, enduring pride in his name. And that name had been violated, torn from the title of the great industry created by his genius through years of intensive planning and unremitting toil."[38]

Was a similar fate in the offing for George Westinghouse? He had successfully, against all odds and expectations, secured the necessary loans and a new pool of capital for the electric company. Nonetheless, the *Electrical Engineer* of February 17, 1892, had reported, "It seems quite reasonable to expect, as many do, and as rumor has it, that absorption of the Westinghouse Company into the proposed new [General Electric] corporation will soon follow. The provision of $16,600,000 of stock—$6 million of which is in preferred shares—remaining to the treasury after taking up the Edison and Thomson-Houston stocks, is thought by many to imply the use of a considerable portion of it in taking over the Westinghouse Company when convenient; but no information of such a plan has been made public." *When convenient.* Even a man as tough as Westinghouse had to read those words with at least a tinge of trepidation. While he had yet to get his Tesla AC system working, no one doubted that he would. Clearly, the people at General Electric, who would now control three-quarters of the nation's electrical business, would want those AC patents. And J. P. Morgan, who was now a firm GE backer, was a man who preferred a tidy monopoly above all other industrial arrangements.

CHAPTER 10

The World's Fair: "The Electrician's Ideal City"

In mid-May of 1892, George Westinghouse was hurtling along in his private railcar, the *Glen Eyre,* crossing the spring Indiana prairie bright with wildflowers. Ahead lay the sprawling metropolis of Chicago, the booming city that Pullman porters admiringly tagged the "Boss Town of America." The Pennsylvania Railroad's powerful locomotive began to clang its bells urgently, for it was speeding into "an industrial amphitheater bigger and blacker than Pittsburgh—endless reaches of factories, marshaling yards, slaughterhouses, grain elevators, and iron mills, and slag heaps and coal piles that looked like small mountains. Soot-covered cable cars and long lines of freight wagons waited at crossings for the train to shoot by, and everywhere, covering everything, were wind-driven clouds of black and gray smoke."[1] As the train slowed, Chicago's already fabled two dozen skyscrapers came briefly into view, modern architectural temples to the city's determined energy and imagination. This fast-moving commercial New World crossroads and hub of three dozen railroads had already created more than two hundred millionaires. Now the most ambitious of them were organizing the next great World's Fair. And George Westinghouse wanted its biggest electrical contract.

A night view of the Court of Honor, World's Columbian Exposition, Chicago, 1893

George Westinghouse had in the spring of 1892 become the darling of the Chicago newspapers, hailed as the electrical white knight sallying boldly forth at the last moment from distant Pittsburgh to joust with the contemptible knaves of the Thomson-Houston and Edison "electrical trust." These arrogant princes of eastern commerce had submitted one extortionate bid after another to light Chicago's upcoming World's Fair. Charles Coffin of the new General Electric Company had thoroughly misjudged the hard-eyed Chicago dreamers, the city's business elite hell-bent on putting their raw but mesmerizing metropolis on the international map, a metropolis hailed by English journalist George Warrington Steevens as "Chicago, queen and guttersnipe of cities, cynosure and cesspool of the world! Not if I had a hundred tongues, everyone shouting a different language in a different key, could I do justice to her splendid chaos."[2] Now its great men were building at lightning speed the most spectacular of all World's Fairs—the Columbian Exposition, set to open May 1 of 1893 (a year late in deference to the presidential election) as a celebration of the four hundredth anniversary of the discovery of America. The fair would showcase the industrial and cultural might of marvelous Chicago, the United States, and foreign nations as disparate as Germany, Brazil, Egypt, and Samoa.

The Chicago fair directors were not amused to be mistaken for rubes and crooks by the hard-charging Coffin, who, seeing that Westinghouse was not in the bidding, demanded sky-high rates. In mid-March, General Electric had submitted to the fair a bid of $38.50 per arc light for six thousand lights. In vivid contrast, just the previous October, World's Fair officials had paid a third that rate, $11 per arc light, to Chicago Edison (before Morgan's "trustification") to light the fair construction site for night work. The fair committee, not liking to be gouged, promptly sidestepped Coffin and worked out cheaper arrangements—$20 an arc light—with several smaller firms that were not part of the new "electrical combine." The local papers gloated: CANNOT ROB THE FAIR, and CUT A HIGH BID IN TWO; ELECTRIC LIGHT COMBINE HUMBLED, and EUCRED THE TRUST; EXPOSITION DIRECTORS AHEAD. But Coffin did not blink, and he next submitted an equally outrageous bid for dynamos—$15.78 per horsepower. Again the irritated fair directors simply improvised and went around him, contracting with a smaller local concern for $2.50 a horsepower.

By early April, the committee was ready to award the most important of the fair's electrical contracts—the ninety-two thousand outdoor incandescent lamps that would set aglow the fairgrounds for six months. All the big lighting companies, save for Westinghouse, were now wrapped up with the trust. So on April 2, when the fair directors opened the big iron bid box, there lay only two bids—General Electric's bid of $18.50 per lamp, or $1,720,000, and that of the little South Side Machine and Metal Works, for $6.80 a lamp, or $625,600. Southside's proprietor was an unknown Chicago businessman named Charles F. Locksteadt. "The big concerns stood aghast," recalled Westinghouse biographer Francis Leupp. "Who was this intruder? Could anyone of consequence vouch for his responsibility? Who would manufacture the apparatus for him?" The answer was not long in coming. "Mr. Locksteadt approached Mr. Westinghouse, hoping to interest him in the situation, and in due course the Westinghouse Electric and Manufacturing Company advised the officials that it would undertake to carry out the Locksteadt bid."[3] The officials, seeing a possible savings of almost $1 million, naturally welcomed Westinghouse into the fray.

And so was launched, in this roundabout way, the next great battle in the War of the Electric Currents. From the very beginning, George Westinghouse had envisioned an entire nation, and ultimately a whole globe, powered by the potential of cheap alternating current. Thomas Edison's malign campaign to banish AC had thus far come to naught, save to cause Edison's own undoing and the loss of his beloved electric company to the fast-talking but talented former shoe salesman Charles Coffin. During the past year and a half, George Westinghouse had been too preoccupied with simply preserving his own electric company to contemplate bidding upon the fair. But now the fiscal storms had cleared, his coffers were flush with Boston and New York capital, and a most interesting electrical proposition was at his door: Join forces with the feisty Locksteadt and provide ninety-two thousand incandescent lights for what promised to be the grandest showcase in American history. The stakes were not immediately about money, but about the unparalleled opportunity to display to an unsuspecting world for six months the true glories and possibilities of electricity, specifically alternating current. The competition was no longer with Thomas Edison, but with Charles Coffin and General Electric.

At the grimy Garrison Alley Works in Pittsburgh, said draftsman E. S. McClelland, "Mr. Westinghouse startled us by informing us that he was going to Chicago, Illinois, to get the contract for the lighting of the Columbian Exposition to be held in that city in 1893. No one took him seriously in this venture. None of us dreamed that he would be successful in this mission."[4] But Westinghouse was utterly serious. Later, the boss returned to the Westinghouse Building downtown and summoned his public relations man, Ernest H. Heinrichs, to his spacious office. Westinghouse was sitting as usual at his immense table desk. Six magnificent upholstered chairs were arrayed around its sides, while a valuable Persian carpet covered the floor. Off in one corner stood a big bookcase containing copies of the *Patent Office Gazette* and various engineering journals. A plain wooden clock ticked the time above the fireplace. Westinghouse looked up at Heinrichs, the former industrial reporter for Pittsburgh's *Chronicle Telegraph,* and asked, "Do you know any newspapermen in Chicago well?" Heinrichs said he did not, but Westinghouse told him to go to Chicago anyway, connect up with the local reporters, and press their side of the story.

Heinrichs cogitated and then called the Pittsburgh Associated Press man, an old friend named Colonel W. C. Connelly. The colonel just happened to be traveling to Chicago himself, and the two took the train together to the sprawling nerve center of the Midwest. "For the next three days the Colonel remained with me," wrote Heinrichs, "taking me from one newspaper office to another and securing for me an introduction and a hearing. . . . In the meantime, Mr. Westinghouse himself arrived at the Auditorium Hotel, [and] no time was lost in taking the newspapermen to see him. His magnetic personality, his affability, his genial manner and straightforwardness, completely won the entire press."[5] And so, the Chicago reporters, already well steeped in their dislike of the unregenerate "electrical trust," embraced George Westinghouse and hailed him as the city's most excellent knight and champion.

The moment Westinghouse charged gaudily onto the World's Fair field to offer a lower bid, GE too returned to the lists, shamelessly slashing its original bid of $18.50 a light down to $6. Coffin always sent his surrogates, men whose entitled, moping manner only further annoyed the hostile Chicago press. This particular electrical tournament entailed numerous jousts spread over many spring weeks, accompanied by much hue and cry and plaints of foul play. Fair president Harlow Higinbotham, an upstanding partner in the palatial Marshall Field's department store, was proud of both his probity and his mercantile ability to drive a good bargain. He elected to allow an entirely new and final round of bids for early May. The *Chicago Times* rejoiced at this news, declaring, WILL UNDERBID THE TRUST: MR. WESTINGHOUSE PROMISES TO MAKE ELECTRICAL FUR FLY.[6] So it was that in mid-May, during a brief spell of lovely weather, Westinghouse was arriving once again in the great smoky hurly-burly of Chicago. Tall, commanding, wearing his usual formal dark vested suit and carrying the ever-present umbrella, his muttonchop whiskers and walrus mustache bristling, Westinghouse stepped off the plush quiet of the *Glen Eyre* into the cacophony of the Pennsylvania Railroad's cavernous Chicago depot with its vaulting steel-and-glass roof. There the arc lights blazed and hissed, gleaming locomotives belched great billows of steam, porters trundled steamer trunks and suitcases, and newsboys in knickers hawked one or the other of the city's twenty-seven newspapers. Westinghouse and his longtime friend and counsel Charles Terry threaded their way toward the street. They had been summoned by telegram to discuss their new bid to furnish and light all ninety-two thousand incandescent lights needed for the exposition's outdoor illumination. The Thomson-Houston and Edison men of General Electric would also be present to argue their case, even as the Chicago press jeered at the precipitous drop in their original greedy bid.

While the World's Fair directors had been thrashing through the complexities of launching this world-class exposition, including the important and costly electricity contracts, seven thousand men had been laboring away for more than a year out at the desolate fairgrounds of Jackson Park, a bleak, swampy bog seven miles south of the city on Lake Michigan. Through bone-chilling gales and snowstorms, stifling heat and dust, the men and great armies of mules had wrested from this forbidding six-hundred-acre swamp a fairy-tale transformation: an utterly unlikely, elegant Venetian-style landscape of canals and lagoons whimsically envisioned (for a $15 million fee) by the venerable landscape architect Frederick Law Olmsted. All around the long rectangular Great Basin, the most central of the placid and lovely waterways, was being built the neoclassical White City, a fantastical Court of Honor of colossal and chimerical palaces, each to serve as a splendiferous exhibition hall of modern wonders. Passengers and crew on passing Lake Michigan steamships could only exclaim at these dreamy, surreal palaces they saw rising from the dust, appearing half-hewn from the most ancient creamy Parian marble, all linked by their mythic nobility and the simple device of common roof pediment lines.

Yet these classical structures with their Ionic columns, slender arches, grand domes, towers, turrets, and spires were but skillful dreams and illusions, the "marble" exteriors being nothing but "staff." This dextrous combination of plaster of Paris and hemp was clad sturdily but ever so beautifully onto gigantic iron-and-steel structures, the exteriors ornamented with the most elaborate of architectural details. Chicagoans were so fascinated by the spectacle that thousands slogged out daily, undeterred by winter's frigid mud or summer's sweltering dust, to gawk at the swarms of fast-moving construction crews building faux ancient palaces. The shrewd fair directors quickly turned it into a tidy tourist business, charging entry.

If there was one man in Chicago who could claim credit for this strange, ethereal classical dream arising from the mire of Jackson Park, it was Daniel H. Burnham, one of the city's preeminent architects and designers. Burnham, forty-four, was a handsome, vital man who, with his partner, John Wellborn Root, had designed and built several of the city's most admired skyscrapers. Burnham was the fair's director of works, the dynamo with the jutting jaw and steely will who pushed mercilessly to complete what many said could not be done. In February of 1891, Burnham had installed himself out in the great wastes of Jackson Park, his command post a rustic log cottage with a massive stone fireplace and an excellent store of wine and aged Madeira, consolation against the howling winds whipping off the lake and the sheer brute grind. There the man known to his construction troops as the "Commander in Chief" spent part of each week.

Most of the engineers and the great army of workers lived in huge barracks. The whole construction site was fenced off with barbed wire to keep out labor organizers, the gates manned with guards. The weather was relentless, but ice or blistering heat, Burnham pushed the men and himself to the brink, seven days a week. Said one observer, "The life of the Director of Works and his staff was like that of soldiers on the field. They seldom went home, their entire energies were put into the work, and there was no cessation day or night."[7] Starting the previous October, another shift of men had begun working the cold wintry nights, something possible because the Edison electric plant now powered bright arc lights. The Edison electricity also ran fifty motors that were speeding work along by operating dredges, crushers, tool sharpeners, sawmills, pumps, the hoists for the heavy beams and trusses, and eventually the electric spray painters that made the "staff" so like pearly white marble.

On Monday, May 16, George Westinghouse and Charles Terry arrived at the World's Fair's offices in the Rookery, a Burnham building on LaSalle Street, where almost two dozen men had assembled. Some were already nervously smoking cigars. Chief Burnham and the other members of the Committee on Grounds and Buildings were installed at a long table, where sat the sealed iron bid box. Sitting in other chairs were Captain Eugene Griffin, a second vice president for General Electric, and the firm's two local managers. Once Westinghouse and Terry were also seated, all waited expectantly as the bid box was unlocked and unsealed. Cigar smoke drifted about amid the low whispering. The final electrical joust was about to begin. Once again there were only two bids. General Electric's was read aloud first. Their new DC-only bid was $577,485, their new AC-only bid $480,694. As every man sitting in that Rookery office was well aware, these new General Electric bids were a shameless one-third the company's original $1,720,000 bid. There were various murmurings and men glancing at one another. Then, as the new Westinghouse bids were taken up to be read, silence descended, only the street noises filtering through. Westinghouse's bid for a combination of DC and AC was $499,559. The GE men squirmed and looked unhappy. Westinghouse was undercutting the trust. Westinghouse's AC offer for all ninety-two thousand lights was $399,000, $80,000 below the trust's best bid. Daniel H. Burnham, fair construction director and a forceful man of definite opinions, said promptly that the contract should go to Westinghouse. But the other committee members balked. One later memoir claimed that some were "stockholders in the General Electric Company" still determined to see Coffin prevail.[8] The committee retreated to a locked office. Hour after hour passed. The lights in the nearby buildings had gone dark. At 7:00 P.M., when the janitors began cleaning the halls and offices, the exhausted committee men agreed to reconvene the next morning. As they all left, Westinghouse said to a *Daily Interocean* reporter, "There is not much money in the work at the figures I have made, but the advertisement will be a valuable one and I want it."[9]

And so, Tuesday morning, as the lovely spring weather held, the joust resumed at the Rookery. Captain Griffin of GE sulkily insisted that George Westinghouse could not possibly carry out the contract, because, reported the *Chicago Tribune,* "his patents . . . were involved in litigation. [GE] had injunction proceedings instituted against the use of the lamp which Mr. Westinghouse proposed to furnish." Westinghouse laughed affably, responding, "There wasn't the slightest question about his ability to furnish the lamps desired." Another fair director complained in exasperation to the *Daily Interocean,* "For many years the Edison Company contented itself with flooding the country with circulars trying to ridicule the Westinghouse system. One morning it suddenly awoke to find it had a competitor. Now it says that if the contract is given to Westinghouse an injunction will head him off. . . . One moment's thought will show how great a bluff is made."[10]

However breezy Westinghouse might be, the issue of the Edison light bulb patent was deadly serious. The bitter legal war was in its final appeals, and every well-informed electrician, including Westinghouse, fully expected GE to win. What no one could know yet was whether GE would be *judicially* obliged to sell Edison bulbs to anyone but their own customers. Burnham agreed they should put off awarding the contract for a few days while they consulted the fair lawyers.

In this Chicago battle of the War of the Electric Currents, Westinghouse was motivated not just by his longtime electrical dreams, but by his great dislike of GE's chief, Charles Coffin, an animosity that was completely mutual. One could easily imagine that Westinghouse would take great pleasure in denying Charles Coffin and GE the World's Fair contract, even at the lowball bids they were both now making. After all, if Coffin had been less grasping and gouging, the whole huge contract would have been his back in early April.

On May 22, the committee once again summoned Westinghouse by telegram, and once again he climbed aboard the *Glen Eyre* for the twelve-hour journey across Ohio and Indiana. Many friends urged him not to undertake this vast and perilous job, one where failure was a high and public probability. But sanguine as ever and wreathed in his electrical dreams, Westinghouse arrived at the Rookery the next afternoon. The lawyers had ruled in his favor, but now the committee was proposing to split the contract in two. George Westinghouse "said he was the lowest bidder," reported the *Chicago Tribune,* "has first-class apparatus and should get the entire job." He was not a man of half measures. When the truculent Captain Griffin again raised the light bulb patent issue, the committee asked if Westinghouse would ease their minds by providing a $1 million bond guaranteeing the contract. Certainly, Westinghouse said genially.

Once more, the committee withdrew to wrangle. Hour after hour ticked by. Outside, the sky grew dark, and the noise of LaSalle Street down below subsided. Many cigars were smoked, and the room grew stale and stuffy. The electric lights came on in the Rookery. Finally, the committee agreed at 7:30 P.M. to vote. Quickly and unanimously, they bestowed their shimmering prize on George Westinghouse and his Westinghouse Electric & Manufacturing Company. A sore loser, Captain Griffin responded angrily, immediately threatening that when the light bulb patent ruling went their way, Westinghouse would "be entirely in our power. He will not be able to make his own lamps and he can only buy from us. We will not injure the fair, but we will not let him continue his contract."[11]

As for Westinghouse, he did not gloat but picked up his black umbrella and prepared to leave, telling the newsmen only, "I shall put in ten or twelve dynamos of 12,000 lamp capacity and furnish a clean-cut, first-class system. I have about 100,000 lamps, either completed or partly so, at the works, and there will be no difficulty in furnishing material. I am required to have between 5,000 and 10,000 lamps installed by the 1st of October. This is an easy task. There will be no difficulty in furnishing the entire plant by the time of the opening of the Exposition."[12] An easy task, Westinghouse said with his signature insouciance. *Electrical Engineer* wondered if Westinghouse would really go through with this latest reckless endeavor. After all, he first needed to ante up another $500,000 for his $1 million bond. "Mr. Westinghouse may not care to put up so large a bond, and the amount does seem rather heavy, but he is not the kind of man to stop short after having gone so far."[13] Amen.

The morning after George Westinghouse's return from Chicago, the "Old Man," as his workers called him, went straight to the company's machine shop and summoned E. S. McClelland, a top draftsman, to the front office. His employee of a dozen years appeared armed with a pad and pencil. McClelland, who had been amazed when his boss sailed off in the *Glen Eyre* for Chicago to seek this great electrical prize, now learned for the first time that his boss had—to everyone's astonishment—won the great contract. He also was learning what would be wanted for this gigantic Chicago World's Fair contract:

"Mr. Westinghouse: 'I want an engine.'

"Reply: 'Yes, sir.'

" '1,200 brake horse power.'

" 'Yes, sir,' with considerable trepidation.

" '200 revolutions per minute.' (Engines of that size usually run about 75 R. P. M.)

" 'Yes, sir,' with considerable consternation.

" '150 pounds per square inch boiler pressure, non condensing.'

" 'Yes, sir.'

" 'Splash lubrication.'

" 'Yes, sir.'

" 'Must go in such and such space.'

" 'Yes, sir.'

" 'I will be in again at 2:00 o'clock to see what you have.'

"Exit Mr. Westinghouse," said Mr. McClelland, "leaving me in a daze. It is hard to describe the feeling of consternation that request caused. We were building 250 horse power engines [then]. . . . A 1200 brake horse power engine to operate at 200 revolutions per minute seemed to me to be entirely out of all reason. Yet this was the task set before us. Mr. Westinghouse needed such an engine." In fact, he needed many. The response of McClelland's boss in the drafting department was that Westinghouse was "asking for the impossible and he just won't get it." Yet all the draftsmen knew somehow they must produce some kind of an engine design by the afternoon, so they set to work. At 2:00 P.M., Westinghouse called to say he would not be by till the next morning. McClelland and another man stayed on in the drafting room till 2:00 A.M., striving to devise something they could show Westinghouse. The next day at dawn, very early, they all reassembled to look at this drawing, turning the drawing board on its side for a better view. "This setting of the board on end, strange as it may seem, gave us the solution to the problem. . . . As a vertical engine there was space to spare. When this solution flashed upon our minds the leading engine draftsman seemed to be electrified and became wildly enthusiastic." Soon George Westinghouse strode in "with his usual good-natured alertness and expectancy," viewed the work with approval, and said, "How soon may I have four of them?" So the work began on "new and untried electrical machines and steam engines of a totally new design."[14]

During the bidding in Chicago, George Westinghouse had laughed in Captain Griffin's face at the Rookery when the GE vice president predicted Westinghouse would have no light bulbs at all, much less ninety-two thousand to light up the fairy-tale White City. But that was a very, very real possibility. In truth, "matters were critical, not to say dangerous," says his friend and biographer Henry Prout. "The Westinghouse Company was committed to the contract for lighting the Chicago's World's Fair. It had already equipped many plants which must have lamps for renewals. Unless a non-patent-infringing lamp could be furnished, the company could sell no more incandescent-lighting material. The need for such a lamp was immediate and urgent."[15]

Westinghouse knew as well as the GE men that he desperately needed a noninfringing light bulb. What they did not know was that he was actually making steady progress on that dire problem. Back in the smoky precincts of his Pittsburgh factories, Westinghouse had been revisiting his patents for an old Sawyer-Man "stopper" light, personally working and tinkering on it for some months, striving to reach the point where he could mass-produce it in the gigantic quantities needed. Unlike Edison's one-piece bulb, Westinghouse's had two pieces with a low-resistance filament sitting in an iron-and-glass "stopper" that was fitted like a cork into a glass globe filled with nitrogen and then sealed. The stopper could be removed and burned-out filaments replaced. With the World's Fair looming and GE's hostility palpable, Westinghouse now set up a glass factory in a section of the Westinghouse Air Brake Company in Allegheny and went there daily to teach the operatives running the grinding machines how to make the stoppers perfectly snug for the lamps. His World's Fair manager, E. E. Keller, marveled at Westinghouse's "own enthusiasm at having overcome a great obstacle. He was bubbling over like a boy. He explained the operation of the grinders and I saw that the men . . . seemed imbued with the idea that this was a game to beat an opponent who held all the aces, and that they were having a lot of fun doing it. He had a sort of magnetic influence on the workmen. . . . It certainly was a great delight to realize that, in spite of what seemed a hopeless situation, 'the boss' was going to furnish lamps without paying tribute. He certainly lifted the worry from me."[16]

By the time Edison's light bulb patent was upheld in the federal court of appeals on October 4, 1892, Westinghouse coolly professed to *The New York Times* that while he, the plaintiff in the case, obviously thought the court decision wrong, it was of no import to him. "Having anticipated it, we shall not be hampered by it," he said. "Our business has been arranged with a view to this happening. The patent sustained has almost expired anyway, and furthermore, such developments have been made in the electrical world in the last year or two that the decision is shorn of much of its effect."[17]

The very next issue of *Electrical Engineer* carried the first Westinghouse ad for its "stopper" lamp, and the magazine, in an editorial comment, noted that because the new-style lamp could be almost entirely machine-made, it would be cheaper than an Edison lamp. This was a good thing, because the "stopper" lamp did not last as long as an Edison bulb. As a stopgap it would do until the Edison patent expired. Meanwhile, the regular Westinghouse light bulb factory continued to churn out Edison-style bulbs, for the Edison case was now on appeal to the U.S. Supreme Court.

GE was no longer content to quietly await their final victory, and in mid-November they struck again, asking a federal court to stop Westinghouse from making Edison-style bulbs. George Westinghouse, knowing the growing public fury over the power of the new trusts, a corporate form that had been mushrooming, struck back hard. In court papers, he accused GE of being "a most vicious trust" that was doing their damnedest to drive honest competitors like him out of business. He urged that GE be investigated under the two-year-old Sherman Anti-Trust Act.[18] Westinghouse demanded that the court force GE to sell their competitors light bulbs. The judge declined to do anything for either side. Just over a month later, on December 15, the U.S. Supreme Court upheld the long litigated Edison light bulb patent. The Seven Years' Incandescent Light Bulb War was over. GE's wrath would indeed be forcibly tempered by certain judicially imposed moderations, and much of the electrical fraternity breathed a collective sigh of relief. However, all those who actually manufactured the infringing bulbs were forced out of business.

But this was not the end of the corporate warfare. George Westinghouse and his wife and son were in Manhattan for the 1892 Christmas season. On the afternoon of December 23, Westinghouse had completed some business with his friend and legal counsel Charles Terry, and the two had just boarded the uptown elevated when they encountered longtime chief counsel for the Edison Electric Light Company, Grosvenor P. Lowrey. Despite their epic legal battles, the men were all friends, so the two Pittsburgh men sat down with Edison's lawyer in the swaying cars. As they were chatting, Lowrey jovially mentioned that one of his many co-counsels, Frederick Fish, was away in Pittsburgh. Westinghouse's ears pricked up and he casually tried to find out why. Lowrey realized he'd said too much. At the next stop, 14th Street, Westinghouse indicated to Terry that they should get off. As soon as the doors shut and the train began steaming away from the platform, Westinghouse asked Terry, "What is Fish doing in Pittsburgh?"

As they walked down the station steps into the bundled-up crowds and passing horsecars, Westinghouse said, "I can't conceive what would call him there except to make new trouble for us. We shall have to act quickly to head it off, whatever it is."

They alerted by telegram a Westinghouse lawyer in Pittsburgh to be at the federal court the next day, the morning of Christmas Eve. And there indeed the following morning stood GE's Mr. Frederick Fish in his sober black suit. He eyed the well-known countenance of the Westinghouse lawyer, who was holding a packet of papers. Now that the final decision had come from the U.S. Supreme Court, GE was stealthily striking again. But this time the firm hoped to get a restraining order, not for the old bulbs, but for Westinghouse's new Sawyer-Man "stopper" bulb, claiming that it, too, infringed on their now rock-solid patent. GE knew that if they could shut down Westinghouse's Allegheny "stopper" lamp factory, even for a few weeks, they might well sabotage his whole World's Fair contract. But with the Westinghouse lawyer on hand, Judge Acheson was not inclined to make any drastic rulings. After the New Year, the judge quickly concluded that the "stopper" lamp was "no infringement of the Edison lamp patents." Wrote Westinghouse biographer Francis Leupp, "Although more or less harassing warfare was kept up afterward, this unexpected proceeding in court so far cleared the way for Westinghouse that he was able to proceed with the manufacture of his lamps and carry out his great undertaking at Chicago."[19]

The Westinghouse engineers were now building from scratch the biggest AC central station yet installed in America, a plant capable of powering 160,000 lights as well as many motors. Up until now, a big-city AC plant powered at most 10,000 lights and, of course, no motors. But at long last, the electric company was making real headway and expected to exhibit a whole Tesla AC system at the fair, including an AC motor. The fashionably dressed Nikola Tesla came forth from his Manhattan laboratory, where the eccentric inventor was steadily advancing into the uncharted terrain of high-frequency electricity, to consult with the Westinghouse engineers at the noisy Allegheny works. Westinghouse wanted to promote two-phase AC, with an eye to showing the glories of his soon-to-be-available Tesla induction motors. Longtime Westinghouse engineer Benjamin Lamme, who did much of the engine work, recalls, "It was at Mr. Westinghouse's suggestion that the machines for the lighting plant at Chicago were each made with two single phase alternators, side by side, with their armature windings staggered 90 degrees."[20] Two of these were combined to create Tesla's two-phase current. Each double unit could light thirty thousand "stopper" lamps. Westinghouse was building in plenty of insurance. If a generator shorted, no fairgoer would ever notice, for another would kick in immediately. They would be powered by one great 2,000-horsepower Allis-Chalmers engine, as well as numerous 1,000-horsepower engines, all fueled with oil (supplied by Standard Oil) rather than coal. The White City would have no smoky pall.

In late January of 1893, the *Electrical Engineer* could report that many electrical visitors had traveled to Pittsburgh to see the twelve almost completed towering generators of seventy-five tons each. The rotating armatures alone weighed twenty-one tons. "With the 12 1,000 h.p. engines required to drive them they will constitute the largest single exhibit of operating machinery ever made at any exposition, and probably the most extensive exhibit in the Fair."[21] Despite Westinghouse's airy talk of "easy tasks," these huge Westinghouse machines arrived in Chicago only weeks before opening day. They were being installed in the south nave of the vast interior steel-and-iron spaces of Machinery Hall, one of the Court of Honor palaces, at the end of April. Opening day was Monday, May 1.

The afternoon before, on Sunday, the very earnest Englishman Reverend F. Herbert Stead ventured forth to the not-yet-opened fair to report what he saw for the highly respected *Review of Reviews*. He took the tramway out as windblown torrents beat down. At Jackson Park, his umbrella open against the chilly, gusting downpour and his high-button shoes sinking in squishy cold mud, the Reverend Stead said, "I found the World's Fair *en deshabille*. . . . The roads within the gates were even more miry than those without. . . . The *disjecta membra* of a whole host of statues lay about . . . helmeted heads, bare arms, and greaved legs of heroes in profusion . . . swarms of workmen gave the same impression of gross incompleteness."

But that first impression was soon swept away. Just as Chicago's cacophonous energy and dirty ugliness stunned visitors, so did the White City, but as the antithesis of all things Chicago. From a balcony high on the majestic gold-domed Administration Building, Reverend Stead forgot his cold, soaked clothes and gazed wonder-struck down through the sheets of rain and drifting fog. There he beheld the classical Court of Honor around the Great Basin, its placid waters patterned by pelting rain, and beyond that, the lovely meandering waterways and all that Burnham and his troops had wrought. Reverend Stead was driven, as so many would be, to wax horribly poetic: "It was a poem entablatured in fairy palaces. . . . It was a dream of beauty which blended the memory of classic greatness with the sense of Alpine snows. . . . It was a vision of the ideal, enhaloed with mystery."[22] The rain eased, and trailing clouds wisped in and around these astonishing alabaster palaces and glistening waters, rendering them even more mutely ethereal.

Mirabile dictu, on opening day the leaden blanket of rain lifted and by midmorning sun blazed down through the vanishing clouds. Reverend Stead arrived with the multitudes squeezing off the packed trams, trains, and Lake Michigan ferries. The good Englishman made his way to the Court of Honor and installed himself with the restless crowd of sweating journalists on benches below the speaker's platform. Soon all was an undulating sea of dark derbies and broad-brimmed millinery, brightened only by the occasional light blue dab of the uniformed fair police. The sun sweltered. Up on the balcony of the gold-domed Administration Building seventy-five American Indians in full war paint and headdresses watched this celebration of the conquest of their lands. Yells and cheers rippled through the crowd as at last President Grover Cleveland, fully three hundred pounds, ascended to the elevated stage, followed by a retinue of dignitaries. Stead approved President Cleveland's "simple morning dress of the ordinary civilian, without ribbon or medal" and his speech, for "he alone of all the speakers made himself heard by any portion of the crowd. His person, which boasts a somewhat extensive periphery, claimed attention. His office commanded it. His voice retained it."[23] Then, the huge choir burst into the "Hallelujah Chorus," and Cleveland's large presidential finger pressed firmly down on a gold-and-ivory telegraph key.

A thousand feet away in the vastness of Machinery Hall, an anxious group of Westinghouse engineers crowded about on the wood-plank floor, holding their collective breath as the 2,000-horsepower Allis-Chalmers steam engine slowly roared to life at this touch from the president. The great machine powered the Westinghouse generators, which now pulsed electricity out to the fairgrounds. The engineers knew all was well when they heard the crowd unleash a delighted roar—the three huge fountains at the Court of Honor were working, sending splashing plumes of water soaring a hundred feet. A giant Stars and Stripes slowly unfurled and was caught up by a breeze, followed by a rainbow of other flapping flags and pennants, brilliant banners of flashing color. As the water sprayed, the crowd, cannon, boat whistles, and jangling fog bells combined into a jubilant caterwauling that rolled through the shimmering heat and across the lake.

Chicago's Columbian Exposition was formally launched. All told, twenty-seven million visitors (half Americans, half foreigners) would pay fifty cents to enter the great fair and experience its astonishments. Above all else, visitors came to marvel at the fair's electrical astonishments. Just as George Westinghouse had hoped, this fair showcased as nothing else ever had the new age of electricity, with its wonders, both startling and prosaic.

Day after day, during the first official week of the 1893 World's Fair, the cold, rainy deluge returned. This was probably just as well, for despite Burnham's best efforts, much of the six-hundred-acre fair was still unready. But even as the naysayers gloated over the inauspicious start and prophesied failure, the weather warmed, the exhibits were whipped into shape, sultry spring zephyrs wafted in off Lake Michigan, and Olmsted's million trees, shrubs, and plants, planted so artfully about the giant lagoon and sinuous canals, unfurled their buds. The people began to arrive, and from the first, they were truly astounded. Everything about this fair was the biggest, the most astonishing, the most exotic, the most glorious. Over on the lively Midway Plaisance, the fair's section for rides and less rarefied pleasures, George Washington Ferris's great wheel loomed 250 feet high, the engineering marvel Burnham had hoped would eclipse Monsieur Eiffel's tower at the 1889 Paris Exposition. Altogether different, the Ferris wheel was equally enchanting. At night, the great wheel glittered in the sky, its gradual (electrically powered) turning outlined with three thousand light bulbs. More than a million fair visitors would pay fifty cents just to ride in one of the thirty-six shiny Pullman cabins and enjoy two stately revolutions with their incredible views of the nearby smoke-enshrouded Chicago and the fair's fabulous grounds. As they circled slowly around in the air, the Ferris wheel riders could see right below them the exotic Turkish village, the delightful German beer garden, and the Cairo street with its camels, minareted mosque, and Egyptian tombs.

Woven through it all was electricity. In one of the many guidebooks, the chief electrician boasted, "The Columbian Exposition is a magnificent triumph of the age of Electricity . . . all the exhibits in all the buildings are operated by electrical transmission. The Intramural Elevated Railway, the launches that ply the Lagoons, the Sliding Railway on the thousand foot pier, the great Ferris Wheel, the machinery of the Libby Glass Company on the Midway, all are operated by electrically transmitted energy . . . everything pulsates with quickening influence of the subtle and vivifying current."[24] There was even an electric kitchen! Here in Chicago's Jackson Park, in this planned, short-lived paradise, ordinary men and women still living in rural worlds untouched by any electrical wonders could see for the very first time abundant electricity naturally employed in modern daily life, an utterly invisible but supremely powerful phenomenon put to the most pleasing, startling, and useful effects. "The World's Fair probably comes as near being the electrician's ideal city as any spot on the globe," declared the *Review of Reviews*.[25] Writer Hamlin Garland urgently wrote his father, "Sell the cook stove if necessary and come. You *must* see this fair."[26] The fair would ultimately generate and use three times as much electricity as the whole city of Chicago. But the White City was not plagued with the dangerous webs of electric wires that lurked overhead in most big U.S. cities. Burnham had sunk miles of big "subways," roomy enough for men to walk in, reachable by 1,560 manholes. There the electrical wires were tidily and safely ensconced, easy to inspect, and unable to shock (or kill) any unsuspecting souls. Just four years earlier, the 1889 Paris Exposition had used 1,150 arc lights and 10,000 incandescent lamps; the Chicago Exposition had ten times that number in its buildings and grounds. Paris had generated a grand total of 3,000 horsepower, Chicago 29,000.

The White City illuminated at night was a radiant electrical vision long remembered by all who witnessed it, acclaimed as the most fabulous spectacle in a fair brimming with fabulous spectacles. On Monday evening, May 8, as twilight gathered, the Reverend F. Herbert Stead was once again among the crowds ringing the Court of Honor. The sky darkened to a deep indigo over the lake, and the air grew perceptibly chillier. Suddenly, the gold-domed Administration Building came brilliantly to pulsating electrical life, provoking a prolonged sigh of pleasure from the crowd. Next, the long classic sweep of the peristyle on the far end of the Great Basin burst forth from twilight's shadow, wondrously luminous with the tens of thousands of Westinghouse "stopper" lamps glowing softly up in the cornices and along the pediment, highlighting the hundreds of statues. The crowd clapped wildly at this bravura burst of electricity. Next, all the white palaces glowed to electric life, radiant visions of an imagined past, followed swiftly by the thousands of lights encircling the dark waters of the 1,500-foot Great Basin, its rippling surface now a-shimmer. Hundreds of arc lights lining the walkways came on, spreading their clear, blue white coronas. Again, the crowd roared its approval. Ghostly gondolas and long electric craft looked like a fairy fleet. Then, from the highest roofs swung to luminous life four massive searchlights, each raking the night sky, even as it shone white, then bloodred, then green, then blue. The people, who had never seen such concentrated, artistic electrical luminosity, let out a steady, breathless chorus of "Oh!" and "Ah!"

Then all went dramatically, suddenly dark, and the White City loomed only as a spectral vision. The murmur of the crowd mingled with the sound of the chilly spring winds off the lake. Out of the total blackness, with a rush and a roar, "the great electric fountains lifted their gushing and gleaming waters. There were two of these fountains, one on either side of the MacMonnies fountain, and through all their many changes each was the counterpart of the other, alike in color and form."[27] The gushing colored waters formed beautiful shapes and hues. Finally, at 9:30, the lights were quelled and the electrically dazed and sated crowds drifted slowly home, awed by the transformation of the dark into something so magically new.

Yet for all its front-page rhapsodizing about the White City's first sublime nighttime illumination, the biggest local story in the *Chicago Tribune* issue of May 9 was not the long-awaited triumph of the fair, but the abrupt closing of the local Chemical National Bank. It was the first of the fast-widening ripples of the Panic of 1893. Forty-eight hours after the celebratory opening day of Chicago's World's Fair, the nation's brittle but booming economy began to collapse in horrifying slow motion. Already that spring, there had been rising uneasiness about the continuing flight of European capital. Then the Philadelphia & Reading Railroad failed, along with the bloated National Cordage trust, setting off the May 5 Wall Street panic known as "Industrial Black Friday." Three days later, the vice president of

Chicago's Chemical National Bank said blandly, "Yes, the rumor is true. We found today our cash was running so low that we decided it would be best to quit." He claimed depositors would lose nothing. The Panic of 1893 was gaining ominous momentum. The *Review of Reviews* wrote, "Not since 1873 has there been such a critical time. Reports of bank suspensions and of the failure of old and established financial and commercial enterprises have been crowding the newspapers. The worst aspect of these failures is the seeming needlessness of so many of them. . . . If people would only think that nothing is wrong, nineteen-twentieths of the problem would disappear at once." The magazine urged President Cleveland to move up to August the extraordinary and early session of Congress he was already calling for September.[28]

Inside the enchanted precincts of the White City during that first month, visitors became well aware that all around them electricity was at work, dazzling with a multitude of lights, powering all variety of engines, and operating that most amusing of conceits—a long "moving sidewalk" along the steamboat pier that fairgoers could leap onto and off of at will. But it was the easy transmission of such great volumes of electricity with alternating current that so clearly distinguished this Chicago fair. Here at the Columbian Exposition, Harold Brown's "executioner's current" quietly shed that fearful sobriquet to become the "subtle and vivifying current." George Westinghouse had, as promised, installed all his AC behemoths in the south nave of Machinery Hall, a vast cavernous space alive with the deafening mechanical clanking and whirring of hundreds of huge machines and unpleasantly redolent of fumes and oil and grease. Great engines in the Westinghouse nave ran even greater generators, which in turn flashed 2,000 volts of AC from each double Tesla machine forth through the subways. Once out in the park, these high voltages came down for use via "converters [transformers] placed in fire-proof and water-proof pits outside the buildings, and the secondary wires were led into the buildings in vitrified-tile ducts. The largest converter used had a capacity of two-hundred lights, and nearly all were of that size."[29]

Westinghouse had initially contracted for ninety-two thousand incandescent lights. But no one quite knew how many lights would really be needed or how many motors. So from the start he had factored in great excess capacity, one of the strengths of the flexible AC system. Delivering additional power was not a great trouble. By the time the fair was fully opened and up and running, the Westinghouse Electric Company had installed almost triple the number of lights—250,000 "stopper" bulbs—called for in the original contract. But each night only 180,000 of those came aglow, leaving 70,000 as a sufficient electrical cushion when others burned out. (There were Westinghouse employees who did nothing but rush about high in the fair rooftops changing dead light bulbs.) Many motors were run also. The nerve center for this biggest of AC central stations was the highly visible Westinghouse switchboard in Machinery Hall, the control panel monitoring the 15,000 horsepower of electricity flashing out to the fair. The switchboard, made out of (noncombustible) marble, was a hundred feet long, ten feet high, and divided into three sections. Accessible by spiral iron staircases and walkways, the switchboard's many plugs, levers, and wires operated forty circuits so arranged that if one broke, another could be substituted instantly.

"What astonished visitors most, perhaps," said Westinghouse biographer Francis Leupp, "was to see this elaborate mechanism handled by one man, who was constantly in touch, by telephone or messenger, with every part of the grounds, and responded to requests of all sorts by the mere turning of a switch."[30] And that young man maintained a cool, easygoing demeanor, as if it were all very simple. Here, finally, at the World's Fair, George Westinghouse and Nikola Tesla could beguile the American public with their shared dream of cheap power, sketching the outlines of the radiant electrical world taking shape, where electricity was cheap and universal, changing forever—in ways almost too momentous to imagine—how people managed the physical world, how they spent their evening hours, the very nature of work and leisure. Here for the first time, millions would see the electrical motors that would take over the burdensome physical tasks long performed by man or his animals, and the lamps that would light their houses.

By the time the Electricity Building was finally ready for its belated opening on the rainy evening of Thursday, June 1, another twenty national banks had gone belly-up and the newspapers were filled with stories of business failures due to "financial stringency." But the White City served as an enchanted retreat from the impending hard times, and those approaching the Electricity Building through the stubborn drizzle that night had their hearts lightened by the loveliness of the looming delicate and airy neo-Renaissance white palace. A colossal statue of Benjamin Franklin attired in colonial frock coat and knee breeches, famous kite in hand, welcomed all who approached. Those walking into the monumental three-acre exhibition hall found their eyes briefly overwhelmed by intense electric glare: the combined candlepower of thirty thousand incandescent and arc lights. Once eyes adjusted, visitors could see the gray evening light filtering in softly through the skylighted vaulted roof high above and many flags and tricolor bunting draped from the second-story balcony. What could not be missed was General Electric's large and dominating corporate presence. Most noticeably, there was front and center a towering eighty-foot Columbian column, set atop a charming round Greek pavilion, its top covered with a huge concealing drapery. As visitors came closer to the foot of the tower, they saw that among the pillars of its base, the adorable colonnaded temple, hung hundreds of lovely and highly artistic Edison light fixtures.

Charles Coffin of GE, who had had little use for Thomas Edison as an actual business partner, made sure the public would still associate the new GE with the beloved inventor and his world-famous name. Therefore, all things Edison were given full and extravagant play: Edison's newest and most amazing invention, the Kinetoscope, forerunner of the motion picture, ran a short film of English prime minister William Gladstone delivering a speech in the House of Commons. On the Kinetoscope's screen, he was as real to visitors as if they were watching him through a window. A tasteful arena of palms displayed 2,500 different kinds of Edison incandescent bulbs. Much of the exhibit catered to nostalgia for Edison's historic lighting breakthrough, thus establishing the primacy of all Edison inventions. Yet Edison's epochal direct current central station was fast becoming a technology of the past. The future, of course, was what the Wizard would never have countenanced in an Edison display—the alternating current central station. But because Charles Coffin now ran GE, the firm showcased variations of its own alternating current apparatus.

One suspects that Westinghouse—who avoided such ceremonies—would have enjoyed the scene on that wet June night when the cavernous Electricity Building formally opened. For as soon as the aisles were suitably full of electrical visitors and the hour struck 8:15, John Philip Sousa's band began to smartly play his "Picadors March." All eyes now turned curiously to the towering top of the drapery-shrouded GE column. Slowly and ceremoniously, the concealing cloth was pulled away. Great hooting, laughter, and clapping erupted, echoing clamorously off the curved rooftop skylights. For there perched atop the sturdy column rose an eight-foot-tall, half-ton Edison incandescent light bulb, shimmering gorgeously through five thousand laboriously installed prisms. It was nothing less than General Electric's triumphal monument to the Seven Years' Incandescent Light Bulb War, the corporate victor's visible and sculptural declaration that the superior light bulb was theirs exclusively. (In keeping with this triumphant theme, GE displayed the full seven-thousand-page, seven-volume "Filament Case" testimony, a happy sight for any lawyer.) Lovely and legally instructive as the gargantuan Edison bulb might be, Westinghouse would have appreciated the irony that it was also one of the few Edison bulbs in the whole vast fair, for everywhere it was *his* "stopper" lamps that lit the night.

Edison's column turned out to be almost alive. It could dance! The audience gasped to see the eighty-foot column come pulsatingly to life. Reported the *Chicago Daily Tribune,* "Electricity danced up and down and all about its circumference in time with the rhythm of the music. First, slender lines of purple fire ran straight up and down its great height. Between them, at the next measure, came waves of crimson flame, and then, cutting up the spaces between in geometric figures and circling the column from top to bottom, clear dazzling whites."[31] One man yelled, "Edison!" in thrilled tribute. Others joined, and soon the whole building resounded to the delighted chant, "E-di-son, E-di-son, E-di-son." When at last the show ended, the crowds returned to roaming the aisles and alcoves. On the main floor were the major inventors, men like Edison and Tesla and Elisha Gray, whose teleautography machine was a protofax. Upstairs was the purview of lesser mortals flogging all kinds of electrical gadgets, many dubious—charged belts for a better sex life, body invigorators, electrical hairbrushes.

With such dazzlements at hand, it was perhaps natural that few appreciated the exhibit that was key to the electrical revolution slowly unfolding. This was, of course, the alternating current system of Nikola Tesla set up within the Westinghouse exhibit in the Electricity Building's north end. Only the most informed understood the extraordinary significance of the humble working model sitting on a long sturdy wooden table. It merited barely any mention in the many written popular accounts of electricity at the World's Fair. Yet here was the technology that would soon change the whole world. It included "a generating station, a high tension transmission circuit about 30 feet long and a receiving and distributing station. The first contains a 500 h.p. two-phase alternating current generator, a 5 h.p. direct current exciter, a marble switchboard and the necessary step-up transformers. In practice the generator and exciter would naturally be driven by water power. . . . Both generator and wheel are driven by a 500 h.p. Tesla polyphase motor with a rotating field, and the exciter by a 5 h.p. motor of the same type, both operated by current from the large two-phase alternators in Machinery Hall."[32] This collection of machines signified little to all but the cognoscenti. Here was power at its most versatile: a single source sending electricity a distance, where it then could run incandescent or arc lights, street railways, motors for factories.

While only the best-informed electricians appreciated the supreme importance of the small working model of Tesla's AC system, thousands of fair visitors breathlessly pressed in around Tesla's other crowd-pleasing exhibits in the Westinghouse section. There was the whirling Egg of Columbus (this being a large ostrich-size copper egg) such as had so captivated the reluctant Mr. Peck but six years earlier. The egg vividly demonstrated the rotating fields created by polyphase currents. The World's Fair version also included many smaller copper "planets." All were set to twirling until slowly but surely they all spun to the outer ring of their small electrified universe. Great crowds watched spellbound as seemingly inexplicable forces moved the copper balls inexorably about. Far more flashy but equally inexplicable to the general public was the eerie and almost occult darkened Tesla room, a tiny claustrophobic place. Over its entry door, a Westinghouse sign emitted brilliant and startling miniature crackling lightning, followed by a thunderous boom that echoed throughout the noisy hall. Inside, above people's heads could be seen "suspended two hard-rubber plates covered with tin foil. These were about fifteen feet apart, and served as terminals of the wires leading from the transformers. When the current was turned on, the vacuum bulbs or tubes [arrayed about the room], which had no wires connected to them . . . were made luminous. . . . Shown by Mr. Tesla in London about two years ago . . . they produced so much wonder and astonishment."[33] Glass tubing formed the names of famous electricians, all hauntingly aglow. As high-frequency currents crackled across the exhibit, beautiful long fingers of white sparks played across the flat, aluminum-covered surfaces. It was ethereal and magical, as if small pieces of lightning had finally been captured from the heavens and tamed.

On the hot Friday of August 25—Colored People's Day at Festival Hall—the great abolitionist and leader Frederick Douglass wearily entreated in this era of Jim Crow, "All we beg is to receive as honest treatment as those who love only part of the country."[34] Over on the South Side of Chicago, a ferocious fire, fueled by two huge coal yards, was roaring its way through 131 buildings, leaving two hundred families homeless. In Freeport, Illinois, yet another bank failed. Seven U.S. deputy marshals traveling the Rock Island Railway were bound for the Sub-Treasury Building on Wall Street, escorting twenty tons of California gold intended to shore up government finances, while at Manhattan's docks New York police were forcibly dispersing striking longshoremen.

But in Chicago on that very warm, sultry Friday evening, a thousand electrical engineers and scientists were streaming excitedly into the already hot Assembly Hall in the Agricultural Building, interested in but one thing: Nikola Tesla. "The great majority of those who came," reported the *Chicago Tribune,* "came with the expectation of seeing Tesla pass a current of 250,000 volts through his body and perform the marvelous feat with lamps lighted through his body that set Paris wild." Those without seats clamored at the entrance door, vainly offering as much as $10 to anyone who would sell them a ticket. Of course, those who really knew something of electricity now understood that deadliness was measured not in volts, but in amps. Rumor had it that Tesla was on retainer again to Westinghouse for $5,000 a month.

As the electricians in town for their electrical congress waited in the heat for Tesla, they could see up on a platform small cylinders of heavy steel mounted on steel pedestals, all with insulated wooden bases. To the right, a wooden table was stacked with strange mechanical appliances. Even the most eminent confessed they had no idea what many of these objects were. As the crowd settled down, white-haired Elisha Gray escorted Tesla in to a tumult of applause, saying: "I give you the Wizard of Physics!"

Though Tesla (always working long hours) looked gaunt and exhausted, his cheeks hollow, his dark eyes sunken, he smiled demurely and joked that many electricians had promised to speak, but "when the programme was sifted down I was the only healthy man left." So he was there to lecture on "Mechanical and Electrical Oscillators." This dry topic encompassed many wonders. Ever the dandy, Tesla was wearing a beautifully tailored gray brown four-button cutaway suit. For those who noticed, his shoes were unusual—thick soled with what looked like cork. The oscillators, he demonstrated, could generate very "precise frequencies that could be used to transmit information or electrical energy. When the oscillator was pulsating at the frequency of light, he could manifest luminescence as well. And mechanically he could create [and send precisely oscillating] pulsations through metal bars, or pipes, and test for harmonic frequencies and standing waves."[35] He had also designed steam generators so tiny, they could fit in the crown of a derby hat. Tesla made objects whirl, flashed huge sparks, lit up all sizes and shapes of protofluorescent lights, and, finally, lit up himself, until he was engulfed in dazzling streams of light. As the streamers of electricity subsided and Tesla was just fine—a living refutation of Edison's calumnies against his beloved AC—his brown suit continued to emit "fine glimmers or halos of splintered light."[36] The amazed audience burst into furious applause. Tesla had come and conquered!

Despite such transcendent moments within the magical precincts of the wondrous Chicago World's Fair, outside in the harsh and sobering light of the real world, the Panic of 1893 was turning into the worst economic disaster in decades. The fair certainly felt the stringency, and its directors were even more grateful to George Westinghouse for the $500,000 he had saved them. They always made sure he was paid first, ahead of all other fair creditors. Wrote Francis Leupp, "When the panic was passing through its most acute stage, and the banks were refusing to cash checks because they had nothing to cash them with, the treasury of the Fair handed over . . . to Westinghouse Electric and Manufacturing Company large quantities of dollars and half-dollars and quarters, which were shipped directly to Pittsburgh, and used to pay off the workmen in the shops at a time when currency was commanding five percent premium."[37] To everyone's surprise and delight, the Westinghouse Electric Company actually turned a small profit of $19,000 on its fair contract. The publicity for AC electricity was, as Westinghouse had anticipated, incalculable.

The millions upon millions of fair visitors who had flooded into town had shielded Chicago through the summer and early fall from the worst blows of the panic. But by November, Chicago reporter Ray Stannard Baker was writing, "What a spectacle! What a human downfall after the magnificence and prodigality of the World's Fair. . . . Heights of splendor, pride, exaltation, in one month; depths of wretchedness, suffering, hunger, cold, in the next." On the last day of the fair, the city's beloved four-term roguish mayor, Carter Henry Harrison Sr., came into the hall of his home to greet a disgruntled young man seeking a job. Instead of a handshake, he found himself fatally shot in broad daylight. On the heels of that jarring tragedy came the grim, ever growing army of unemployed, swelled by those who had built and then run the fair. Writes Chicago historian Donald L. Miller, "Thousands of them roamed the streets, unable, on some nights, to afford a urine-soaked mattress in a ten-cent flophouse."[38] When Baker the reporter stopped in late at City Hall one night, he was stunned to see every foot of cold stone floor space covered by men sleeping on newspapers, wet, worn shoes serving as pillows. By the end of that terrible year, 500 banks across Gilded Age America had failed, as had 150 railroads and 16,000 businesses. In Chicago, it had been the best of times. Now, as was true across America, it was the worst of times.

CHAPTER 11

Niagara Power: "What a Fall of Bright-Green Water!"

As the nineteenth century edged toward its final years, the Niagara Cataract was still one of the world's most celebrated natural wonders, a required destination for every genteel and self-regarding tourist, American or foreign. Visitors to Niagara Falls marveled at the powerful torrents thundering so violently down a 160-foot cliff that the very earth trembled while rainbows danced high above. The Canadians had the more spectacular, broadly curved Horseshoe Falls on their side of Goat Island, while the Americans had to content themselves with the narrower American Falls on their side. The earliest visitors had been uniformly rapturous in praise of all this tumbling, terrifying, awesome beauty. Charles Dickens arrived on the train straight from Buffalo on a raw April morning in 1842. From below the Canadian side, he watched "stunned and unable to comprehend the vastness of the scene.... Great Heaven, on what a Fall of bright-green water!... Then I felt how near to my Creator I was standing.... Peace of Mind, tranquility, calm recollections of the Dead, great thoughts of Eternal Rest and Happiness... Niagara was at once stamped upon my heart, an Image of Beauty."[1] He stayed ten glorious days, watching the oxygenated glass-green Great Lakes water crash over the falls on its journey to Lake

Niagara Falls, Prospect Point, 1890

Ontario, a man blissfully immersed in the sheer natural immensity of water, rumbling earth, mist, and dancing rainbows.

Fifteen years later, American landscape artist Frederic Church painted a vast and idyllic vision of the falls, rendered as if one were wading perilously in at their very top from the Canadian side. Beneath a luminous but glowering sky, Church caught the broad and boiling thunder of the stupendous green falls as it roared into the chasm. The combination of the immediacy and the majestic mesmerized the critics and the public. When Church's *Niagara* went on display in New York City in May of 1857, in just two weeks one hundred thousand people waited patiently in line to see it.

This sense of the sacred amid the wilderness that had so enthralled early visitors and that Church had captured so vividly on canvas was, in the years leading up to the Civil War, largely gone from Niagara Falls. The railroads made travel there easy, thus setting off the building of hotels, museums, stables, icehouses, bathhouses, laundries, and curiosity shops catering to the tourist dollar. A tawdry and aggressive commercialism engulfed both sides of the falls—tacky tea gardens, curiosity shops, huge and unlovely hostels, taverns, and viewing towers. The venal Niagara hackmen, vying loutishly for fares, quickly dispelled any pilgrim's spiritual frame of mind. Where a Charles Dickens had been content simply to *be* at Niagara, to literally soak up its sacred aura, others now sought more practical glory there. Most notably were those known as "funambulists," a Latin word meaning "rope walker." In the summer of 1859, the Great Blondin in his pink tights had attracted crowds of twenty-five thousand—not to admire the booming and foaming great cataract, but to watch him cross a tightrope strung up below the roaring falls. With each crossing, Blondin added more heart-stopping stunts—carrying his manager across on his back, then a small stove on which he nonchalantly cooked not one but two omelets a hundred feet above the chasm, and finally a table to hold his champagne and cake, as he casually sipped, munched, and balanced on his tightrope.

The next summer, as the tourist hordes again swarmed north to escape the heat and enjoy the delights of Niagara, Blondin was weekly challenged by a Monsieur Farini on *his* tightrope—performing headstands above the roaring waters of the gorge, hanging by toes as onlookers gasped on the *Maid of the Mist* excursion boat below, crossing the wide gorge while covered by a bulky bag, or taking a washtub, lowering it to gather water, and then studiously washing hankies! Finally, the Great Blondin did what even Farini did not dare: He sidled out onto a tightrope high above the gorge balancing on stilts—inspired, one suspects, by the presence of the Prince of Wales. His Royal Highness could only gasp, "Thank God it's over!" when Blondin returned to the bank to leap gracefully down.[2] Stuntmen and daredevils of all sorts would follow, making the falls a mere backdrop. Yet such was Niagara's reputation that all felt compelled if possible to visit, hoping to recapture the rapturous awe of its great waters.

Then there were the businessmen, practical sorts who from the start had admired the roaring waters but fretted that all that pounding waterpower was going to waste, just tumbling greenly and gorgeously over a 160-foot cliff. One early prospectus in 1857 for water wheels lamented "a power almost illimitable, constantly wasted, yet never diminished—constantly exerted, yet never exhausted—gazed upon, wondered at, but never hitherto controlled."[3] One local entrepreneur diverted Niagara water down a canal and by 1882 had as water wheel customers seven small industries, including pulp and flour mills and an Oneida silver-plating factory. This nascent industrial activity was the galvanizing event in the 1885 creation of the Niagara Reservation, a New York State government preserve that forbade all man-made excrescences on its four hundred acres of state land (three-quarters of it submerged) around Niagara Falls.

The very restrictions imposed by the new Niagara Reservation inspired Erie Canal engineer Thomas Evershed in the winter of 1886 to propose a wildly ambitious dream to harness Niagara Falls: a canal water wheel power system whose intakes would be more than a *mile* above the falls, well out of sight. There Niagara River water would be siphoned off and surge into a long, broad canal that would diverge off at factory and mill sites to feed some two hundred water wheels.

The Niagara water turning each wheel would then discharge straight down one hundred feet into a *two-and-a-half-mile-long* sloped tailrace tunnel deep below the ramshackle tourist town of Niagara. All the water wheel flow would return to the river through this tailrace tunnel just below the falls. By June of that year, a dozen influential upstate men promised to subscribe to $200,000 in stock in the Niagara River Hydraulic Tunnel, Power and Sewer Company and secured necessary state charters. In early September, the village of Niagara Falls gave permission for the tailrace discharge tunnel to pass far below its streets. Over the next several years, the Evershed group, which never anted up the actual cash for their stock holdings, tempted various investors but failed to convince any to actually finance their grandiose plan.

But the Niagara men did not give up. They drew in New York attorney William Rankine, who had studied law locally before making a name in Manhattan legal circles. On July 5, 1889, the handsome and debonair Rankine delivered, bringing north to the river-cooled Niagara Reservation two highly influential New York money men. The first was Edward A. Wickes, a rather fleshy, porcine Vanderbilt representative. The second man was none other than Francis Lynde Stetson, one of the most powerful attorneys in America. Scion of a distinguished New York legal and political family, Stetson served as a political confidant to New York governor Grover Cleveland in his whirlwind rise to the presidency. Out as president after one term, "the Big One" was now one of Stetson's law partners. So brilliantly had Stetson represented the railroad interests of the Vanderbilts that since 1887, J. P. Morgan had Stetson's firm on permanent retainer. So Francis Lynde Stetson was much courted and much feared. Sandy hair side parted, thick mustache topping a skeptical mouth, Stetson always carried about him an aura of intimidating power. His hard-looking eyes appeared as if they easily sized up adversaries.

By that September of 1889, with summer gone and both Stetson and Wickes asking again and again for postponements and concessions as they considered financing the Evershed plan, the enterprising William Rankine went himself to see J. Pierpont Morgan. The Niagara project would require very large infusions of capital. It had been more than a decade since Thomas A. Edison had installed the first incandescent lights in Morgan's cigar smoke–wreathed Wall Street office. Since those days, Morgan had been steadily amassing financial power and personal girth. In the office, he was as ferocious as ever, and his famed glare dared visitors to notice an unfortunate development of middle age: a reddened, pitted nose, progressively more inflamed by acne rosacea. Morgan, who liked to be surrounded by the aesthetically pleasing, showed a marked partiality for hiring young, handsome partners willing to work at a killing pace. Rankine came to Morgan's office and tried—to no avail—to entice the participation of the great Wall Street financier in this "floundering power enterprise." But Rankine was not one to be deterred. The next day, he reappeared and sought Morgan out again. "Mr. Morgan, you seemed interested in our project until we invited you to join us. Is there some feature we could change which would make it satisfactory?"

Said Morgan, "Your scheme is all right but you have no man to run it."

Replied Rankine, "Whom would you suggest?"

"Well, there is Adams. If you can get him, I'll join you."[4]

So it was that Edward Dean Adams came to Niagara.

Edward Dean Adams was a low-key New York investment banker renowned for his ability to revive and reconfigure crippled railroad companies. Naturally, Morgan, whose financial forte was just this, viewed Adams as a most valuable asset. A short man, Adams looked somewhat like a wise old frog with his heavily hooded eyes, weak chin, and mouth lost in a huge dark clump of mustache merging into an odd little chin tuft. While unimpressive in appearance, Adams was a full-fledged Boston Brahmin and descendant of two presidents. He had received a scientific and military education at Norwich College, then attended MIT. In 1878, he had become a partner at the exclusive Winslow, Lanier & Company in New York, where he was much admired as a brilliantly effective executive and attorney, a man who knew how to make things happen and get things done. In later years, the grateful stockholders of the American Cotton Oil trust bestowed a solid gold vase upon him as their rescuer from corporate ruin.

Adams had been a major stockholder in the Edison companies since 1884, and he was indeed quite interested in Rankine's proposition to take over the stymied Niagara power project. He, like others, wondered whether the new technology of electricity might not play an important role. Adams, who lived with his wife and children at 455 Madison Avenue in a mansion not far from Morgan, immediately consulted one of America's grand old mechanical engineers, Coleman Sellers of Philadelphia. On the last day of September 1889, Adams wrote to Sellers that while he knew that electricity had moved forward "with great rapidity," he was quite uncertain if it warranted "the investment of a large amount of money," and "I desire your advice."[5] He sent Sellers the Evershed prospectus, and by October 5, Sellers was writing that this huge project based on powering hundreds of time-honored mill wheels did indeed seem financially feasible, but he cautioned Adams to seek further advice from engineering experts in various fields. As for electricity, Sellers lamented the paucity of experience with long-distance transmission, telling Adams, "Large amounts of power have been cheaply sent short distances say two or three miles, and small amounts of power have been sent very much farther than is now under consideration." But nothing like the 100,000 horsepower envisioned at Niagara had been sent as far as the distance between Niagara Falls (population 5,000) and Buffalo (population 256,000)—twenty-six miles.[6] For surely Buffalo and its thriving industries were the likely consumers of Niagara power if its form was to be electrical.

By December, Coleman Sellers, a vigorous sixty-two-year-old with a cropped snow white beard and big handlebar mustache framing a jolly countenance, had journeyed up to Niagara Falls himself to survey the actual site and to confer with many of the Niagara men promoting the plan. He gave particular attention to the quality of the rock near the falls, for that would be paramount in excavating the long tunnel under the town. On December 17, 1889, he sent Adams a seventeen-page report that included actual figures and various proposals to reduce costs, concluding that the project was indeed feasible. He ended by noting it had been "one of the most interesting engineering problems ever given me to consider."[7] Sellers's enthusiasm sold Adams, and with Adams sold, so were the other New York financiers. A syndicate of 103 men, "one of the most powerful combinations of New York capitalists . . . ever . . . formed,"[8] invested a total of $2,630,000 in the Cataract Construction Company. This formidable new firm would transform Niagara's sublime and mighty flow of tumbling water into the riches of usable power. While the original Niagara men had little to offer monetarily, it was considered good local politics to retain them all on Cataract's board.

Adams, president of the newly formed and richly financed Cataract Construction Company, now began his organized and methodical search for a solution to one overarching question: "How can the power at Niagara best be utilized?" Adams hired the jovial and levelheaded Sellers as his main engineer, along with a handful of other prominent scientists and engineers to proffer expert advice. Adams and Sellers then sailed to Europe in early 1890 to seek whatever enlightenment was to be had there, for hydropower was far more familiar to the Swiss, French, and Italians, with their Alpine waterfalls. By May 1890, the two Americans had crossed the English Channel to travel through all manner of remote and lovely mountainous regions, examining water wheels and power transmission. Once back in England, Edward Dean Adams persuaded the eminent Sir William Thomson, a craggy-faced Scottish mathematician and physicist with a wild bushy beard, to formalize the quest as head of an International Niagara Commission. Thomson had earned his knighthood some decades earlier by making a reality of the transatlantic cable and had amassed an enviable fortune serving as the engineer for numerous other submarine cables in places as remote as Brazil. The other Niagara Commission members were a *Who's Who* of continental engineering. The whole raison d'être of the Niagara Commission was to invite engineers from around the world to submit their solutions to harnessing Niagara and then transmitting its bountiful energy. The best would be awarded prizes. It was an ingenious way to gather the collective knowledge of the scientific elite.

Adams, the wise old frog, and Sellers, the genial dean of mechanical engineers, held court (when not out reviewing Alpine hydro projects) at an elegant chandeliered salon in the posh Brown's Hotel on Dover Street in London. It was there in that June that young Westinghouse engineer Lewis B. Stillwell, also in Europe to survey the state of the electrical art, ventured to first meet Adams and Sellers. He was accompanied by Reginald Belfield, William Stanley's onetime assistant in Great Barrington and now chief electrician for British Westinghouse. They emerged from this one meeting in Edward Dean Adams's electrical salon burning to submit an entry to the Niagara competition. When Stillwell returned in the fall of 1890 to Pittsburgh and tried to convince his boss, George Westinghouse famously growled at him, "These people are trying to secure $100,000 worth of information by offering prizes, the largest of which is $3,000. When they are ready to do business, we will show them how to do it."[9] Fourteen proposals were received and various prizes dispersed, but the commission merely learned that there was no obvious solution. Sellers tried to put the best gloss on the situation by reassuring a rather discouraged Adams that it had "brought the scheme before the world with a prestige that cannot be measured by dollars . . . and won for its management respect, as wise, far-seeing cautious business men and not followers of any one visionary schemer or inventor."[10]

Even as Adams and Sellers and their International Niagara Commission picked the best engineering brains of Europe, the Cataract Construction Company pressed boldly on with its plans to begin excavating the great tunnel. On October 4, 1890, a mellow fall morning, Cataract secretary William Rankine, stylish in his well-cut dark vested suit, high collar, and clipped mustache, returned to Niagara Falls to attend the groundbreaking for the tailrace tunnel at the edge of the New York Central rail yards at Falls Street and Erie Avenue. When the first ceremonial shovel of dirt was dug, the warm autumn air resonated with tolling bells of nearby churches and shrieking mill whistles. At long last, the great Niagara power project was actually launched.

That lovely October morning marked the start of nonstop work, as 1,300 workmen dynamited, sledgehammered, steam-shoveled, and pounded round the clock to excavate the tailrace tunnel. Teams of mules strained as they moved out wagonloads of rock and debris. But the tunnel the men began digging was not the same design they ultimately built, for Edward Dean Adams and Coleman Sellers boldly abandoned the original plan rooted in the age of steam, a plan that relied largely on 238 water wheels strung along the long canal to collectively provide almost 120,000 horsepower. Instead, by the next summer, in July 1891, the two had committed themselves to a radically revised plan, one that extracted all the hydropower from two massive central stations on each side of a long intake canal right off the river. The water was still drawn off above the falls and still returned to the river via the tailrace tunnel deep under the town. But now it was only a mile long, a third the length of Evershed's scheme. Edward Dean Adams and Coleman Sellers were deeply influenced in this decision by a Swiss-born Englishman named Charles E. L. Brown, who was working for the Swiss firm Oerlikon but would soon leave to begin his own firm, Brown, Boveri & Company. On February 9, 1891, Brown delivered in Frankfurt a seminal talk titled "High Tension Currents," describing his highly successful experiments transmitting 100 horsepower of electricity several miles using 30,000 volts. Brown, with a high-domed forehead and rimless glasses, was utterly confident as he declared, "The transmission of electrical energy by means of current tensions of, for example, 30,000 volts is possible, the distribution of energy to great distances by electrical methods is a fact."[11]

Shortly thereafter, in mid-December of that year, Adams and Sellers bravely embraced the age of electricity and invited bids from six companies, three Swiss, including Brown, Boveri, and three American, those being Westinghouse, Edison General Electric, and Thomson-Houston. The Niagara central station powerhouses would no longer turn water wheels attached to shafts and pulleys. Instead, Cataract now planned for ten 5,000-horsepower turbines set deep in each of two central stations, each water-powered turbine to run an electrical generator. The two Niagara stations would generate a mind-boggling 100,000 horsepower, equal to all the central stations then operating in America. Never had electricity been generated on such a scale. Initially, only one powerhouse would be built and only the first three turbines and three generators installed. As the demand for electricity rose beyond 15,000 horsepower, more turbines and generators would be added.

Under the new Cataract plan, Niagara's mighty green waters were to be diverted into the powerhouse, funneled into eight-foot-wide penstocks (giant pipes), fall 140 feet straight down, rush around a crooked "elbow," and then roar into the double wheels of gigantic twenty-nine-ton turbines, the world's largest. These perpetually whirling turbines in the deepest basements of the central stations would turn attached vertical steel shafts that would twirl the electrical generators up on the main floor. Having powered the turbines, Niagara's waters would then begin their three-minute journey back to the river, *whoosh*ing along at twenty miles an hour through the 6,800-foot sloping tailrace tunnel. The new plan was extraordinary in its simplicity.

More than two hundred feet under the town of Niagara Falls, 1,300 men and their machines worked steadily, excavating day and night in the dank, close spaces. The tunnel itself was to be a spacious and elegant marvel, twenty-one feet high, with a gently curved roof, and eighteen feet across. Yet in some sections, water was visibly squirting and seeping through the raw tunnel cuts. Eight months and two fatal cave-ins after work began, Coleman Sellers ordered the tunnel shored up with sturdy pine and oak. The tunnel walls would—contrary to initial assumptions—have to be lined. Skilled bricklayers appeared and began to line the tunnel, fitting four layers of hard-burned Buffalo bricks into imported Portland cement. By the time the 6,800-foot tailrace tunnel was triumphantly completed on December 20, 1892, six hundred thousand tons of rock had been hauled out, sixteen million bricks lined the walls, and twenty-eight workers had died. Once the great dynamos were operating, all this would be eternally filled with racing river water. For the last two hundred feet, the curved brick walls gave way to cast iron, and every step brought closer the roar and thunder of American Falls. This immense water tunnel was the largest in the world.

Adams, Rankine, and the New York money men, in their methodical way, were covering all bases. To ensure that there would be adequate customers for their hydropower (if they could not figure out how to get it cheaply to Buffalo), they had quietly purchased almost two miles of land along the river and then inland, an L-shaped 1,500 acres where they envisioned dozens of factories powered by smokeless hydropower. And to house a portion of the future workers, the New York men were already planning to construct a small model workers' town, complete with the electricity that most viewed as a luxury. Adams chose the name Echoata, meaning something akin to "refuge" in Cherokee.

Season after season, as the tunnel had been pushing steadily forward under the village of Niagara Falls, Adams and Sellers had continued their methodical consideration of the electrical state of the art. At any inkling of advance, Cataract dispatched experts to report back. Meanwhile, the fight between AC and DC had been raging year after year. No one group—besides the antagonists themselves—could have followed this bitter electrical battle more avidly than Adams and Sellers and their experts. For who else had so many millions of dollars already riding on the future of electricity and, especially, the future of electrical transmission? The initial financing of $2.6 million had expanded steadily as expenses relentlessly mounted. Within their own Cataract ranks, the AC-DC debate was also being fought. Sir William Thomson remained an ardent and determined foe of AC. As of January 1891, so were most of the experts on the International Niagara Commission. Coleman Sellers would later describe them as "strongly antagonistic to an alternating current transmission."[12] Yet they also knew that despite all the arguments and accusations, alternating current was winning in the lighting marketplace. A February 1891 issue of *Electrical World* showed there were only 202 Edison central stations, all DC, naturally, versus almost *1,000* AC central stations installed by Westinghouse and Thomson-Houston.[13] Coleman Sellers retained an open mind and a propensity toward AC. The truth was, Sellers said in a July 1891 lecture at the Franklin Institute, "the progress of invention is going on so rapidly that we are at a loss to know what particular line should be pursued."[14]

In April 1892, having failed to persuade Charles Brown to set up in Niagara Falls, Adams and Sellers engaged George Forbes, a rather supercilious Scottish electrical engineer. Tellingly, his was the one entry in the original International Niagara Commission competition that proposed using alternating current. Professor Forbes's first official act was to dismiss the two DC designs submitted to Niagara by Edison GE and Thomson-Houston. In his report a month later, Forbes wrote, "I do not consider that these designs have sufficient merit to induce you to accept any delay in the hopes of getting something more perfect in this direction."[15]

Certainly, one of the most potent arguments *against* AC was the lack of a workable motor. In February of 1892, William Stanley, who had six years earlier set up the first working AC system in Great Barrington, Massachusetts, reminded others rather snidely that "a commercial AC motor . . . is a thing unknown to the practical engineer."[16] Or as Edward Dean Adams put it, at this juncture "the Tesla motor was still a prophecy rather than a completely demonstrated reality."[17] How could Cataract profess to be planning a *power* plant—as opposed to a *lighting* plant—if its electricity could not run motors? But this greatest of stumbling blocks was about to be surmounted. Very quietly and out of the limelight, the Tesla single-phase AC motor was finally distinguishing itself as a working *commercial* reality.

The proving ground for the Tesla AC motor was notably tough: the rugged and arcticlike San Juan Mountains of Telluride, Colorado. There, the owners of the financially ailing Gold King Mine had to find cheap energy or close. Having heard of Tesla and his AC system, they inquired of the Westinghouse Company in the spring of 1891 if electricity could be generated three miles away down in the valley, where a 320-foot waterfall could operate a water wheel. If a water-powered generator could indeed transmit electricity twelve thousand feet up the mountain to run a motor in the stamping and crushing mill above the timberline, the mine might be saved. The mine was quickly running out of nearby wood, and importing coal was too expensive. Westinghouse sold Gold King Mine a single-phase alternator (the phrase then for an AC generator) early that summer, and by June it was housed in a wooden shack next to the falls and its new water wheel. From this shack, 3,000 volts of electricity traveled on $700 worth of copper wire strung up and across three miles of steep, hardscrabble mountain terrain. At the mill the electricity was stepped down to power a 100-horsepower single-phase Tesla motor. The question was—would it run reliably? So all that summer, fall, and winter, and all through 1892, the Westinghouse engineers were delighted as the simple power system and the sturdy motor hummed steadily along, surviving the mountain's usual severe electrical storms, high winds, blizzards, and avalanches. The Tesla AC system had passed its first real job with flying colors.

In June of 1892, a jubilant Charles Scott, the young Westinghouse engineer who had assisted Tesla when he'd first come to Pittsburgh, announced this first successful *commercial* use of the whole Tesla system, including the long problematic induction motor, in the *Electrical Engineer.* "The aggregate time lost . . . was, by actual count, less than 48 hours during three-fourths of a year. . . . This success is confirmed in a substantial way by the immediate extension [expansion] of the plant. A 50 h.p. motor is now being installed at a mill a few miles from the Gold King . . . work in this field is fast passing from experimental investigation into practical electrical engineering."[18] At long last, the Tesla motor had arrived, operating off a Tesla-designed generator. In the same triumphant article, Scott also described how for two years a forty-foot waterfall on the Willamette River had powered a big Tesla AC generator, which sent electricity *thirteen* miles to their electric lighting central station in Portland, Oregon. Here at last the Tesla polyphase system had begun in the smallest of ways to fulfill its long touted promise. But above all, there was now a working Tesla AC induction motor, albeit single-phase. The last great AC hurdle had been overcome.

As the Cataract Construction Company pushed steadily ahead upstate at Niagara Falls, New Yorkers in Manhattan were reveling in what was shaping up to be one of the city's juiciest, messiest scandals ever. It had all begun the previous month when the thickly bewhiskered Reverend Charles Parkhurst stood in his pulpit at the Madison Square Presbyterian Church and accused the city's Tammany mayor and police department of being part of an "official and administrative criminality that is filthifying our entire municipal life, making New York a very hotbed of knavery, debauchery and bestiality." Hauled by the Tammany DA before a grand jury to prove these intemperate charges of vice run amok amid police complicity, Reverend Parkhurst had to admit he could offer none. The good reverend emerged from this public humbling determined to gather the necessary evidence. He hired Charles Gardner, a private detective, who agreed to lead Parkhurst and two others on a tour of Manhattan's darkest dives.

Disguised as "toughs," the foursome rode the Third Avenue el from 18th Street down to Franklin Square and began their education in Manhattan lowlife, starting off with "a whisky saloon, an opium den in Chinatown, a stale beer dive in the Italian section, and other horrors.... In a Tenderloin house, five girls stripped and did a 'dance of nature.'... At Scotch Ann's Golden Rule Pleasure Club on West 3rd... they found it subdivided by partitions into cubicles... in which sat a boy with a painted face."[19] As an appalled Parkhurst and his two companions toured every kind of beer-soaked dance hall, cheap saloon, and sleazy brothel, they became astonished, genteel witnesses to a large and rowdy underworld thriving on illicit gambling, cheap liquor and beer, and organized, commercial sex catering to every kind of appetite. Gardner also had his own detectives out gathering affidavits. So when Reverend Parkhurst mounted his pulpit a mere month later on Sunday, March 13, 1892, he possessed documented evidence that Tammany was "rotten with a rottenness that is unspeakable and indescribable." He had proof that 254 saloons and 30 brothels had been roaring with business just the previous Sunday. In the ensuing months, a grand jury handed down a few placating indictments, but little would change immediately. The urban poor (and a certain number of their better-off confreres) wanted jollity and dazed forgetfulness, be it craps, bawdy dance halls, or cheap, quick sex, for theirs were hardscrabble lives. And there were great sums to be made, as Tammany showed, in organized vice. A few outraged citizens were, for the moment, but an annoyance.

While Reverend Parkhurst was roiling the murk of the Manhattan underworld to little avail, George Westinghouse was informing Edward Dean Adams that he would indeed bid on the Niagara contract. That done, Westinghouse turned his attention to his ongoing crusade to best GE for the Chicago World's Fair lighting contract, "the largest electrical project ever undertaken" up to that time. While Parkhurst achieved little but frustration, Westinghouse emerged from his battle in mid-May with his lowball contract in hand. Westinghouse then concentrated full bore throughout the second half of 1892 on building the fair system, knowing that if it worked, it would well serve his pursuit of the Niagara contract. Since most of the Chicago World's Fair AC electricity was for lighting, the young Charles Scott proposed they stick with the tried-and-true single-phase for the tens of thousands of incandescent "stopper" lights. But Westinghouse said, "No, they have been telling me that the two-phase is the right system and I want to find out whether it is or not."[20] So the Westinghouse engineers dedicated themselves to creating commercially viable two-phase, something possible only because they finally conceded what Nikola Tesla had insisted on in 1888—AC induction motors could not function if the frequency was too high. As Charles Scott so dryly put it, "Commercial circuits were single-phase at a frequency of 133 cycles. Strenuous efforts to adapt the Tesla motor to this circuit were in vain. The little motor insisted on getting what it wanted, and the mountain came to Mahomet. Lower-frequency polyphase generators inflicted obsolescence on their predecessors in a thousand central stations—such was the potency of the Tesla motor."[21] Without these new AC generators, the new AC motors would not be an option for Westinghouse customers.

The Westinghouse Electric Company now set standard AC frequencies of 60 cycles per second for lighting and 30 cycles for motors. Nikola Tesla once again joined the Westinghouse effort. In late September of 1892, he was urgently ordering various apparatus from Pittsburgh "to carry out some improvements on my motor," noting that "time is very precious."[22] He asked that all be delivered to his fourth-floor laboratory on South Fifth Avenue, a loft building in the ramshackle commercial and industrial area. By December of 1892, Westinghouse had submitted its two-phase AC design for Niagara to the Cataract Company. GE's entry came in soon thereafter, with a design similar in all ways but that it employed three-phase AC.

Up at Niagara Falls, the gargantuan Cataract construction project blasted and rumbled onward. In December 1892, the tailrace tunnel had been completed. By January of 1893, Professor Forbes was writing in the *Electrical Engineer* that the big intake canal "about a mile and a half above the American Fall . . . has been dug out 500 feet wide, and 1,500 feet long, with a depth of 12 feet. Along the edge of this canal wheel pits are being dug 160 feet deep, at the bottoms of which the turbines will be placed. The water is admitted to the penstocks by lateral passages (or head-races) which can be closed by gates."[23] The other notable event at Niagara Falls that January was "the most ample and substantial" ice bridge in almost three decades, since the winter of 1855. Reported *The New York Times,* "The steady zero weather of the past week has filled the upper river with ice which is pouring over the falls in vast quantities and adding each hour to the jam which is called the 'bridge.' . . . The heavy fall of snow and the clouds of mist from the Falls, which settle on top, freezing as fast as they fall, form a natural cement."[24]

Day after day that January, the river ice crashed over the falls, until the gorge below became a fantastical arctic landscape, a massive shifting, grinding ice field, dazzling in the winter sun and gorgeous with its towering white ice mountains. Hundreds of tourists bundled in dark winter coats and cloaks, women and girls with fanciful fur hats and muffs, streamed out onto the broad ridges of the ice bridge, tiny dark figures laughing and shouting as they gazed up at the beauteous sight of the falls frozen into hundreds of gigantic craggy icicles. Professor Forbes, who was at Niagara later that winter, marveled that "the precipices are concealed behind icicles 60 feet long. Every rock in the river is the nucleus for a dome of frozen spray rising 150 feet."[25] On cold wintry mornings, all the trees and bushes near the river were encased in a shimmering coat of ice, and visitors felt they had entered a wondrous fairyland. But the river ice had a prosaic and serious side, and keeping it out of the power plant machinery would, some winters, prove a brutal struggle.

While the ice bridge and the thrilling danger of walking across its ever shifting, dazzling landscape was attracting winter tourists to Niagara, Cataract consultants were visiting the sprawling, noisy

Westinghouse electrical plant in Pittsburgh. From January 9 to 13, 1893, Coleman Sellers and Johns Hopkins physics professor Henry Rowland, another Cataract consultant, observed and subjected to many tests the new Westinghouse AC generators and transformers, assessing if they were suitable to this biggest of all electrical projects. They watched the working of a new rotary converter that could turn AC into DC (important for street trolleys), measured safety and switching apparatus, observed how too low frequencies would make lights flicker, and tested motors. Sellers came away very favorably impressed and in his report noted, "A careful examination of the work done in this establishment showed excellent workmanship and correct engineering design in all the machinery examined. . . . The workmanship is beyond criticism in quality."[26] In his report, Professor Rowland concluded that Westinghouse had "the greatest experience in the practical use of the alternating system and they seem to control the most important patents."[27]

The next month, the two men visited the General Electric plant in Lynn, Massachusetts, and Sellers noticed how similar—but by no means equal—the GE apparatus was to the Westinghouse equipment. He noted that "very considerable change would have to be made to make it mechanically the equal." Moreover, he was very leery of GE's proposed use of three-phase AC, saying, "I should incline to the biphase on account of its greater simplicity and its adaptability to a broader field of usefulness." Knowing that Professor Forbes, who was annoyingly away in England during these crucial visits to Westinghouse and GE, favored the AC design of the Swiss firm Brown, Boveri, Coleman Sellers ended his twenty-five-page report sent from his Baring Street office in Philadelphia thus: "I do most earnestly protest against the purchase of the foreign plant if as good electrical results can be anticipated from the home made machine, even if the first cost is seemingly greater."[28] Aside from his patriotic chauvinism, Sellers the engineer could not see how a foreign firm could fix and maintain the inevitable problems in a timely fashion.

Not surprisingly, the AC patent issue was looming larger and larger. As Coleman Sellers bluntly told Edward Dean Adams, "Until the contrary is proved by the courts, [Westinghouse] claims control of what is most important for our purpose at the present time in America. I am not aware of any claim to ownership in this country of what can stop the owners of the Tesla patents from commanding the market. . . . My present opinion is that no foreign company can secure the Cataract Construction Co. against all losses from patent litigation."[29] The previous month, February, Edward Dean Adams, from his office on the fourth floor of the prestigious Mills Building, had begun a private correspondence with Nikola Tesla, seeking his opinion on a variety of electrical matters, often about whether reports sent to Adams were technically correct. Other times, he was grappling to better understand the new AC technology, such as Tesla's synchronous and multiphase motors. Tesla, acutely aware that George Westinghouse was vigorously competing for the Niagara power contract, used his private letters to Adams (scrawled on stationery from his new residence, the Gerlach Hotel) to press home again and again the broad scope of his AC patents. The inventor's message was clear: If any other company said it could provide a multiphase AC generator and, above all, AC motors that would then power factories, they were infringing on his Westinghouse patents. In a February 2 letter, Tesla wrote, "I have not heard from Germany yet, but I have not the slightest doubt that all companies except Helios,—who have acquired the rights from my Company,—will have to stop manufacture of phase motors. Proceedings against the infringers have been taken in the most energetic way by the Helios Co. It is for this reason that our enemies are driven to the single phase system and rapid changes of opinion."

Modesty played no part in Tesla's self-confident missives to Edward Dean Adams, a financier so influential, so utterly critical in deciding whether Nikola Tesla would live out one of his oldest, most cherished electrical dreams. While in high school, Tesla would later recall, "I was fascinated by a description of Niagara Falls I had perused, and pictured in my imagination a big wheel run by the Falls. I told my uncle I would go to America and carry out this scheme."[30]

In fact, Tesla had devised something far more original than a water wheel, and now the decision about the contract hung in the balance. He lost no opportunity to trumpet his advantages. When Adams inquired whether a Thomson-Houston patent might be comparable to Tesla's, the inventor asserted in a March 12 letter that this patent had "absolutely nothing to do with my discovery of the rotating magnetic field and the radically novel features of my system of transmission of power disclosed in my foundation patents of 1888. All the elements shown in the Thomson patent were well known and had been used long before." When Adams sought his opinion on a DC system, Tesla wrote on March 23, barely able to contain his horror at "how disadvantageous, if not fatal, to your enterprise such a plan would be, but I do not think it possible that your engineers could consider seriously such a proposition of this kind."[31]

And in fact, on May 6, 1893, the officers of the Niagara Falls Power Company declared unequivocally that polyphase alternating current would be their choice. This was, at the time, still a very bold and highly controversial stance. The eminent Sir William Thomson, chairman of the International Niagara Committee and just elevated to become Lord Kelvin by Queen Victoria, cabled Adams on May 1 to head off the announcement. He proposed an ambitious DC plan, urging, "Trust you avoid gigantic mistake of adoption of alternate current."[32] Edward Dean Adams, in his two-volume history of the Niagara Falls Power Company, noted how much that momentous decision was based on "faith and hope that electrical engineers could produce apparatus much larger in size than ever had been built and that new types which were then hardly beyond the stage of experiment would prove successful."[33] After all, on what did they truly base this leap of faith, this major commitment of many (steadily mounting) millions of their own hard cash? The "outstanding actual achievement in power transmission was at Telluride," where the gold mine stamping mill had been running for two years now, sending, in truth, piddling amounts of power up through rugged mountains to run a small motor. Then there had been a highly successful—but nonetheless experimental—demonstration at the 1891 Frankfurt Exposition, where power was transmitted one hundred miles.

Then, of course, there was the triumph of the just opened Chicago World's Fair, with the nighttime Court of Honor awash in dazzling electricity. Wrote Adams, "The construction of twelve polyphase alternators of a thousand horsepower each and the electrical illuminations of a great White City for the first time in history were great events, but they were overshadowed in real significance by a more important though less spectacular exhibit."[34] That was, of course, the working model of Nikola Tesla's universal AC power system, with its AC generator, transformers, transmission lines, working induction motors, synchronous motor, and Westinghouse-invented rotary converter that supplied direct current for the railway motor. George Westinghouse and Nikola Tesla had finally convinced the Niagara engineers and millionaires, many of whom had been deeply skeptical, some outright hostile, that alternating current truly was the ideal for creating and distributing power in the coming age of electricity. And so, in the spring of 1893, as the Chicago World's Fair opened, Westinghouse and Tesla verged on complete electrical triumph, with the great Niagara dream tantalizingly within their grasp.

Their chief American rival for this sought-after prize, this ultimate showcase of electrical power, was, of course, GE, the much maligned electrical trust. For some time, George Westinghouse had been wrathfully suspicious that GE was stealing his company's hard-won, highly valuable mechanical and electrical knowledge. Even though all Niagara submissions were supposed to be completely confidential, he had discovered the incredible similarity between his and GE's Niagara plans, something mentioned by Sellers in his report to Adams. (GE had varied its design only by making it triple-phase.) In early May, one of the Westinghouse engineers learned that the Westinghouse Company's blueprints and many documents about prices, labor costs, and other privileged information were indeed at GE's Lynn plant. George Westinghouse immediately sought a search warrant, and GE was caught red-handed. Westinghouse had one of his draftsmen arrested for secretly selling the firm's World's Fair and Niagara blueprints for thousands of dollars to two GE men, one a general superintendent in the GE railway department. GE insisted they were only trying to see if Westinghouse was infringing on *their* patents. The Pittsburgh district attorney happily announced the discovery of this "conspiracy" May 8, 1893, and his intention to seek grand jury indictments of not just the underlings on both sides, but the eminent Charles Coffin, GE's president and top executive.[35]

Coffin was furious and quickly wrote an assuaging letter to his financiers, saying to Vanderbilt son-in-law Hamilton McK. Twombly, "I hope you are not disturbed . . . while it is altogether probable that some of their blue prints may have been in our possession, it was absolutely without my knowledge or sanction. . . . If there be any similarity between their [Niagara] plans and ours . . . it is purely accidental. Be that as it may, there is an implied charge against the Niagara Co. of very bad faith [not keeping each submission confidential] in the statements of the Westinghouse Co. . . . It is part of the bitter and vituperative work of the Westinghouse people. . . . [They] will distinctly lose prestige and business as the result of their ridiculous behavior in connection with this matter."[36] (When the case went to trial that fall, Coffin was no longer a defendant and the Pittsburgh jury deadlocked.)

But galling as the spying case was to George Westinghouse, it paled in comparison with the perfidious deed that came next. On May 11, Edward Dean Adams and the Cataract Construction Company dropped a large, unexpected, and outrageous bombshell. Adams wrote a one-page missive to each of the four electrical companies still competing—GE and Westinghouse in America and Brown, Boveri and Maschinenfabrik Oerlikon in Europe—coolly informing them that their services were no longer needed. The Niagara dynamo contract, this most glittering, sought-after, and prestigious of electrical prizes, was not going to any of them. Instead, the Cataract Company, having had the privilege of examining every aspect of these four firms' proprietary electrical designs, having availed themselves of the companies' best men and technical advances, was now appointing their *own* electrical consultant, Professor George Forbes, to design a generator to accommodate its 5,000-horsepower water turbines.

Most outrageous, these designs by Professor Forbes, said Adams, "are well advanced," which meant that even as Westinghouse, GE,

Oerlikon, and Brown, Boveri engineers were struggling to solve and overcome all manner of obstacles for Niagara and discussing these with Sellers and Rowland and Forbes, Cataract well knew that Professor Forbes was already working on a generator. Adams had the gall to inform the rejected American bidders that "we expect to submit the same to you, as well as to others, for proposals for construction within a brief period. Please accept our sincere thanks for the response you have made to our invitations for proposals."[37] One of the world's most eminent electricians, Silvanus Thompson, later expressed the profession's collective outrage, denouncing Cataract's blatant and "ungenerous picking of the brains of others." He deemed this "contemptible collaring of rival plans . . . the one discreditable episode the savour of which will ever cling about the undertaking."[38]

Adams, ever the gentleman even as he dished out such bitter soup, wrote Tesla to let him know that Professor Forbes would design the dynamo, not Westinghouse or GE, and also indicated his own Brahmin ire over the sordid public airing of GE-Westinghouse spying charges. Tesla, himself ever the gentleman, replied that same day, dismissing the GE spying brouhaha as "a trifle, to be sure, not worthy of any consideration." The elegant Serbian inventor was the soul of cordiality, writing from his comfortable suite at the Gerlach Hotel on West 27th Street (conveniently around the corner from Delmonico's, the gastronome's delight), "I can assure you that your decision does not in the slightest affect my sympathies and my sincere wishes that your magnificent enterprise may meet with the success it deserves."

But after these pleasantries, Tesla got down to business, which was to strongly warn Adams he "could not help seeing difficulty ahead." The obvious problems were that Cataract had been shown "in good faith" Westinghouse plans based on "long continued experience and items on the subject not found in any treatise on engineering," and of course, there were Tesla's patents and the many Westinghouse improvements developed since to make them commercially ready. This uncomfortable fact would, Tesla indicated, make it very difficult for Professor Forbes to design a noninfringing alternating current system. This was sufficiently worrying to Adams that he passed it along to Coleman Sellers in Philadelphia. Sellers advised letting Tesla know that it was Cataract's intention to redesign the generators to better suit the turbines and then go back to the "competitors, with the expectation that they can find it to their advantage to aid us in the development."[39] And there the matter rested, leaving a residue of bitter ill will.

THE NIAGARA FALLS POWER CO.

CHAPTER 12

"Yoked to the Cataract!"

All the dappled summer of 1893, Professor George Forbes lived at Niagara and worked on the Cataract dynamo design. "I had a lovely house in parklike grounds . . . on the banks of the placid river above the upper rapids," he later wrote. "I went to bed early and rose at five or six in the morning, and I shall never forget the delights of those glorious summer mornings at one of the most beautiful sites in the whole neighborhood." From time to time, Professor Forbes squired around visiting electricians en route to or from the World's Fair who also wished to see the marvels of the Great Falls and Cataract's progress on its mammoth power project. The gigantic wheel pits at Power House No. 1 were well advanced, and soon the three mighty turbines would be installed deep in its bowels. Highly favored visitors were treated to the clammy stygian walk through the completed mile-and-a-quarter tailrace tunnel from the powerhouse down to the roaring river gorge.

Professor Forbes, a tall, supercilious Scotsman with fair hair, a large nose, and a mustache, had very few kind words for his Cataract engineering colleagues. His attitude toward most things American could be summed up in one word: condescension. He preferred to live on the Canadian side of the falls and disdained the nearby town

Niagara Falls Power Company horse and wagon

of Niagara Falls as "dirty . . . [and full of] cheap restaurants, merry-go-rounds, itinerant photographers, and museums of Indians and other curiosities."[1] Forbes clearly gloried in his high-profile, plum assignment as Cataract's consulting electrical engineer, a man central to the world's most ambitious, expensive, and closely watched electric power project. Yet he petulantly complained that designing the Niagara dynamo that summer and generally carrying out the technical aspects of his work were made far more difficult by "politics," which he described as "intriguing, underhand dealing and jobbery."[2] The haughty Forbes was quick to take sole credit for work that was highly collaborative or others' entirely.

As the August heat clamped down on the eastern seaboard that summer of 1893, Coleman Sellers and Edward Dean Adams felt growing pressure to move ahead. The Cataract investors had anted up $4 million thus far for this vast enterprise, all of which now hinged on a workable AC generator. Niagara was a huge investment of purely private capital, a fact acutely felt as the American economic system was slowly and disastrously imploding. No day of the gloriously lovely summer passed without more terrifying news—bank failures rolled from region to region, farmers in the far West could not ship their crops for lack of credit, railroads slid into receivership. Even the mightiest millionaires had to look closely to their accounts. One associate described how "every morning while in New York Mr. Rankine has submitted to him a statement showing the exact balance in the bank of the Cataract Construction Company, Niagara Falls Power Company, Niagara Falls Water Works Company, Niagara Development Company, and the Niagara Junction Railway Company. When he is here at the Falls this statement is mailed to him."[3] J. P. Morgan wrote a friend in late July that summer, "Everything here continues blue as indigo. Hope we shall soon have some change for the better, for it is very depressing and very exhausting."[4] As the financial markets teetered on the brink, the U.S. Congress convened in emergency session on August 7, and Grover Cleveland, the only president to return to office for a second term after being defeated following his first, urged the legislators to repeal the parity of silver with gold.

Days later, on August 10, 1893, Coleman Sellers, now president of the Niagara Falls Power Company and its chief engineer, wrote both GE and Westinghouse to announce that Professor Forbes had designed a suitable dynamo and transformers. Therefore Cataract was again looking for a firm to manufacture and install its generating equipment. George Westinghouse, still seething from the earlier dismissal, wrote Sellers a stiff reply a week or so later from Pittsburgh, reminding this genial engineer who had seemed to so favor his firm, "We have given several years time to the development of power transmission, and have spent an immense sum of money working out various plans, and we believe we are fully entitled to all the commercial advantages that can accrue to us . . . we do not feel that your company can ask us to put that knowledge at your disposal so that you may in any manner use it to our disadvantage."[5] Nonetheless, Westinghouse relented. After all, this was still the great Niagara Falls power project, his dearest electrical dream, a world showcase for alternating current. He had pushed forward through far worse. Then, of course, there was the work this would guarantee for his men, at this grim time when every factory was watching its orders plummet.

So George Westinghouse dispatched a couple of his top engineers, one of these being Lewis Stillwell, to Niagara Falls on August 21 to see just what Professor Forbes had come up with. As the two men came off the train, the resort was jammed with tourists in a holiday mood, enjoying the merry-go-rounds, clambering about the falls and the Cave of the Winds, riding the *Maid of the Mist.* The engineers' mood was nowhere as jolly. The Scotsman's dynamo design, they would soon conclude, was so hopelessly flawed that they could not consider constructing it. After studying the blueprints thoroughly, they told Coleman Sellers that "mechanically the proposed generators embodied good ideas . . . [but] electrically it was defective and if built as designed . . . would not operate."[6] Historian Harold Passer summed up the flaws in Forbes's generator as these: such low frequency—16⅔ cycles a second—that it would cause noticeable flickering in lights; worse yet—for a project aimed at providing industrial power—it would be "too low for satisfactory operation of most polyphase power equipment [notably the all-important rotary converter to change AC to DC]. The Westinghouse engineers also sharply criticized the high-generating voltage [an unheard-of 22,000]. The insulation problems would be difficult to solve and perhaps impossible."[7]

The haughty Professor Forbes seemed to have forgotten during his delightful summer sojourn in Niagara the whole point of AC. Letters were exchanged and a negative report made to George Westinghouse. Then on September 15 two top Westinghouse engineers, one again being Lewis Stillwell, returned once more to the ramshackle precincts of Niagara, where the last tourists still lingered on in the chill of early fall. The engineers met first with Sellers and other Cataract experts and once again talked about the shortcomings of the Forbes dynamo. They then proceeded to Professor Forbes's office, where, Sellers later related, "Professor Forbes discussed some of the questions raised and declined to take up others, stating that he had fully considered the subject and was sure he was right."[8]

Coleman Sellers and Edward Dean Adams well knew that they could not move forward without George Westinghouse and his patents and know-how. (Adams seemed to view GE's bid merely as a means to keep the overall price down.) And now the top Westinghouse engineers had returned from repeated Niagara inspections of Forbes's dynamo blueprints oozing contempt for Cataract's design. Soothing ruffled corporate feathers was, however, Adams's forte. He had made his name as an attorney bringing together angry and fractious railroad investors and railroad officers and convincing these hissing rivals to agree on their joint salvation. So in early October he proposed a dinner in a comfortable private dining room at the venerable Union League Club, one of Manhattan's most exclusive men's clubs, formed during the Civil War to support the North. That evening, the millionaires and engineers gathered in their formal dinner clothes and over many sumptuous courses reviewed every contentious aspect of the proposed Cataract dynamo contract with Westinghouse.

By the time the cigars and brandy were served, the two sides had come to terms on everything but frequency, with Cataract clinging loyally to Forbes's too low frequency of 16⅔ and Westinghouse insisting it could not guarantee any dynamo of less than 30 cycles. Decades later, Westinghouse engineer Benjamin Lamme wrote, "There was more or less of a deadlock on the question of frequency. . . . [We] did not wish to build such a machine, due to the great probability of complete failure from the operative standpoint."[9] As the dinner broke up, Adams pulled aside Lewis Stillwell, the chief Westinghouse engineer. Could they compromise at 25 cycles? The eventual answer turned out to be yes. On October 27, 1893, three days before the Chicago World's Fair was coming to a triumphant end, George Westinghouse finally had in hand the coveted contract that had slipped away from him earlier in the spring. He and Nikola Tesla would finally show the world what real electrical power could be.

By the end of that year, 1893, the whole Westinghouse camp was united on two matters: perfecting the generators for Niagara Power House No. 1 and visceral dislike and distrust of Professor George Forbes. In the wake of the professor's misguided dynamo design, they doubted his electrical competence and viewed him as a decided impediment to their job. Nor did they appreciate his condescension. The ever genial Coleman Sellers found himself uncomfortably in the middle, for in December of 1893, George Westinghouse announced in his usual definite and decisive way that he and his men simply would not work with Professor Forbes (soon to return from a Christmas dash across to Great Britain). Westinghouse now saw Forbes as "a possible rival in dynamo design" based on a lecture Forbes had delivered, and he had no intention of providing the professor any edge. Sellers wrote a rather anguished private memo to Adams about this "very delicate matter" right after Christmas and blamed Forbes for being away "when the most important measures are to be decided." He wanted Adams to be fully aware of the "absolute unwillingness" of Westinghouse to have dealings with Forbes.[10]

After a February 6, 1894, meeting in Manhattan with George Westinghouse and two of his engineers, Edward Wickes reported to Adams that the Pittsburgh magnate was unyielding on Forbes. It created, he conceded, "considerable difficulty. We must get along the best way we can."[11] The upshot was that Coleman Sellers largely sidestepped Forbes from that time forward. Professor Forbes, whose dynamo design had been thoroughly lambasted by American engineers when he presented it at various engineering forums, did not appreciate his diminished Cataract status. (Charles E. L. Brown, head of the Swiss firm Brown, Boveri, formally accused the Scotsman of copying the unique umbrella-style-design dynamo *he* had submitted to Cataract in late 1892.)[12]

In a parting shot at Cataract for trifling with him, Professor Forbes penned a subtly poisonous article on Niagara for *Blackwood's*, an English magazine, the next year. Predictably, he portrayed himself as the presiding engineering genius behind the great power project and the Americans largely as annoying ignoramuses. Forbes wreaked some measure of revenge when he airily explained, "I had at times great difficulty in keeping the president and vice presidents in hand.... Most of them began to think they knew something about the subject.... All this was generally amusing enough, but became almost tragic at times when I found them endangering the whole work. On such occasions I would write to my millionaires and tell them that if they did not do what I told them they would be personally answerable to the directors and shareholders for any disaster that might occur."[13]

All through 1894, the engineers at the Westinghouse Electric & Manufacturing Company were engaged in the mammoth yet delicate task of fine-tuning the designs and starting to build the first two of Niagara's 5,000-horsepower generators, completely new kinds of machines five times bigger than those at the World's Fair, which themselves at 1,000 horsepower had been considered behemoths. The third would be finished only when the first two had been shown to work properly. In its report on the contract, Westinghouse emphasized the great novelty of almost all the machinery: "The switching devices, indicating and measuring instruments, bus-bars and other auxiliary apparatus, have been designed and constructed on lines departing radically from our usual practice. The conditions of the problem presented, especially as regards the amount of power to be dealt with, have been so far beyond all precedent that it has been necessary to devise a considerable amount of new apparatus.... Nearly every device used differs from what has hitherto been our standard practice."[14] The original intended size and scale of the generators had to be reduced to ensure that they could be hoisted onto a railroad flatcar and transported safely to Niagara.

COURTESY OF THE CHICAGO HISTORICAL SOCIETY

Westinghouse Tesla Polyphase System exhibited in the Electricity Building at the Chicago World's Fair of 1893. While General Electric dominated the exhibits, Westinghouse won the big lighting contract.

COURTESY OF THE GEORGE WESTINGHOUSE MUSEUM

The Tesla polyphase exhibit at the Chicago World's Fair of 1893 showed how alternating current worked and included the twirling Egg of Columbus. Few appreciated its epochal importance.

COURTESY OF NIAGARA MOHAWK, NATIONAL GRID USA SERVICE COMPANY, INC.

This 1893 photo shows a few of the thousands who labored to excavate the Niagara Falls Power Company's mile-long tailrace tunnel under the town of Niagara Falls. A mile above the famous falls, river water was drawn off to power the Westinghouse AC dynamos, and then roared out through the tunnel just below the falls.

COURTESY OF NIAGARA MOHAWK, NATIONAL GRID USA SERVICE COMPANY, INC.

One of first Westinghouse Niagara Falls Power Company generators being built in Pittsburgh in 1894.

COURTESY OF NIAGARA MOHAWK, NATIONAL GRID USA SERVICE COMPANY, INC.

The first three Westinghouse dynamos in Stanford White's "Cathedral of Power" at Niagara Falls in a photo taken April 6, 1896.

COURTESY OF NIAGARA MOHAWK, NATIONAL GRID USA SERVICE COMPANY, INC.

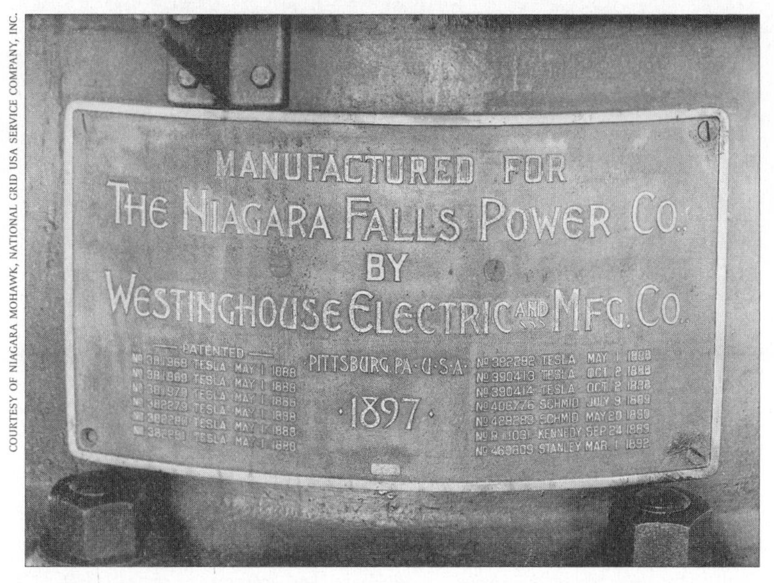

The plaque on one of the Westinghouse dynamos shows the tremendous importance of the Nikola Tesla AC patents.

COURTESY OF NIAGARA FALLS PUBLIC LIBRARY

A 1903 photo of the famous Niagara Falls ice bridge, when the falls and the river below froze into a strange and beautiful arctic landscape.

COURTESY OF THE NIKOLA TESLA MUSEUM

Nikola Tesla in the lab he set up in Colorado Springs in 1899 to study electric energy by generating millions of volts. This is a double exposure.

COURTESY OF THE GEORGE WESTINGHOUSE MUSEUM

George Westinghouse (*left*) with Lord Kelvin, long in the anti-AC camp, when he visited the Westinghouse Company in August 1897.

COURTESY OF THE GEORGE WESTINGHOUSE MUSEUM

The Westinghouse country estate, Erskine Manor, in the Berkshires. They always liked to have family and friends in residence.

COURTESY OF THE NEW-YORK HISTORICAL SOCIETY

Starting in the 1890s, Nikola Tesla lived in the Waldorf Astoria, Gotham's most luxurious and socially elite hotel.

COURTESY OF THE NIKOLA TESLA MUSEUM

Nikola Tesla's Wardenclyffe Tower and power plant in Shoreham, Long Island, was to serve as the heart of his World System of Power. Lacking sufficient money, he never finished it and got it operating.

COURTESY OF THE NIKOLA TESLA MUSEUM

The white pigeon that was Nikola Tesla's great love in his final years.

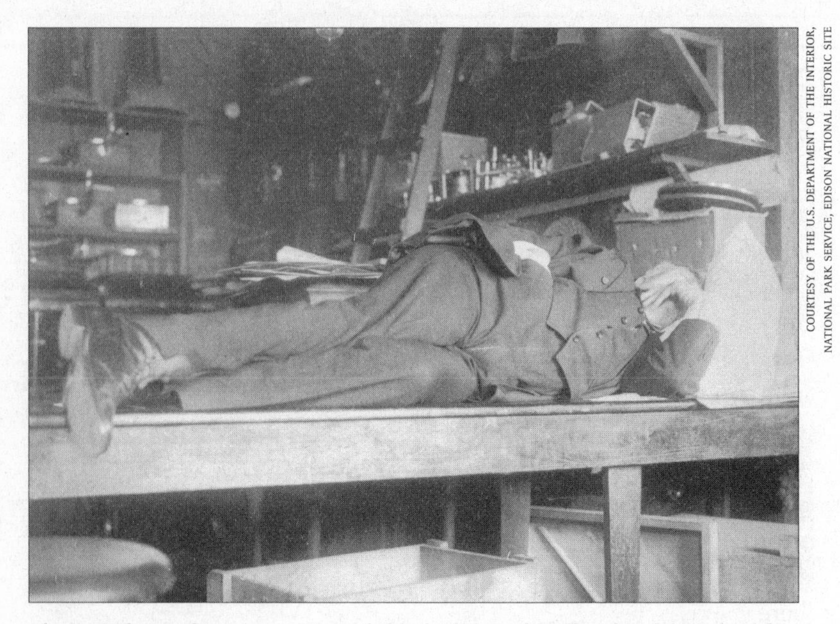

COURTESY OF THE U.S. DEPARTMENT OF THE INTERIOR, NATIONAL PARK SERVICE, EDISON NATIONAL HISTORIC SITE

Thomas Edison taking a nap. He needed little sleep and preferred to doze in his laboratory.

COURTESY OF THE U.S. DEPARTMENT OF THE INTERIOR, NATIONAL PARK SERVICE, EDISON NATIONAL HISTORIC SITE

After his second marriage, Edison lived with his family in Glenmont, a beautiful estate near his West Orange laboratory in New Jersey.

As the hard times sank in across the nation, unemployed men who had tramped from place to place upon rumors of work began to form "industrial armies," groups intent on pressuring the government to create work for them. Without a job, how were they to live? And anyone could look around and see only crushing unemployment. All spring the coalfields had been racked by strikes as 170,000 angry miners whose wages had been slashed closed most of the nation's bituminous mines. When some state militias marched in to reopen the mines, there were murderous clashes. Public backlash doomed the strikers. While those coal miners returned sullenly to the pits for lower wages, trouble flared in Chicago. The great Pullman Palace Car Company, which manufactured every railroad's sleeping cars, fired half its workforce and cut wages by a fourth. But the company did not reduce rents comparably on its overpriced company housing, where all operatives were required to live. When three unionized workers protested, George Pullman fired them, laid off *all* his workers, and closed the entire plant. For two months Pullman rejected any arbitration. At the end of June, his employees union, the American Railway Union, vowed its members would not serve on any train with a Pullman Palace Car until the company accepted arbitration. With so much anger and misery festering, the strike spread like wildfire, and within a couple of weeks the heartland's rail system shuddered to a halt. Factories went dark for want of coal. A badly ailing economy was dealt a debilitating blow.

President Grover Cleveland, who abhorred anything that smacked of "socialism," had resisted fashioning any federal solutions to nationwide want and misery. But he did not hesitate to call out the militia to break the strike. On July 5, as the soldiers marched in, Chicago, just a year earlier the proud host of the World's Fair, was engulfed in deadly riots and rampant looting. The violence spread elsewhere. With that, even the most liberal sympathy for the strikers evaporated. The Sherman Anti-Trust Act was invoked to jail union leader Eugene Debs and others, and gradually calm returned. But not prosperity. Many Americans had come to view the nation as in a struggle for its very being, pitting the greedy rich against the ordinary folk. The *North American Review* decried the new "plutocracy . . . their octopus grip is extending over every branch of industry; a plutocracy which controls the price of bread we eat, the price of sugar . . . the price of the oil that lights our way, the price of the very coffins in which we are finally buried."[15] It had been a very grim year.

Despite the nation's labor strife and abysmal economic state, Edward Dean Adams had forged resolutely ahead with the complex Niagara Power infrastructure, readying the site for the day the electricity finally began to flow, a date that was being pushed back continually. Cataract's new far-flung industrial domain at this stage included not just the unfinished powerhouse and the smaller transformer building, but a dock for Great Lakes ships to unload and a huge swath of rough reclaimed land created with tunnel debris. All of this was then connected by the company's railroad, seven miles long. Then there was Cataract's workers' village, Echoata, with sixty-seven modest two-family houses with trim lawns designed by renowned New York architect Stanford White of McKim, Mead & White. Adams had pressed White to find time in his overloaded schedule to serve as the architect for Power House No. 1, the gigantic but simple "cathedral of power" constructed with chiseled Queenston quarry limestone, as well as for Echoata. The powerhouse was two hundred feet long, sixty-four feet wide, and forty feet high, topped by a slate-and-iron roof, a building whose plain exterior gratified the Brahmin sensibilities of Edward Dean Adams. Tall, graceful windows kept the powerhouse interior aglow with lambent natural light. Deep in the basement, the three powerful turbines sat unmoving, awaiting Westinghouse's dynamos and Niagara's roaring green waters. As at the World's Fair, the switchboard was a huge marble affair set up on a railed platform.

George Westinghouse's hard-fought triumph at Niagara and the industrial spying charges had done nothing but aggravate his already acrimonious relations with GE. And many, including historian Harold Passer, have wondered why the powerful Morgan forces behind GE had not exerted their considerable influence at Cataract to win GE the coveted dynamo contract. (Early on, Edward Dean Adams had honorably divested himself of all Edison GE stock to remove his own possible conflicts.) Passer concluded that the stakes were just too high, "the financiers were afraid to go against the judgment of their engineering advisers."[16] GE did win a contract for transformers and transmission. Yet so parlous was GE's situation as the post-panic depression settled in that its Wall Street backers decided to take another tack, eyeing a takeover of the unwilling Westinghouse. This would put GE firmly in control, not just of Niagara, but of 90 percent of the electrical market, thus creating J. P. Morgan's favorite industrial arrangement, near or total monopoly. But there was also the paramount issue of who owned the all-important Tesla AC patents. Already that issue had made it impossible for GE to win the Niagara dynamo contract. Westinghouse and GE were reportedly locked in three hundred patent lawsuits, many over AC designs, and a "merger" would save each $1 million a year in legal fees. So the markets and rumors began to bubble and boil.

"General Electric was most anxious to bolster its jerry-built structure with the solid Westinghouse concern," wrote Thomas Lawson in his muckraking Gilded Age classic, *Frenzied Finance,* which examined how Wall Street's robber barons made easy and unscrupulous millions through watered stock, market manipulation, and monopolies. "Suddenly the financial sky became overcast. The stock market grew panicky . . . Wall and State streets [were] full of talk about General Electric's probable absorption of Westinghouse. . . . This was the signal. From all the stock-market sub-cellars and rat-holes of State, Broad, and Wall streets crept those wriggling, slimy snakes of bastard rumors which, seemingly fatherless and motherless, have in reality multiparents who beget them with a deviltry of intention. . . . [Rumors] . . . seeped through the financial haunts of Boston, Philadelphia, and New York, and kept hot the wires into every financial centre in America and Europe, where aid might be sought to relieve the crisis. There came a crash in Westinghouse stocks and their value melted."[17] In this era long before any such regulating body as the Securities and Exchange Commission, Westinghouse fought back with the weapons of the time. He hired Lawson, a specialist in stock market manipulation, and Lawson masterminded a retaliatory attack on GE stock, an attack so ruinous that the Morgan forces retreated. The Westinghouse stock recovered, and George Westinghouse got on with his Niagara power contract. Eventually, a wiser Westinghouse reached an 1896 patent-sharing agreement with GE that allowed the firm use of the all-important Tesla patents, ending such GE takeover attempts.

As George Westinghouse moved the mammoth Niagara Falls Power Company project steadily to completion, Nikola Tesla's aura of fame glowed more brightly. Thomas Commerford Martin, the ever ambitious editor of *Electrical Engineer,* had been faithfully promoting his friend's work and career since Tesla's reluctant but momentous first lecture at Columbia College in mid-1888 on the AC induction motor. In late 1893, Martin and Tesla published a thick book titled *The Inventions, Researches and Writings of Nikola Tesla,* which included all the inventor's major lectures, as well as a brief profile by Martin and many of Tesla's early patent applications. Martin, a distinctive figure in Manhattan social circles with his handsome bald head, soulful eyes, bristling mustache, and energetic ways, had decided the time was ripe to introduce Tesla beyond the confines of the world of electrical science. So Commerford, as his friends called him, had set off in the December cold to the Union Square offices of the genteel and middlebrow *Century* magazine, there to persuade associate editor Robert Underwood Johnson, amid his towering stacks of manuscripts and books, that Nikola Tesla was the next wizard, a figure comparable to Edison. Commerford proposed to write a profile of this fascinating Serb. Johnson, a handsome man with a close-cropped black beard and gold-rimmed glasses, was sufficiently intrigued to tell him to bring Tesla to a dinner party at his Lexington Avenue brownstone, where the editor and his vivacious wife, Katherine, cultivated a wide range of famous and lively luminaries, writers like Mark Twain, the shaggy naturalist John Muir, musician Ignace Paderewski, and popular actors and actresses playing the New York stage. Just before Christmas, Tesla dutifully accompanied Commerford to dine, the tall, slender inventor wearing one of his elegant tailored evening suits. Tesla appeared haggard and wan, but he proved to be a riveting conversationalist.

Both Johnsons were completely enchanted by the charismatic and courtly Tesla. Robert Johnson had met a great many famous and accomplished men and women and was more than familiar with all the usual egotism combined with notable limitations of real learning and intellect. In Tesla, Johnson believed he had found that rare man, one steeped in the most abstract electrical science, but also "widely read in the best literature of Italy, Germany and France as well as much of the Slavic countries to say nothing of Greek and Latin. He is particularly fond of poetry and is always quoting Leopardi . . . or Goethe or the Hungarians or Russians. I know of few men of such diversity of general culture or such accuracy of knowledge." Yet this electrical and cultural prodigy was also a lovely human being. Johnson described his new friend's personality as "one of distinguished sweetness, sincerity, modesty, refinement, generosity and force."[18] So that evening began a long and warm friendship. Commerford, of course, secured his magazine assignment. The Johnsons insisted that the exhausted-looking inventor return a few days hence for a rejuvenating Christmas dinner, a jovial celebration with their two teenage children and others.

Christmas Day in Manhattan dawned wonderfully balmy and springlike, a great blessing for the poor in their cold tenements. The widespread economic misery brought on by the summer's panic had reduced so many to destitution and so many others to outright beggary that *The New York Times* described a muted holiday largely dedicated to alleviating want. Every church gathered clothing and food for the needy, while many served large, filling meals and provided small presents for the ragged, unwashed children of the new poor. A thousand newsboys—abandoned urchins ever present on the city's streets eking out a living selling papers—gobbled down their annual Noel feast of turkey, ham, and mincemeat pie served at the Newsboys' Lodging House on Duane Street. At the Gerlach Hotel, where such want was out of sight, Tesla donned his elegant clothes and strolled forth, joining the fashionable Christmas crowds thronging Broadway, window-shopping the stores bedecked with winter greens. The whole city, rich and poor, was out enjoying the April-like respite from winter. Many were streaming up to Central Park, but Nikola Tesla was going east to Lexington Avenue.

Tesla so enjoyed the Johnsons' gaiety and intelligence that day, and it was such a strangely warm evening, that he invited them after dinner to venture downtown to see his laboratory, the first of many such nocturnal visits. Long after, Robert Johnson recalled those extraordinary forays down to South Fifth Avenue and the tromp up the stairs to Tesla's laboratory loft. These outings eventually included other favored guests like Twain and architect Stanford White, whose Power House No. 1 at Niagara was to shelter all three of Tesla's thirteen-foot-tall AC dynamos. Wrote Johnson of those lab visits, "Lightning-like flashes of the length of fifteen feet were an every-day occurrence, and his tubes of electric light were used to make photographs of friends as souvenir of their visits. He was the first to make use of phosphorescent light for photographic purposes—not a small item of invention in itself."[19]

Through the Johnsons and then White, Tesla became something of a society darling, a sought-after guest swirling through Manhattan's most glittering homes, private salons, and lavish restaurants. Tesla reciprocated by hosting elaborate dinner parties in private rooms at the delectable Delmonico's. At Stanford White's urging, Tesla and Robert Johnson joined the arty Player's Club across from Gramercy Park. Tesla still worked prodigious hours and frequently declined the invitations that showered down. But when he allowed White to cajole him into going sailing late one November, the architect wrote gleefully, "I am so delighted that you have decided to tear yourself away from your laboratory. I would sooner have you on board than the Emperor of Germany or the Queen of England."[20] It was something of an odd pairing, for Tesla, with his priestly devotion to electricity and his phobias for germs, showed no interest in women or sex, while White was a sexual satyr whose lubricious pursuit of delicious young females eventually led to his shooting death by an outraged husband.

In February of 1894, Commerford's *Century* profile of Tesla ran, complete with a moody engraving from a handsome Sarony photograph. The article was Commerford at his most florid, gushing that "Mr. Tesla has been held a visionary, deceived by the flash of casual shooting stars; but the growing conviction of his professional brethren is that because he saw farther he saw first the low lights flickering on tangible new continents of science."[21] This admiring (if overwrought) article in a major national magazine naturally caused quite a stir, prompting the New York press to become interested. A few months later on Sunday, July 22, Joseph Pulitzer's *New York World,* Manhattan's biggest daily, ran a long and prominent profile. Written by popular columnist Arthur Brisbane, the headline and subheads ran OUR FOREMOST ELECTRICIAN, "Greater Even Than Edison," "The Electricity of the Future," and so forth. Brisbane, unlike Commerford, did not understand anything about electricity, nor did he pretend to. "Every scientist knows his work," wrote Brisbane of Tesla, "and every foolish person included in the category of New York society knows his face. He dines at Delmonico's every day. He sits each night at a table near the window . . . with his head buried in an evening paper."

Brisbane's article was illustrated with a full-length drawing of Nikola Tesla resplendent in formal cutaway coat and striped dress pants and radiating "the Effulgent Glory of Myriad Tongues of Electric Flame After He has Saturated Himself with Electricity." This was, of course, Tesla's most famous lecture stunt, running many thousands of volts through himself until electrical flames licked all about his person. He confessed to Brisbane, "My idea of letting this current go through me was to demonstrate conclusively the folly of popular impressions concerning the alternating current. The experiment has no value for scientific men. A great deal of nonsense is talked and believed about 'volts' AC. . . . You see voltage has nothing to do with the size and power of the current." Brisbane, like most who met Tesla, found him mesmerizing, and the two sat up all that hot, still night at Delmonico's talking, leaving only when the dawn broke and the scrub ladies appeared with their mops and pails to swab the restaurant's marble floor. Brisbane told the *New York World*'s 280,000 Sunday readers: "When Mr. Tesla talks about the electrical problems upon which he is really working he becomes a most fascinating person. Not a single word that he says can be understood. He divides time up into billionths of seconds and supplies power enough from nothing apparently to do all the work in the United States. He believes that electricity will solve the labor problem. That is something for Mr. [Eugene] Debs to ponder while he languishes in his dungeon. It is certain, according to Mr. Tesla's theories, that the hard work of the future will be the pressing of electric buttons."

That fall, *The New York Times* weighed in on Sunday, September 30, 1894, with multiple columns on NIKOLA TESLA AND HIS WORK, with the subhead "Advancing with Certainty to Greatest Triumphs." In contrast with Brisbane's lighthearted romp, *The New York Times* made an exhaustive attempt to explain Tesla's work in high frequency and the science behind his wireless lights. Oddly, Niagara was not even mentioned. The truth was that even as work on the Westinghouse dynamos was advancing steadily, Tesla was passionately and utterly absorbed with a new and more abstruse electrical frontier. Daily he labored away in his laboratory, oblivious to the commercial cacophony wafting up from the busy street below. Nikola Tesla was deep into a completely new electrical dream, one that he had confessed to *New York World* reporter Arthur Brisbane in their long evening at Delmonico's: "I look forward with absolute confidence to sending messages through the earth without any wires. I also have great hopes of transmitting electric force in the same way without waste."[22]

Wireless transmission of power. While Tesla would speak openly of his work in a general way, he was highly secretive about the specifics of his research into what we know today as radio. By mid-1894, Tesla had built a small portable radio transmitting station, and all that year he continually tested and improved it. Many an afternoon and evening he climbed the stairs with one of his draftsmen to the broad roof above his laboratory and set up the transmitter. Then he would take his receiver and head out to high-up places progressively farther away, testing to see how far his wireless radio signals could travel. By winter Tesla was setting up on top of where he himself lived, the Gerlach Hotel (a "Strictly Fireproof Family Hotel"), thirty blocks uptown, a mile and a half north of his laboratory. Up on the Gerlach roof, ten stories above the stylish and exclusive shops of Broadway, Tesla would carefully send up a tethered balloon filled either with helium, hot air, or hydrogen. The balloon and its string were holding aloft, as high as was practical, an aerial. A cable was attached to the hotel's water main. Tesla would then tune in his receiver to successfully receive his draftsman's broadcast signals from the lab roof downtown. All that winter, Tesla fine-tuned his primitive radio, knowing that as soon as the Hudson River was clear of ice in the spring, he would take a river steamer north and see just how far he could sail toward Albany and still receive transmissions. So as the year ended, Tesla was in fine fettle, with every reason to feel triumphant.

Certainly the year 1895 promised yet more glory for Nikola Tesla, for even as he perfected his new radio, he could anticipate the turning on of the electric power at Niagara Falls. At long last, seven years after the unknown electrician had first sold his patents to George Westinghouse, four years after he gave up his royalties to help save the company, Tesla's Westinghouse AC dynamos would be installed in Stanford White's cathedral of power. The great moment was on the horizon, when the inlet gates to the Cataract canal would open and Niagara's glass-green waters would flood through. Funneled into the three giant penstocks, the cold river water would fall with a tumbling roar toward the water turbines. As those great sunken wheels revolved in a frothing, whirling blur, each steel shaft, too, would whirl. Up in the glowing riverine light of the powerhouse, atop the gleaming, whirling steel shafts, Tesla's trio of great dark dynamos would also whirl, creating magnetic fields of power. And from those humming dynamos would flow an invisible river of electricity, quietly crossing the bridge to the transformers, there to become high voltage and mighty, flashing out into the world, where individual transformers would again lower its invisible and silent flow to light tens of thousands of bulbs, to power great industries, to run the Buffalo streetcars, to brighten man's nights and lighten his load. They had called him a dreamer, but this was no dream.

And so great glory would rightly be Tesla's. Others in time surely would have solved the problem, for the conundrum of the AC motor and the polyphase generator had been very much in the air. But he, Nikola Tesla, had been there first. His place in history was secure. Yet it was not just glory Tesla sought, but the financial means to work as he would. He fervently believed AC was just the beginning, an interim stage in a far more sophisticated yet simpler system of power he was already working on: wireless transmission. And no one less than Edward Dean Adams, a much admired judge of men, money, and prospects, wished to be his financier. Adams had, it turned out, been quietly courting Nikola Tesla for some time. Adams, that wise old frog, understood far more than almost any other Wall Street money man the enormous value of Tesla's AC patents. And Tesla was quite forthright in asserting that these were just the first of his revolutionary inventions to be developed and commercialized. In mid-February 1895, a one-paragraph story in *Electrical Engineer* announced the formation of the Nikola Tesla Company, which would "manufacture and sell machinery, generators, motors, electrical apparatus, etc."[23] The company had a gilt-edged group of directors—Edward Dean Adams and his son Ernest, the hardworking and ambitious William Rankine, Tesla's long-ago rescuer Alfred S. Brown, one Charles Coaney of New Jersey, and Tesla. The story spoke of $5,000 in capital, but this seems a laughable sum, and Tesla later said Adams alone invested $100,000. So as the spring of 1895 neared, Nikola Tesla was an inventor to be much envied.

Then calamity struck. At 2:30 in the morning on March 13, 1895, Nikola Tesla's laboratory burned to the ground in a fire so intense, the whole loft building imploded and his whole floor collapsed into the inferno, obliterating every single piece of his electrical apparatus. When the cold gray dawn shed its light, all that was to be seen at 33-35 South Fifth Avenue, reported the *New York Sun,* were "two tottering brick walls and the yawning jaws of a somber cavity aswim with black water and oil." Charles Dana, one of the most revered newspaper editors of his time, wrote, "The destruction of Nikola Tesla's workshop with its wonderful contents, is something more than a private calamity. It is a misfortune to the whole world. It is not in any degree an exaggeration to say that the men living at this time who are more important to the human race than this young gentleman can be counted on the fingers of one hand; perhaps on the thumb of one hand."[24] Fortunately, Tesla had not been toiling away late that night or he might have been trapped in the conflagration and incinerated.

Instead, he strolled down for work at 10:00 A.M. as usual, only to be greeted by the horrifying sight. "It cannot be true," he repeated again and again as he paced up and down before the charred, smoking ruin. His fifteen employees, who had arrived quite a bit earlier, stood there disconsolate. They had not had the heart to summon him from the Gerlach Hotel to such a tragedy. When *The New York Times* reporter approached him, Tesla waved him away, saying, "I am in too much grief to talk. What can I say? The work of half my lifetime, very nearly; all my mechanical instruments and scientific apparatus, that it has taken years to perfect, swept away in a fire that lasted only an hour or two. How can I estimate the loss in mere dollars and cents? Everything is gone. I must begin over again."[25] Tears filled his eyes. All his specially designed dynamos, oscillators, motors, and vacuum bulbs, not to mention all his records and papers and correspondence, his World's Fair exhibit, all his newly developed radio transmitters and receivers, the many years of work and thought, all gone up in raging flames.

While one might readily suspect the fire had arisen in his laboratory with all its electrical wonders, the nightwatchman reported it had begun on the ground floor. Another tenant, a steam-fitting manufacturer, had over time saturated the loft building with oil, and "it burned like a tinder-box." The watchman's buckets of water were futile. All the firemen could do, though they battled the fire for three hours, was prevent its spread to an adjacent box factory and the nearby elevated railroad. In a heartsick daze, Tesla slipped away and wandered the city streets. Robert and Katherine Johnson looked for him everywhere, wanting to help him at this moment of "irreparable loss." Some of his apparatus existed in similar form elsewhere—his dynamos and oscillators and motors—but his radio work was unique and would all have to be completely rebuilt. Nothing in his laboratory was insured. The financial loss was complete and devastating. Tesla had made huge sums in recent years, but he had poured almost all of it into the now smoldering ruin.

There was, of course, nothing to do but start again. Encouraged by Commerford, the Johnsons, and his many Manhattan friends and acquaintances, Tesla regathered his broken spirits and secured a new laboratory at 46 East Houston Street. Later he would tell a reporter, "I was so blue and discouraged in those days that I don't believe I could have borne up but for regular electric treatment which I administered to myself. You see, electricity puts into the tired body just what it most needs—life force, nerve force. It's a great doctor, I can tell you, perhaps the greatest of all doctors."[26] By March 22, Tesla was sufficiently recovered to write one of the top Westinghouse managers with orders for new equipment. "You have, no doubt," Tesla wrote, "learned through the papers of the unfortunate accident which has deprived me of all my apparatus, and of some results of my recent work. I must now rebuild my laboratory." Within the month, needed machines began to arrive. Tesla also wrote Westinghouse engineer Charles Scott, who had supervised AC at the Telluride Gold King Mine, asking him to help push through his many orders. "This kind of work is almost essential to my health," he explained.[27] In the interim, Tesla found refuge in the most unlikely of places—Thomas Edison's gigantic laboratory in West Orange, where once Harold Brown had electrocuted dogs, calves, and horses. The press had taken to portraying Edison and Tesla as rivals for the title of America's greatest wizard, but in such a time of loss, Edison could set aside his competitive instincts to offer temporary working shelter to the grieving Tesla.

When first Edward Dean Adams considered the financial prospects of a hydropower plant at Niagara Falls, he (and all others) had assumed that transmission to thriving big-city Buffalo was *the* key to success. Now, as the first of the eighty-five-ton dynamos was fitted into place around the thick steel shaft of the giant turbine, the Cataract officers discovered that entire new industries were prepared to move to the firm's industrial acres and contract for large amounts of cheap Niagara power. The first industrialist was Chester Martin Hall, who had announced in 1893 that he would be moving his Pittsburgh Reduction Company to the falls. Until the enterprising and energetic Hall, aluminum was a highly sought after strong, light metal whose high price—$15 a pound—prohibited its widespread use. While Hall was a student at Oberlin College in 1880, his professor had told his class that whoever could cheaply manufacture the wonderfully useful metal would make a fortune. Hall had whispered to a classmate, "I'm going for that metal."[28] And he'd kept at it doggedly from that time until 1884. Back in his woodshed laboratory, Hall finally, after much trial and error, discovered that double fluorides would extract aluminum from the clay where it was most abundant. When he energized that with electric current, he ended up with pure aluminum. Soon, Hall's Pittsburgh company (later renamed Alcoa) had the price down to less than a dollar a pound. But to cut costs further, he needed plenty of cheap electricity. At Niagara he hoped to get it.

Hall's daring commitment to cheap power that did not yet exist was matched by that of Edward Goodrich Acheson, a chemical genius who had acquired his electrical training with Edison in Menlo Park and then Europe. He was a sufficiently talented inventor that George Westinghouse had bought some of his patents. Acheson decided that what the emerging industrial world needed was a cheap abrasive, something to replace the $1,000 plus it cost for a pound of diamond dust. Eventually he devised an electrochemical process that created what he called Carborundum, a substance hard enough to cut glass. At his factory in Monongahela, just outside Pittsburgh, he was already selling twenty pounds a day at $576 a pound, but he could have sold twice that, were the price not so prohibitive. Like Hall, Acheson needed massive amounts of cheap electricity. In his case, electricity would fire new arc furnaces capable of reaching unheard-of temperatures. He, too, sought electricity at Niagara. When Acheson informed his board that he had signed a contract for 1,000 horsepower a day (with an option for 10,000 more) from the Niagara Falls Power Company, which had yet to transmit so much as a single horsepower, they resigned en masse. Hall with his aluminum and Acheson with Carborundum would soon be followed by many other entrepreneurs starting or expanding electrochemical or electrometallurgical firms producing "acetylene, alkalis, sodium, bleaches, caustic soda, chlorine."[29] So it was that Edward Dean Adams and William Rankine discovered, as their fixed costs mounted and the dynamos were not yet running, they could probably sell all their first 15,000 horsepower of electricity locally.

Finally, on August 26, 1895, almost a year later than predicted by the engineering journals, Niagara power was harnessed for full-time commercial use. For nine months, the engineers had been testing and calibrating and retesting all aspects of the system, especially the behemoth Westinghouse dynamos. Lead engineer B. J. Lamme described what happened during one early test in Pittsburgh of a giant dynamo when numerous little temporary steel bolts had "loosened up under vibration, and finally shook into contact with each other, thus forming a short circuit. . . . In a moment there was one tremendous [electric] arc around the end of the windings of the entire machine. . . . It looked, at first glance, as though the whole infernal regions had broken loose. Everybody jumped for cover." One man managed to shut down the machine, and gradually the huge flaming electrical arc that had engulfed the dynamo subsided. Peering forth from their shelters, the engineers then rushed back and "someone climbed underneath to see what had become of our man inside . . . expecting him to be badly scorched. . . . He said the fire came in all around him but did not touch him."[30] No one present had ever seen such a sight.

But now, at long last, the first Niagara dynamo was ready. At 7:30 A.M. that late-summer morning, the inlet gates at the canal opened, the river water flooded into one of the penstocks, the turbine whirled, and so did Dynamo No. 2, flashing alternating current off to the Pittsburgh Reduction Plant. The specific electrical needs of this voracious first customer made for delicious irony. For, as *The New York Times* noted in its small story buried back on the ninth page, "The power from the power house is sent over copper cables laid in a conduit to the aluminum works. The current sent is an alternating one, and before it can be used in the making of aluminum it must be transformed to a direct current. This is done by passing through four of the largest rotary transformers ever built. These are 2,100 horsepower each, and three of them are running. Everything was found to work perfectly and great satisfaction was expressed by the officers."[31] The Niagara Falls Power Company had spent several arduous and expensive years determining the best means of sending large amounts of power the long distance to Buffalo. Adams and Sellers had audaciously chosen AC, and now they had all the customers they needed within easy reach of DC and the power they needed was DC!

The city of Buffalo, the original intended market, the booming metropolis that had proudly declared itself the City of Light, had been bogged down month after month about what sort of franchise they should grant. Its Common Council and then its Board of Public Works wrangled on and on about how to proceed. Should the city itself manage the electrical power? William Rankine, a man whose brutal schedule for six wearing years had brought on a serious case of heart disease, handled the negotiations. Rankine, Niagara Falls Power Company secretary, had first approached the Buffalo city fathers in October of 1894 to secure the necessary franchise, explaining that his company would like a commitment for 10,000 horsepower before Niagara began excavating new wheel pits, ordering more dynamos, and installing transmission lines twenty-six miles across the farms and forests. A full year later, Buffalo's Common Council and Board of Public Works were still dithering and the Niagara Falls Power Company had no franchise. They were separated by such serious issues as the city's wish to have the right to revoke the franchise on ten days' notice or the right to order all wires underground at any time.

But at Niagara Falls, Power House No. 1 was at last up and running, and it had been successfully flashing electricity to the Pittsburgh Reduction Plant for more than a month. Now both Westinghouse dynamos were humming quietly away, which meant the Niagara Falls Power Company was finally earning income. Edward Dean Adams, president of the Cataract Construction Company, felt the time propitious for a full-scale, formal tour of the $4 million investment, a small and private celebratory savoring of their mighty work. So on September 30, 1895, Adams assembled his board of directors at Power House No. 1. All the men (with the exception of William Rankine) were well-known Manhattan millionaires, long admired for their financial prowess but increasingly resented during hard times as plutocrats running Wall Street and government as their own private club.

When the Cataract directors stood for their photograph inside the luminous powerhouse, all wore the standard uniform of Gilded Age gentlemen: somber vested suit, dark overcoat, respectable black bowler, and black umbrella in hand, powerful men dwarfed by the gargantuan dark dynamos brooding behind them. The tall, slender John Jacob Astor, not yet thirty, was a scion of the gigantic New York real estate fortune. An inventor with numerous patents, he interested himself in all manner of odd schemes and adventures. Darius Ogden Mills had made his first money in the craziest days of the California gold rush before settling in New York, where he became a famously shrewd investor. His purchase of Edison stock convinced J. P. Morgan to invest more also. His nine-story Mills Building on Broad Street was a prestige address for stockbrokers and lawyers and was the first office building in the city to have its own generator and electricity. It housed, not incidentally, the offices of the Cataract Construction Company. Edward Wickes represented the interests of the Vanderbilts, who, as the major stockholders in the almighty New York Central, remained one of America's richest and most socially prominent families. Wickes was also a vice president, with Francis Lynde Stetson, of Cataract Construction. Charles Lanier was an old friend of J. P. Morgan's whose family's Wall Street investment firm (where Adams was a partner) specialized in railroad finance. George S. Bowdoin was a blue-blooded Morgan partner who often joined Morgan on the *Corsair,* Morgan's luxurious yacht (with its six staterooms and working fireplaces), where he entertained lovely actresses and desirable widows. John Crosby Brown was a white-bearded partner of the Wall Street firm Brown Brothers.

Last, but by no means least, was Francis Lynde Stetson, who had by now acquired the sobriquet of "attorney general" to J. P. Morgan. Earlier that year, in February, he and Morgan had rushed down to the White House in a private railcar to meet with President Cleveland, as the government, hemorrhaging gold from the U.S. Treasury, teetered on the verge of bankruptcy. Stetson found some legal loophole that allowed August Belmont and Morgan to arrange a huge bond sale and thus replenish the U.S. coffers with gold, despite opposition from the Republican Congress. Republicans, Populists, and newspapers across the land howled in outrage, especially as the Wall Street bankers proceeded to make a profit on it all. But with no federal banking system in place, only men like Morgan had the financial clout to steady the markets. Later, Morgan, who viewed himself as a staunch patriot, wrote a friend, "The dangers were so great scarcely anyone dared whisper them."[32] So Francis Lynde Stetson's aura of power shone ever brighter. A millionaire in his own right, Stetson was a hard-nosed legal adviser to many of his own ilk. Rumor had it Morgan paid him a $50,000 annual retainer just to be on call.

We have no record of what this mighty assemblage of Cataract directors thought of their tour. They had arrived, as was the way with the rich, on private railroad cars, no doubt ignoring the ever-present mélange of hard-selling hackmen scouting late-season fares and touts promoting tourist hotels. Coming out that late September morning to the powerhouse, they would have passed Erie Avenue's clapboard houses with their wandering chickens. Our eloquent witness to the wonders of the new Niagara power is, instead, Englishman H. G. Wells, science fiction writer turned social observer:

> These dynamos and turbines of the Niagara Falls Power Company impressed me far more profoundly than the Cave of the Winds; are indeed, to my mind, greater and more beautiful than accidental eddying of air beside a downpour. They are will made visible, thought translated into easy and commanding things. They are clean, noiseless, starkly powerful. All the clatter and tumult of the early age of machinery is past and gone here; there is no smoke, no coal grit, no dirt at all. The wheel pit into which one descends has an almost cloistered quiet about its softly humming turbines. These are altogether noble masses of machinery, huge black slumbering monsters, great sleeping tops that engineer irresistible forces in their sleep. . . . A man goes to and fro quietly in the long, clean hall of the dynamos. There is no clangor, no racket. . . . All these great things are as silent, as wonderfully made, as the heart in a living body, and stouter and stronger than that. . . . I fell into a daydream of the coming power of men, and how that power may be used by them."[33]

One suspects that the New York millionaire Cataract directors, observing these very same sights on this, their first formal inspection of their company's electrical industrial tour de force, were also beset with reveries of what these first monster dynamos augured. Two months later, steelmaker extraordinaire Andrew Carnegie, his reputation for benevolence tattered by Pinkertons unleashed on strikers at his Homestead works, had come to see this latest industrial marvel. He wrote in the official guest book, "No visitor can have been more deeply impressed nor more certain of the triumphant success of this sublime undertaking."[34] J. P. Morgan made his pilgrimage that fall also, bringing along his wife, Frances, and several other women. As was his wont, he left no comment.

The one person who had not yet set foot in Niagara Falls, the man who had never seen the world-famous cataract, much less Power House No. 1, was Nikola Tesla, the very electrical dreamer whose system of AC (above all his induction motor) had made it all possible. Again and again, the inventor had been invited to make the journey to Niagara and experience two astounding sights: the thundering Great Falls with their towering clouds of mist and rainbows, and the herculean power enterprise under construction, with its astonishing subterranean horseshoe-shaped tunnel or the gigantic turbines sunken in the bowels of Stanford White's cathedral of power. But for four years Tesla had turned down all invitations. Not until the summer of 1896 did he finally agree to come. First, he would spend a day in Pittsburgh with George Westinghouse, where they would visit the firm's splendid new twenty-acre electrical works out in Turtle Valley. That evening, Edward Dean Adams and several others would join them as they traveled overnight in the comforts of *Glen Eyre,* Westinghouse's sumptuous private railcar.

The next morning, at the height of the tourist season, Nikola Tesla came for the first time to Niagara Falls, debarking with the rest of his travel companions around 9:00 A.M. on July 19, 1896, at the small but always bustling Pennsylvania Railroad depot. Also present for this momentous and historic encounter that lovely Sunday were George Westinghouse, the electrical magnate whose steely determination, tremendous courage, and considerable charm had prevailed in the War of the Electric Currents. Accompanying Tesla and Westinghouse were Edward Dean Adams, who as president of Cataract had advanced resolutely toward the bold choice of AC; William Rankine, the upstate lawyer whose incessant dedication to this gigantic venture had wisely shepherded its physical construction; Westinghouse attorney Paul Cravath, who had steered his friend and client through so many legal and financial perils; and Westinghouse's thirteen-year-old son, George junior. The group made their way through the holidaymakers, fathers in their straw boaters herding gaggles of excited children and wives in cool white lawn dresses with parasols ready against the sun. All the Westinghouse party boarded a trolley, which trundled them a mile southeast on Erie Avenue toward the edge of town, the lush green foliage of deep-summer trees spreading cooling shade along the dusty tracks. Ahead loomed Stanford White's handsome limestone Power House No. 1, fronted by a broad lawn. This many-windowed cathedral of power was situated on one side of the broad inlet canal, where the diverted river water sparkled in the sunlight as it flowed steadily into the powerhouse.

Across the canal stood the limestone transformer building, also by White, a much smaller, faithful echo of the powerhouse. A limestone bridge spanned the canal, waiting to carry the electrical conduits from the powerhouse to the transformer building. Once the men stepped down from the trolley, the wide Niagara River's powerful presence was palpable with its rushing, tumbling rapids and cooling breezes. Here at last it all was, the river, the canal, the cathedral of power, its quiet austerity belying the gigantic achievement. Inside, the luminous morning light poured in from the ceiling-high windows. It was all so unexpected, so unlike any other industrial venue, so silent, so pristine. Yet the dark mammoth metal dynamos were tall, brooding presences, their vast, ominous power silently controlled by the switchboard. Only one dynamo was in operation that quiet Sunday morning, but Tesla and the others inspected it enthusiastically, clambering up and around on the special walkways. And there was all the ancillary apparatus on the ground floor to be seen and discussed. Next they descended in the ornate elevator to the wheel pits, that cloistered space where one could hear the river water passing through the penstocks and then hear and see the turbines whirling. When the group came back up, they walked out of the big powerhouse and crossed over the canal to the transformer building, which as yet had no transformers. The GE machines were still being manufactured. Tesla closely considered all the complex's machinery, delightedly asking many questions. As the morning waned, Rankine escorted everyone back to his favorite hotel and dining spot, the Cataract Hotel overlooking American Falls. There they lunched.

When the Westinghouse-Tesla party emerged from their repast, a crowd of newspapermen was lying in wait. After demurring at first, the famous Tesla finally agreed to answer a few questions. "I came to Niagara Falls," he said, "to inspect the great power plant and because I thought the change would bring me needed rest. I have been for some time in poor health, almost worn out." And what did he think of the power plant? Tesla's face lit up. "It is all and more than I anticipated it would be. It is fully all that was promised. It is one of the wonders of the century . . . a marvel in its completeness and in its superiority of construction. . . . In its entirety, in connection with the possibilities of the future, the plant and the prospect of future development in electrical science, and the more ordinary uses of electricity, are my ideals. They are what I have long anticipated and have labored, in an insignificant way, to contribute toward bringing about." And what of Niagara Falls? the local reporters urgently asked. Tesla replied without hesitation: "The result of this great development of electric power will be that the falls and Buffalo will reach out their arms and will join each other and become one great city. United, they will form the greatest city in the world." Was it truly possible, asked another reporter, that this was Tesla's first visit to this world-famous cataract? To the cathedral of power? "Yes," he said, "I came purposely to see it [the plant]. But and it is a curious thing about me. I cannot stay about big machinery a great while. It affects me very much. The jar of the machinery curiously affects my spine and I cannot stand the strain."

Of course, the Niagara Falls Power Company's greatest challenge still remained unmet, and that was the long-distance transmission of AC power to Buffalo. The reporters now asked Tesla what he thought of all that. Was it an assured undertaking? There was good reason to wonder. Just six months earlier, on December 16, 1896, after fourteen months of indecision and politicking, the city of Buffalo had finally come to a franchise agreement with William Rankine that obliged the Niagara Falls Power Company to deliver to a newly formed Buffalo entity, the Cataract Power and Conduit Company, 10,000 horsepower of electricity on or before June 1, 1897. The first declared customer was the Buffalo Street Railway Company, which contracted for 1,000 horsepower (DC, of course). The Conduit Company cautiously sought no further contracts until such time as all went smoothly with this. Tesla's eyes flashed as he responded to the question of transmission. "Its success is certain. The transmission of electricity is one of the simplest of propositions. It is but the application of pronounced and accepted rules which are as firmly established as the air itself."[35] As he became enthusiastic, his long, thin hands gestured about his face, trembling noticeably.

While Nikola Tesla was the star of the show, the reporters were also interested to hear the thoughts of George Westinghouse, one of the nation's great industrialists and the man who had taken Nikola Tesla's brilliant AC inventions and made them a commercial reality. Westinghouse was feeling quite jovial and listened to Tesla enthuse "with much interest and good nature." Did he believe the Niagara Falls Power Company could really sell so much electricity, 100,000 horsepower ultimately? Many people said there would never be that much demand. "This talk is ridiculous," said Westinghouse, his walrus mustache bristling. "When you think that a single ocean steamship like the *Campania* uses 25,000 horse power, it is easy to be seen that there will be no surplus here. All the power here can and will be used."[36] And what would electricity do? The first and greatest benefits would obviously accrue to Niagara Falls, for Cataract had 1,500 acres to fill with industries, which were coming. "But Buffalo's possibilities are to be made marvelous as well," he said.

First, however, the power had to reach Buffalo. And until such time as electricity was conveyed successfully from Power House No. 1 to the Queen City of Lake Erie, the whole great Niagara power enterprise and effort would not have surmounted its greatest challenge. After all, Edward Dean Adams, Coleman Sellers, William Rankine, Nikola Tesla, George Westinghouse, and his excellent team of engineers had not been toiling all these years to provide direct current to local factories. Where was the glory in that? Their electrical dream, what they had been striving to show the world, was the revolutionary delivery of electricity to entire cities and regions, electricity that was generated cheaply and abundantly in one locale. From there it would flash forth over far distances to satisfy every imaginable need—incandescent lighting in offices and homes, arc lighting for the avenues, DC for street railways, and motive power for factories and mills. And this was just the beginning. Once electricity was cheaply, reliably available, who knew what inventions would draw on its energy? Yes, so these men had stilled the doubters by showing they could harness large quantities of hydropower using massive turbines and dynamos. But could they send big volumes of electricity big distances? And indeed, as William Rankine told the reporters that day with Tesla, the Niagara Falls Power Company was even now arranging the contract for erection of the towering wooden transmission poles, modeled on those of the telegraph companies. But not until early November were the long-awaited GE transformers finally installed in the transformer house. They were testament to the continuing rapid advance of the electrical arts: The dynamos' two-phase AC was to be stepped up to even more efficient three-phase AC for transmission. By mid-November of 1896, the Niagara Falls Power Company was finally ready.

All afternoon on the cold Sunday of November 15, William Rankine and several engineers were at Power House No. 1 testing the delivery of the 1,000 horsepower to the transformer. However, Rankine had promised his father, an Episcopal minister, that the actual transmission of power would not begin on the Sabbath. So late Sunday night he was back at the powerhouse, lit now not by daylight, but by brilliant hissing arc lights. Inside, the mighty dynamos hummed. Outside, the Niagara River flowed swiftly past and the remaining dead leaves rustled in the almost bare trees. Rankine and one Westinghouse engineer mounted to the switchboard platform. Over at the transformer house, a GE engineer had been supervising and testing all day. They waited for midnight. As soon as Monday came, at 12:01, William Rankine pulled down three switches. Twenty-six miles away in Buffalo, a small group huddled in the southwest corner of the Buffalo Railway Company powerhouse. One man's eye was on his watch, and when it read 12:01, he pulled down three knife-blade switches on their two rotary transformers. "Perhaps two seconds elapsed," reported a jubilant *Buffalo Enquirer.* "Electrical experts say the time was incapable of computation. It was the journey of God's own lightning bound over to the employ of man."[37] The Niagara Falls Power Company's alternating current electricity had flashed out at a pressure of 2,200 volts, been swiftly stepped up in the GE transformer to 10,700 volts, flashed over the twenty-six miles of cable to the Cataract Power and Conduit Company's transformer, been stepped down to 440 volts, and flashed on to the Buffalo Railway Company, where rotary transformers—delivered and tested just that day—brought it up further to 550 volts of DC. The Buffalo streetcars were soon jogging along their rails with Niagara hydropower, mundane yet wondrous proof of this miraculous new and invisible reality—long-distance transmission of AC energy.

YOKED TO THE CATARACT! proclaimed that day's *Buffalo Enquirer,* with the subheads "Niagara's Energy Ready to Stir the Wheels of Buffalo's Great Industries—Power Transmitted Successfully at Midnight Last Night" and "Now for Prosperity for Greater Buffalo." In truth, after an hour of transmission the power had been shut off and all involved had gone off to celebrate. Buffalo was indeed now a city of dazzling prospects. Already it was the world's sixth largest commercial center, storing mountains of Great Plains cereals in its fifty-two grain elevators before they were shipped out to feed the world. Five million head of livestock passed through each year. Buffalo boasted the world's largest coal trestle; twenty-six railroads had seven hundred miles of track and depots; every day 250 passenger trains alone came and went. Almost six thousand vessels docked at its port each year.[38] And soon the city would boast the nation's most abundant, cheapest electricity.

Edward Dean Adams had taken Nikola Tesla's counsel against any formal dedication when the cathedral of power began its mighty work at Niagara Falls. The Buffalo men were of no such reticent nature. Though only a paltry 1,000 horsepower was theirs as yet, the men running the new Cataract Power and Conduit Company were determined to celebrate in grand banquet style. They set a date in mid-January of 1897 and invited Nikola Tesla, who graciously agreed to be the guest of honor. So, for the second time in six months, Tesla found himself journeying up to Niagara, traveling overnight in a private railcar from New York City with Edward Dean Adams, Francis Lynde Stetson, Edward Wickes, several other millionaire directors, and two top Westinghouse engineers, one being Lewis Stillwell. On Tuesday, January 12, at 9:00 A.M. the group debarked at the New York Central depot in Niagara Falls. It was very cold and had been snowing lightly all morning. The town was at its most lovely, its usual ramshackle appearance hidden by the snow and tumbling flakes, the trees shimmering with snow and icicles. The gentlemen climbed into waiting horse-drawn carriages and made the short trip to the elegant Prospect House Hotel. There they met William Rankine and sat down to a hearty breakfast in the marble octagonal dining room, the gentle winter light diffusing softly through its famous domed ceiling of stained glass. Once breakfast had been finished, the group declined to speak to waiting reporters, ascended into the carriages, and headed toward Erie Avenue and Power House No. 1. This time, Nikola Tesla would see the transformers at work as well as several new factories powered by Niagara. In the afternoon, they visited the falls.

That evening, Tesla and the other men, now in formal evening clothes, all returned to Buffalo by train to attend the Cataract Power and Conduit Company's lavish electrical banquet at the new Ellicott Square Building. Designed by Daniel H. Burnham, the architect and mastermind of the Chicago World's Fair, this handsome ten-story neo-Renaissance behemoth with its beautiful, airy, glass-roofed interior was said to be the world's largest office building, with six hundred suites. On the top floor was the Ellicott Club, jammed and noisy with hundreds of arriving guests. Each man was handed a souvenir menu and seating list, indicating his table, bound in engraved aluminum covers made with Niagara power. Outside the club's dark windows, the snow swirled thickly down on the city below. The din of excited talk rose louder and louder. Nikola Tesla and the New Yorkers were cloistered in a small office away from the local hoi polloi until it was time to be seated.

Three hundred of Buffalo's leading citizens, from the mayor to the major merchants, were present for this gala occasion, celebrating the long-awaited arrival into the City of Light of the first 1,000 horsepower of Niagara electricity, that mystical, amazing energy. Some fifty eminent scientists and "electricians" had also come north to pay homage, including such important names as GE's Charles Coffin and

Elihu Thomson, Charles Brush of arc light fame, and T. Commerford Martin of the *Electrical Engineer.* George Westinghouse, never one for ceremonies and occasions, had sent in his stead chief engineer Lewis Stillwell and longtime friend and attorney Charles Terry. Then, of course, there were the rich and powerful Manhattanites, fifty strong, the capitalists whose dollars had underwritten it all.

At 8:00 P.M., the chattering multitudes streamed into the gold-and-white dining room, scintillating like a starry night with hundreds of electrical fairy lights. A giant silver Neptune vase on the long raised speaker's table displayed crimson roses beribboned with tricolor electric lights. The company settled in at the eight long banquet tables, each decorated with fresh ropes of sweet-smelling smilax, its evergreen bright against the starched white linen cloths, and set with the finest colored china, crystal goblets, and silver. The air warmed and conversation rose, and the hothouse scent of the roses, carnations, palms, and ferns subtly scented the room. The waiters skillfully served course after course of this festive electrical evening: succulent oysters, lobster, tender terrapin, and rare beef filet, all washed down with sherry, Rhine wine, champagne, the palate refreshed by the sorbet electrique. Pontificated the editors of the *Buffalo Morning Express,* "Such a company never sat down in Buffalo before, while such an event had never previously been celebrated in the history of the world."

For three hours, the four hundred guests made merry, the best citizens of Buffalo relishing this rare convivial rubbing of elbows with important and powerful New Yorkers whose names "appear almost daily in newspapers around the country. They not only control events. They create many and strive for more." By 10:00 P.M., the dessert plates with their delectable petit fours were whisked away and the faint blue smoke of hundreds of cigars wisped and curled toward the ceiling. Sandy-haired Francis Lynde Stetson, the first of six toastmasters, responded to the toast to "the Company." He stood up and bluntly launched on a series of complaints: Since 1889, the New York investors had put up more than $6 million to build the great Niagara power plant and the transmission facilities "without thus far receiving one penny of profit or dividends or interest." Moreover, Stetson lectured his suddenly silent audience, the most profitable way to use this new power would be to sell all of it to the firms in their own industrial park. Nonetheless, Stetson declared curtly, the Niagara Falls Power Company intended to honor its less profitable arrangement to supply Buffalo electricity, but not on time. The 9,000 horsepower of electricity due the city by June 1897 would not be available until an unspecified future date. Stetson's ungracious toast received stunned, perfunctory applause. Only a man as hard-nosed as Stetson, a man used to wielding great power and having his way, would possess the nerve to coolly announce such churlish news at a banquet. He earned an editorial rebuke from one local paper the next day.

The mayor and the state comptroller gamely delivered their platitudinous toasts through the pall of cigar smoke. Then it was time to hear from Nikola Tesla, already a legend. All evening eyes had been upon him, for the inventor was an unusual-looking man, so tall and thin, with his coal black wavy hair, his wide forehead, his luminous eyes, his nervous, uncomfortable manner. As soon as he was introduced, the packed room of men rose to their feet and began wildly waving their big linen napkins, cheering for the famous scientist. They would show Stetson who mattered in this world. Smiling shyly, Nikola Tesla stood as the rapturous roars enveloped the room, shaking the china and crystal. When he tried to speak, the audience began clanging their wineglasses with knives and forks, setting up an even greater clamor of approval. Slowly, the pandemonium subsided, and the smiling crowd calmed and took their seats expectantly.

Tesla then began his toast in his high-pitched voice, the model of modesty, saying he "scarcely had courage enough to address them." Then, as the room became completely silent, Tesla the idealist urged his listeners to honor the "spirit which makes men in all stages and position work, not as much for any material benefit or compensation, although reason may dictate this also, but for the sake of success, for the pleasure there is in achieving it and for the good they might do thereby to their fellow men." Finally, here were sentiments fitting to the epochal nature of the night. The crowd's wild clapping sent the cigar smoke swirling, and some had trouble hearing Tesla's words about a "type of man . . . inspired with deep love for their study, men whose chief aim and enjoyment is the acquisition and spread of knowledge, men who look far above earthly things, whose banner is Excelsior!" Tesla and his audience were enjoying each other in this electrical, uplifting moment, as the great creator of Niagara power elevated the occasion above the bitter, mundane concerns of the money men. At Niagara itself, had not Tesla given up already almost $50,000 in royalties?

Unmoved by Tesla's noble sentiments, the ever practical Stetson looked at his watch, stood up, and whispered loudly in his ear, "Mr. Tesla, we will have to leave in three minutes." The private railcar of the New York millionaires was attached to a soon-to-depart train. Always good-natured, Tesla endeared himself with his final kind words: "Let me wish that in no time distant your city will be a worthy neighbor of the great cataract which is one of the great wonders in nature." With that, he bowed. The diners rose as one, their cheers echoing back and forth as Tesla departed from the hot, smoky room redolent of good food and drink, following Stetson, Adams, and the other big New York money men as they made for the elevators.

Those who remained to celebrate into the wee hours, and they were many, felt thrilled. The reporter for the *Buffalo Morning Express* captured their rapture, declaring, "Great are the powers of electricity. . . . It makes millionaires. It paints devils' tails in the sky and floats placidly in the waters of the earth. . . . It creeps into every living thing. . . . Last night it nestled in the sherry. It lurked in the pale Rhine wine. It hid in the claret and sparkled in the champagne. It trembled in the sorbet electrique. . . . Small wonder that the taste was thrilled and that the man who sipped was electrified, if nothing more. Energy begets energy."[39] All felt energized themselves as they emerged from the world's largest office building to a cold gray January dawn and the snow-covered streets. "There was not a man who went home [that morning who] did not feel the wheels of commerce whirring in his head and hear the buzz of the mighty dynamos of the electric dinner surging in his ears, beating about him, driving away all other thoughts and drowning all other contemplation save the glorious knowledge that the introduction of power had been commemorated."[40]

The War of the Electric Currents was over. George Westinghouse, Nikola Tesla, and the alternating current had won. The world was about to be forever changed. Great indeed are the powers of electricity.

CHAPTER 13

Afterward

WESTINGHOUSE

Three titans of America's Gilded Age—Thomas Edison, Nikola Tesla, and George Westinghouse—dreamed of spreading the ethereal power of electricity throughout the world. But in the wake of the War of the Electric Currents, only George Westinghouse among the Promethean three truly stayed the electrical course. He alone of the three completely mastered the new industrial order of gigantic capital, swift and treacherous change, and colossal corporate enterprise. Ever the audacious innovator and empire builder, George Westinghouse continued to make prodigious advances in AC technology and capacity, determined to make AC electricity so cheap, abundant, and versatile that it could power anything, including the heaviest locomotives. He had no doubt that once that was so, the whole world would embrace electricity. Yet such was the new American industrial order that George Westinghouse had had to share his great AC prize with his far bigger rival, General Electric, or risk repeated corporate attacks. That great concession made, Westinghouse also began setting up foreign affiliates in France, Italy, and Russia, the first links in spanning the whole globe with the Westinghouse-generated "ethereal fluid."

George Westinghouse; Thomas Edison (in chair), circa 1890s; Nikola Tesla, ever dapper, in his hotel room in the 1930s

Guido Pantaleoni, the cultured young Italian engineer who had worked for Westinghouse in the early AC days, returned occasionally to Pittsburgh. He found that while his former boss "would often be highly entertaining . . . very witty, fond of puns, he really had only one thing definitely and always on his mind, and it would be quite disconcerting . . . to be interrupted suddenly by questions such as: How is that machine doing?" Pantaleoni was struck by how even more absorbed Westinghouse now was. "Business matters were gradually coming more and more to the front. . . . In business he was a real Giant; I have never met a human being who in that period when he was at his prime could keep track and direct as many things simultaneously; he had a farsightedness that was almost uncanny to me; every new idea was almost immediately analyzed by him and acted upon before you realized what was in his mind."[1] George Westinghouse, having triumphed at Niagara with his dynamos, saw his electric works flourish while thousands of other businesses collapsed or barely survived the grinding years of the 1890s depression. He was free to plunge forward in the way he loved best—boldly and recklessly, erecting ever bigger works to manufacture ever more gargantuan machines, all in pursuit of cheaper, more efficient power. He delighted in his freedom to operate on such a monumental scale.

Not long after the Niagara plant began operations, Westinghouse was already developing a better version of Englishman Charles Algernon Parsons's revolutionary steam turbine, a completely new technology for America. These machines, far more powerful than steam engines and far smaller, soon became the standard means of running generators in electrical plants and factories. Wrote Westinghouse biographer Henry Prout, "The obsolescence of the engine-type alternator was almost pitiful. Here was a branch of heavy engineering built up at great cost and backed by years of experience. In the coming of the turbogenerator this experience was mostly thrown away, for the engineering . . . was so radically different . . . designers had to start practically anew . . . at enormous expense and through years of effort."[2] Other industrialists might have blanched at such waste, but Westinghouse did not blink an eye. He was always, as Prout said, after progress over profits.

As Westinghouse traveled restlessly among his many businesses, he continued his lifelong practice of seeking out brilliant inventor/engineers, buying their patents, and then collaborating with them to create even better industrial versions. A French physicist, Maurice Leblanc, had invented an air pump that could signally improve the efficiency of the new steam turbines. General Electric helped itself to the invention, and Leblanc swiftly sued. George Westinghouse happened to be in Paris in 1901 and had a friend find Leblanc and bring him to his hotel on the rue de l'Arcade. Westinghouse genially inquired, "So it is you who have sworn to make the fortune of all the lawyers in America. Can we come to terms?" He then proceeded to arrange to buy Leblanc's patents, which he would share with GE. He also hired Leblanc as a consultant for Société Anonyme Westinghouse in France. Leblanc became an enthusiastic collaborator, for he adored Westinghouse: "He was before all things a perfect gentleman and a great-hearted man, and he himself a mechanician beyond compare . . . [as were] his vigor and power of work."[3]

Maurice Coster, who ran the French Westinghouse operation, was returning to Paris on the train with his boss after a "very hectic day in Havre. We had a compartment to ourselves and I said to him, 'Mr. Westinghouse, can you go to sleep after such a busy day?' He replied, 'Coster, I never think of the past. I go to sleep thinking only of what I am going to do tomorrow.' "[4] With this constant forward thinking, Westinghouse squandered no mental energy on what might have been. This forward focus resulted in his receiving a new patent about every six weeks during his career, or almost four hundred patents. These were not speculative inventions, but products of proven commercial value. Westinghouse had learned the hard way the crucial nature of patents in this new industrial order, and ferocious defense of his company's patents was standard corporate strategy.

It was inevitable that Westinghouse would seek to apply the benefits and advantages of electricity to the era's most important infrastructure—trains and streetcars, a form of transportation already so improved by his air brake and automatic signaling inventions. In

1896, he began collaborating with the Baldwin Locomotive Works to incorporate electricity into moving trains. When Westinghouse designed AC systems to power the Manhattan elevated and the New York subway system, the dynamos were so gigantic that they had to be assembled in special shops at the East Pittsburgh plant. But that AC power still had to be converted to DC to run the train engines, a challenge Westinghouse sought aggressively to solve.

Following the pattern of Chicago's White City and Niagara Falls, in 1905 Westinghouse showcased his hugely ambitious new AC locomotive project in the most high-profile, high-wire way. During the first two weeks of May 1905, the top railway men of forty-eight nations convened, a thousand strong, in Washington, D.C., for the International Railway Congress held every five years. Westinghouse had reluctantly agreed to serve as chairman, for he harbored a great dislike of public speaking. For two weeks, the delegates strolled about the one hundred pavilions gaily arrayed about the grounds of the Washington Monument, there to inspect the multitude of technical exhibits. They assembled for lectures and speeches, visited the White House, and were feted in grand style.

On May 16, when the International Railway Congress ended, Westinghouse escorted three hundred of these railroad men on a special train back to Pittsburgh to witness a practical demonstration of his latest triumph, a powerful locomotive operating strictly on AC power. Once in Pittsburgh, the silk-hatted gentlemen of the railroad world assembled in the company's grimy rail yards, gussied up for the occasion with red, white, and blue bunting. Before them stood two locomotives, one the familiar powerful steam engine, the other a strange boxy affair with accordion wires on its flat roof attached to overhead electric lines. The Westinghouse engineers had worked so feverishly to be ready, there had been as yet no trial run of this strange engine. "This was the first electric locomotive of its size, the first alternating-current locomotive, and the first real main-line electric locomotive."

The demonstration was a rousing success. Westinghouse, in his time-tested hard-charging way, then undertook to bring the New York, New Haven, and Hartford Railroad over to the single-phase system. In June 1907, the work was complete and regular service began.

Company historians later saw this project as the "consummation of Westinghouse's great career, for it combined into one masterful work of engineering his revolutionary contributions" in railways and AC.[5]

In the first years of the twentieth century, the Westinghouse Electric & Manufacturing Company was expanding steadily, each passing prosperous year bringing bigger sales and higher dividends. From 1901 to 1907, sales doubled from $16 million to $33 million. By mid-1907, the company stock was yielding a very handsome 10 percent dividend. In his usual fashion, Westinghouse had been audaciously wading into the financial markets to underwrite his tremendous expansions and innovative enterprises, first increasing the amount of stock, then floating collateral trust bonds and debenture bonds. During the spring and summer of 1907, once again financial weakness and jitters overseas began to infect the New York Stock Exchange. On August 10, the jitters escalated into outright terror, and the market began a precipitous downward slide. The nation nervously wondered once again about its banks. "America's obsolete banking system was like an immense tangle of dry brush and timber waiting for a spark," Jean Strouse wrote in her biography of J. P. Morgan. "There were in 1907 nearly twenty-one thousand state and national banks across the country, with no coordinated management or pool of common reserves. Most of them lent their surpluses to correspondent banks in New York, the national money center, and the New Yorkers lent the money out to the Stock Exchange, individuals, and business. Banks beyond the Hudson could call in the loans at a moment's notice."[6] All September, Americans leerily eyed the markets and their fragile financial institutions, waiting for more woe. In a case of monumental bad timing, two wealthy American speculators tried to corner copper in early October, setting off a chain reaction of bankruptcies—including a mining firm, two brokerages, and a bank. Then word surfaced that one speculator was a trustee at Knickerbocker Trust in Manhattan. Depositors, in this era before deposit insurance, sensed a structural wobble and began lining up at the trust's lavish Fifth Avenue offices to extract their dollars.

George Westinghouse was at his company's New York offices at

165 Broadway in Manhattan during these lovely but anxious October days, himself trying to raise money. A recent Westinghouse foray into the stock market to raise $7.5 million for the electric company had yielded barely $2 million. With the company humming with success and prosperity, George Westinghouse had allowed its debt to balloon to $44 million, $30 million of that being in bonds, another $14 million being notes due immediately or soon. Now nervous banks were once again calling in loans. On a cool, bright Friday, October 18, Westinghouse telegraphed to Walter Uptegraff, his longtime financial secretary, to come quickly to Manhattan. Westinghouse and his faithful aide spent a grim Saturday reviewing the books. The foreign companies, launched as part of Westinghouse's great electrical dream of worldwide AC, had been a huge financial drain. The intrepid Pittsburgh magnate needed $4 million in cash right away. Such sums were not, of course, to be found on Wall Street in its present prostrate state. So they ferried across to Jersey City and boarded the *Glen Eyre* for the rail journey back to Pittsburgh. The next day, Westinghouse had the painful task of immediately laying off 1,500 men at the air brake and electrical companies, cushioning the blow by saying it was only temporary.

Westinghouse's public relations man, Ernest Heinrichs, noticed as soon as he walked into the Westinghouse Building that Monday morning of October 21 that something seemed amiss, for there was "an atmosphere of ominous oppression pervading the offices. Conversations were carried on in whispers. Everybody seemed to have a feeling of fearful expectancy, as if some dangerous catastrophe was about to descend upon the place. Although nobody appeared to have any idea what was going to happen."

By Tuesday there could be no question something big was brewing. "The telephone leading into Mr. Westinghouse's office was unusually busy. The 'boss' sat in his chair during the entire day, except for an hour when he went to the Duquesne Club to eat his lunch. Besides the busy telephone, many strange visitors made their appearance, and one after another was ushered into the big office and the door pushed closed behind him." Westinghouse had spent all day Monday and most of Tuesday trying to raise his millions in Pittsburgh. And he had been hopeful until 2:00 P.M. Then word came that back east in Manhattan, the Knickerbocker Trust had given out its last dollars to its desperate depositors and then closed its ornate doors for good at Fifth Avenue and 34th Street. With that, the panic was unleashed. As Heinrichs prepared to leave for the day, Westinghouse stepped into his office and said, "You had better stay close to the telephone this evening. I may have something for you to give to the papers." Nothing more. The company's vice president had no greater inkling of what was impending. At 5:30 that afternoon, as he and Westinghouse were finishing discussions about some routine matter, his boss said pleasantly, "I shall have a new job for you tomorrow."

"What's that?"

"Receiver of the Electric Company." The bankruptcy papers were being prepared as they spoke.[7]

When Heinrichs returned the next morning to the office, he found a newspaper friend from the *Pittsburgh Chronicle Telegraph* waiting for him. The reporter wanted a statement about the stunning news that the Westinghouse Electric & Manufacturing Company, the Westinghouse Machine Company, and the Security Investment Company had all gone belly-up and were in receivership. Heinrichs was shocked, but he maintained his composure, saying, "Oh, yes, you want a statement from Mr. Westinghouse." When he walked through to his boss's office, Heinrichs found Pennsylvania's U.S. senator George T. Oliver and Pittsburgh political boss Chris Magee, both sitting there grim faced. Westinghouse looked as serene as ever and pleasantly said he'd soon have a statement. When it came, it emphasized the phenomenal prosperity of these companies and depicted the bankruptcy as a temporary and unfortunate expedient until the present "financial stringency" was resolved.

Throughout that sad and unnerving day, Wednesday, October 23, streams of friends and well-wishers filed in to see Westinghouse in his office. Western Union boys bounded in and out. The chief appeared unperturbed and said buoyantly to one aide, "By the way, MacFarland, I've got an idea now for our turbine that will make a sensation when we bring it out."[8] Westinghouse emerged to join his office staff for lunch. As they ate sandwiches, he munched on an apple. At one point, he turned to Heinrichs and urged him when talking to his many newspaper friends, "Do not forget to make it very emphatic to them that this receivership is not the end of the company. . . . [The] company is fundamentally as sound and solid as ever, and it will emerge out of this unfortunate situation a greater and more prosperous concern than ever."[9] The bankruptcy was a great shock to both insiders and outsiders, but with money so tight, Westinghouse saw no alternative. He treated it matter-of-factly. " 'I grant you this is not pleasant,' he told one friend. 'But it isn't the biggest thing in the world. All large business has its ups and downs. The crisis through which we are passing is only part of our day's work.' "[10]

Back in Manhattan, as the Panic of 1907 unfolded, the world turned once again to J. P. Morgan, now seventy, a powerful man with a plutocrat's large belly, the legendary ferocious eyes, and a monstrous nose, red and deformed from acne rosacea. In wake of the Panic of 1893, Morgan's firm had restructured so many failed railroads, he now had a hand in half of America's rails, the nation's economic lifeblood. His combining of U.S. Steel from Carnegie's firm and other smaller rivals had created the country's first billion-dollar corporation in 1901. So only J. Pierpont Morgan had the reputation and power to steady the floundering economy. The great financier returned from an Episcopal gathering in Richmond just as the run on Knickerbocker began. Hunkered down in his sumptuous Madison Avenue library, smoking his huge Cuban cigars, he conferred over the next few days with the nation's most powerful bankers and money men, bailing out interests he felt worthy of the effort, advancing huge hunks of millions as needed. He and his associates let certain weak banks, trusts, and firms fail. Working with the U.S. secretary of the Treasury, Morgan and his allies during the next few weeks staved off complete calamity. One of his partners later said, "In the dark days of 1907, he knew no fear, he believed in the country & himself & imparted pluck & spirit to others & infused strength & hope into men 20, 30, 40 years younger than himself. If he had given way, the whole house of financial cards would have fallen."[11]

Others suspected Morgan had gained some great advantage while others suffered disaster. Among them was a highly skeptical President Theodore Roosevelt. TR had deeply angered Wall Street by attacking its entrenched system of power. The day the Knickerbocker collapsed, Roosevelt emerged from a fifteen-day hunting trip deep into the Louisiana bayous, where he had bagged and eaten bear, turkey, squirrel, opossum, and wildcat. In a speech the next day, he acknowledged that many financiers were blaming his policies for Wall Street's woes. But he stated firmly his intention of punishing "successful dishonesty" among wealthy manipulators and swindlers. TR wondered privately in a letter whether "certain malefactors of great wealth" were not shaking things up to their own nefarious purposes "so that they may enjoy unmolested the fruits of their own evil-doing."[12]

As for George Westinghouse, the eternal optimist who had successfully weathered previous financial panics when Westinghouse Electric's prospects were nowhere so rosy, he assumed the best. Five thousand of his own employees put up $600,000 to help in the firm's reorganization by buying "assenting stock," far more expensive stock that would help retire much of the debt. But when the electric company emerged from bankruptcy a little more than a year later in December 1908 (under a plan largely of Westinghouse's own devising), Westinghouse was no longer the supreme ruler. At long last, the New York and Boston banking interests had gained majority control of a company they had long coveted. They promptly elected as chairman of the board the dour Robert Mather, a railroad attorney of no major accomplishments, to rein in the Westinghouse exuberance they found so galling.

The very reckless boldness that made the domineering Pittsburgh magnate a great inventor and entrepreneur was now denounced—his lavish treatment of inventors, his big spending on research, his generosity with his men, and his expensive gambles on new and experimental machines. In short, the usual complaints of the money men. (Andrew Carnegie, whose steel company was famous for its fabulous profits and infamous for its poor labor relations, was very much of that mind-set. He described Westinghouse as "a fine fellow and a great genius, but a poor businessman."[13] These criticisms infuriated Westinghouse, who had created one major pioneering company after another.) In any case, Westinghouse and the new president, Mather, were like oil and water, and bitterness festered. One Friday afternoon at the end of a meeting in Pittsburgh, as Mr. Mather packed up his briefcase, Westinghouse, ever genial, said, "If you are going to New York this evening, I shall be glad to take you in my car. I am going East myself."

There was an uncomfortable silence, punctuated only by Mather's methodical gathering of papers and stacking them into his briefcase. He finally said, "I prefer to go to New York by myself and pay my own fare."[14]

By late 1910, Robert Mather and the new directors had pushed Westinghouse off the electric company board entirely. They made various lukewarm overtures to bring the great industrialist back into some kind of active management, but he disagreed with their whole corporate philosophy. He saw himself as interested in progress and profits and the welfare of his men, while they were interested only in profits. The board, predictably, saw Westinghouse as an utter profligate. Westinghouse also disapproved of the new board's reneging on their promise to pay dividends on the "assenting stock," shares bought at high prices to get the firm out of receivership. Westinghouse felt a special responsibility to these stockholders, who had stepped forward in tough times and seen him through. So as the electric company flourished and the board refused to pay these dividends, Westinghouse launched one last effort to recapture his stolen company.

A short-lived, two-week attempted coup, the Battle of the Proxies, stirred up great excitement in the broiling month of July 1911, when a deadly heat wave enveloped East Coast cities. Those who had come to loathe the Wall Street money men and their hard-eyed greed were rooting for Westinghouse. From his Pittsburgh office, he rallied the troops, proclaiming to the nation's newspapers that proxies were rolling in by the boxload to unseat the tightfisted board. But when the showdown came, the great electrical dreamer could produce but 200,000 votes, compared to management's avalanche of 490,000 shares. His public relations officer, Ernest Heinrichs, watched this poignant last hurrah with sadness. "The loss of the Electric Company was to Mr. Westinghouse a disappointment from which he never recovered. There is no doubt that it broke his spirit."[15] The boss who had once proudly and impatiently informed a subordinate, "Impossible is not in my vocabulary," had discovered that retrieving his beloved company, the corporate chariot that was to spread electricity across the universe, was indeed impossible. It had fallen irrevocably into hostile hands.

The loss of his electric company was a cruel blow, but Westinghouse was at heart an optimist, a doer, and a builder. He still had his four other major American companies and his insatiable desire to improve the world. Although he had always shunned publicity and avoided public speaking, Westinghouse now began to use his renown and prestige to lobby frankly in support of the Progressive agenda. On January 21, 1910, as the after-dinner speaker at a Boston meeting of the American Engineering Societies, he denounced the "evil practices" of the "large and powerful railway and industrial combinations" and their "selfish and unwise course in suppressing competition by methods transparently wrong."[16] He urged his fellow engineers to support government regulation that would rein in the ruthless and predatory capitalism that dominated the American business world. Though Westinghouse never mentioned his own company, it stood as a chilling example of how the Boston and New York "money power" could use an unregulated stock market and fragile banking system to gain control of someone else's valuable property.

At a time when many Americans had come to view the nation's industrialists and financiers as little better than robber barons and "malefactors of great wealth," Westinghouse was a notable exception: an honest industrialist who sold the best product for the best price, who relished competition and valued his workers, and who deserved his hard-earned fortune. His late-life besting by Wall Street was a chastening reminder of the feared and concentrated "money power." During the 1912 congressional Pujo Committee hearings, the full extent of that long whispered, incestuous stranglehold on money and capital was finally revealed: The officers of the five biggest New York City banks also held 341 key directorships in 112 major U.S. companies. The Morgan partners alone occupied seats on 72 boards. Westinghouse always viewed such blatant collusion and power grabbing as wrong.

In the two years leading up to his death in 1914, Westinghouse was in obviously declining health, pining, many believed, for the loss of his great life's work—the Westinghouse Electric & Manufacturing Company. In June of 1912, the American Institute of Electrical Engineers awarded him the Edison Medal for "meritorious achievements in the development of the alternating current system." He and many others savored the exquisite irony of an Edison honor being bestowed upon Westinghouse, once so maligned by Edison for being the stalwart champion of alternating current. Numerous other honors rained down upon him from engineering groups and foreign governments. Through all his struggles and triumphs, he and his wife of fifty years remained deeply devoted to each other and their son. At George Westinghouse's sixty-sixth birthday party luncheon, held on a glorious blazing blue October day at their Berkshires mansion, many toasts were drunk. Then Westinghouse himself arose. "A glass of champagne in his hand, [he] smiled across the table to his wife, and in a voice so tender, so affectionate, so deep in feeling that all were spellbound, said: 'If I have had any success in life it has been due to my wife.' "[17] All rose and lifted their glasses of champagne to toast this devoted couple's lifetime of ardent marital felicity, and to honor a man who had been so dedicated to improving the world.

After the Westinghouse Electric & Manufacturing Company's bankruptcy of 1907, *The New York Times* wished the best for Westinghouse, calling him "an American institution in whom we have patriotic pride."[18] But Westinghouse never really recovered from his loss. It was as if part of his heart and noble spirit had been excised. On March 12, 1914, George Westinghouse died in Manhattan at the Hotel Langham, where he and his wife had stopped en route to their house in Washington, D.C. At his death, fifty thousand people worked for the many companies he had created. The companies were valued at $200 million, and Westinghouse himself was worth $50 million. In the heyday of the robber baron, Westinghouse showed that an honest and honorable entrepreneur could triumph, building admired and valuable companies. The Pittsburgh magnate had provided good and useful work to his legions of employees, as he hoped to do. Probably more important to him, he had daringly brought the wonders of electricity to the world. With each passing year, more industries, more towns and villages, more homes, became electrified. He had, despite his many powerful opponents, achieved his life goals. Wrote one biographer, "Westinghouse was always working for ideals. . . . High spirit flowed down through the Westinghouse companies and left enduring love, loyalty, and enthusiasm. A corporation can have a soul." Money meant little to George Westinghouse but the means to lift up the world around him. One day toward the end of his life, he was on a train whose air brakes helped avert a derailment at a bad track washout. He told an aide, "If some day they say of me that with my work I have contributed something to the welfare and happiness of my fellow men, I shall be satisfied."[19]

THOMAS A. EDISON

After the formation of General Electric in 1892, Thomas A. Edison was little involved in the development of the electric power field. Some months after that bitter day when Morgan combined his company, Edison General Electric, with Thomson-Houston into GE, Edison's secretary, Alfred O. Tate, came over from Wall Street to the West Orange laboratory to confer with Edison about a battery project. The jolly Tate found his boss alone in the magnificent wood-paneled library. " 'Tate,' [Edison] said, and I had never heard him speak with such vehemence, 'I've come to the conclusion that I never did know anything about it [electricity]. I'm going to do something now so different and so much bigger than anything I've ever done before people will forget that my name ever was connected with anything electrical.' "[20]

True to his word, Edison plunged himself and his new fortune of $2 million (his part of the GE merger) wholeheartedly into something "so much bigger": an iron ore separating and concentrating plant on nineteen thousand acres in the bleakest wilds of New Jersey. For some years, Edison, believing (along with many others) that America was running out of iron ore, had concluded there was money to be made extracting iron from played-out East Coast mines via powerful magnets. As early as 1889, he was full of schemes and iron dreams, writing an associate, "If I could get possession of practically all the magnetic ore deposits in the center of the coal and iron district of Penn[sylvania] I would have a monopoly of one of the most valuable sources of national wealth in the U.S."[21] The caveat in all these grandiloquent plans was the need for significant capital investments in giant crushing machines that pulverized the old ore-containing rocks, preparatory to extracting the iron with giant magnets. Experienced mining engineers doubted the feasibility of such (as-yet-unbuilt) giant machines. As Edison biographer Matthew Josephson said, that "was all Thomas A. Edison needed to hear."[22] Edison still loved to be doing "big things," but his understandable mistrust of Wall Street (and their view of him as a poor businessman) would hobble him to some extent, for "big things" in the industrial world of the 1890s generally required big capital.

Edison pressed ahead full bore and designed and built for $750,000 (his and other investors' money) a large ore-crushing plant in the remote highlands of Ogden, New Jersey, the site of an old iron mine. By April 1891, the monstrous machine was clanking away, its huge rollers crushing hundred-pound boulders gouged out from the dynamited hillside by gigantic steam shovels. The powdery grains of extracted ore, removed from the magnets, were then pressed into iron briquettes. But Edison had a hard time finding steel mill customers at first because the briquettes contained too much phosphorus or too little iron ore. Meanwhile, mechanical problems continually rippled through the noisy, dusty works, cropping up here or there, and the issue of damp or wet ore was a constant headache. Mere months after the giant plant roared into production, it had to be closed for an overhaul. Edison threw himself wholeheartedly into this "much bigger thing," often spending six days a week there from 1894 to 1897. During those years, he regularly closed and rebuilt the plant or engaged in major, expensive tinkerings, all with the ultimate goal of producing three hundred thousand tons of iron briquettes yearly. In 1895, when Edison learned his men planned to strike, he shut down the whole unprofitable enterprise for several months.

Through year after year of setbacks and expensive solutions, Edison was his usual cheerful self, glorying in the rough life of the mining operation. He wore a filthy old duster coat, a broken-down hat, and a dust filter mask, which he continually lifted up to spit out his chewing tobacco. Sundays he would clean up and return home to his wife and two small children at the family mansion, Glenmont. Although by 1898 Edison finally had a fully automated plant that required no human labor—to the amazement of the press and his good friend Henry Ford—the endless mechanical problems and breakdowns required a staff of almost two hundred men to keep the automated wonder up and running. The plant was now able to crush four- and five-ton rocks, and the capital investment had reached $2 million, most of this being Edison's money. His early investors, beset by the nation's worst depression, had long since refused to ante up new funds for what many had started to call "Edison's Folly."

In August of 1897, Edison sold off his stock in the Edison Electric Illuminating Company of New York, saying, "I actually needed the money. My Wall Street friends think I cannot make another success, and that I am a back number, hence I cannot raise even $1000 from them, but I am going to show them that they are very much mistaken. I am full of vinegar yet, although I have had to suffer from the neglect of an absentminded Providence in this scheme."[23] The economics were not promising. Edison needed to sell his extracted ore at about $6 or $7 a ton to make a profit, and he anticipated that ore prices—hovering at $4 a ton—would rise. But then in 1899, longtime aide Charles Batchelor, who had declined to get dragged into what Edison fondly called his "Ogden Baby," came out to the dust-covered camp with its gargantuan grinding and clanking machines to show Edison a newspaper story. Of course, the terrible racket didn't bother Edison, for he was deafer than ever. The article described John D. Rockefeller's plan to mine the vast high-grade iron deposits of Minnesota's Mesabi Range.

When Edison read the newspaper clipping, which noted iron ore prices had dropped to below $3, he laughed out loud and said, "Well, we might as well blow the whistle and close up shop."[24] In fact, he lingered on another year, losing money fulfilling existing contracts. In typical Edison fashion, he had no regrets: "I never felt better in my life than during the five years I worked here. Hard work, nothing to divert my thoughts, clear air, simple food made life very pleasant."[25] One suspects crushing rocks out in the sticks with his "boys" was a nice change from the rancorous years of the War of the Electric Currents and the machinations of Wall Street. And so, giving up on rock crushing, Edison went back to his West Orange laboratory. (When he recycled his disastrous "Ogden Baby" into a cement plant, it made excellent Portland cement, but it still was not very profitable.)

Thomas Edison never completely left the West Orange laboratory. During slow periods at the ore mill, he would return and work feverishly on various projects. During those years, his original phonograph was dramatically improved, creating a much needed stream of income as it became good enough to play music. Edison's Kinetoscope had dazzled visitors to the 1893 World's Fair in Chicago, and the whole field of motion pictures had advanced rapidly, with many striving to perfect the technology. When on April 23, 1896, Thomas Edison's Vitascope debuted at Koster & Bial's high-toned Music Hall at Herald Square, the silk-hatted crowd of theatrical promoters and businessmen watched the strange big screen in amazement as across it swept, in full color, "ballet girls dancing with umbrellas, burlesque boxers, some vaudeville skits, and finally so realistic a scene of waves crashing upon a beach and stone pier that some of the viewers in the front row recoiled in fear."[26] At the end, the audience burst forth in wild applause, cheering the gray-haired Edison up in his balcony box. He had the good grace not to come down on the stage, for he knew he was not the inventor, but the promoter, of this projector. Nonetheless, Edison, who tore himself away from his New Jersey iron ore operations to attend this Manhattan premiere, was unwilling ever to acknowledge the important contributions of his longtime laboratory assistant William K. L. Dickson, or the reality that the Vitascope machine showing the "moving pictures" that night was actually patented by inventor Thomas Armat, who was up in the projection room making it run smoothly.

Edison had set up a primitive movie studio out at West Orange, a sweltering structure known as the Black Maria, and there, from mid-

1893, his movie men had been churning out short, hyperlively silent motion pictures—prizefights, funny skits, and such. Only in 1904, with the Edison studio's *The Great Train Robbery,* did early movies begin to move beyond skits to real stories. This was a revelation, soon copied by others flocking to the business. Generally these "movies" ran about fourteen frenetic minutes, during which time trains were spectacularly wrecked, ladies saved from villains, and all manner of calamities and heroics acted out. Audiences could not get enough, and by 1909, there were eight thousand theaters just for movies, a figure that quickly doubled. Once again, Edison, whose laboratory was continually improving the motion picture technology, dispatched his legions of well-trained attorneys into long-drawn-out patent wars, emerging triumphant as the holder of the key motion picture patents in 1907. Like Westinghouse, Edison understood the supreme importance of owning technology.

A year later, Edison's biggest rivals, including the American Mutoscope Company, gathered for a wary December peace lunch in Edison's West Orange library. There they ostensibly celebrated the formation of the Motion Picture Patents Company, essentially a movie trust. It guaranteed fees worth $1 million a year to Edison, who smiled blandly from that day's group photo. When one former colleague expressed concern to Edison about his financial status, he wrote back cheerfully, "My three companies, the Phonograph Works, the National Phonograph Company, and the Edison Manufacturing Company (making motion picture machines and films), are making a great amount of money, which gives me a large income." And so, as he turned sixty, Edison was flourishing. While Westinghouse had become a great industrialist building ever bigger, heavier machines, Edison was almost unwittingly developing a whole other sector of the American economy, one that was far less capital-intensive and far more glamorous—the entertainment industry.

His next big project took him back to a long-standing challenge in the field of electricity—developing a better storage battery. As with the light bulb, the ever optimistic Edison had thoroughly underestimated the technical difficulties. He began work in earnest in 1900 and introduced the battery in 1903 with his usual fanfare and hyperbole, asserting: "I am sure one [battery] will last longer than four or five automobiles."[27] Unfortunately, the battery had two fatal flaws—it began to leak, and it did not recharge as needed. Edison yanked it from the market and soldiered on. He had poured $1.5 million of his own into the effort when at last in 1909 he delivered a new kind of nickel-iron-alkaline battery that proved to have many uses. Initially its greatest market was in electric vehicles, and before the gas engine triumphed, half of American delivery trucks used the Edison battery. When that market disappeared, Edison hustled up many other industrial uses. The Edison Storage Battery Company slowly recouped the huge investment. As Edison once said, "I always invent to obtain money to go on inventing."

Edison reveled in presiding over his numerous companies and remained understandably leery now about ceding any control to outsiders. In 1912, the aging inventor explained to Henry Ford in a letter, "Up to the present time I have only increased the [battery] plant with profits made in other things, and this has a limit. Of course I could go to Wall Street and get more, but my experience over there is as sad as Chopin's 'Funeral March.' I keep away."[28] By having few stockholders besides himself, Edison never had to worry that the money men would sell him down the river again, nor did he have to listen to their carping. But it also limited him at times because he did not have big capital. His legendary stubbornness served him well and ill. It certainly helped him push on toward success when anyone else would have despaired and quit. But it also still prevented him from recognizing important new technologies that he did not invent. Just as he had refused to see the importance of AC in the 1880s, in the 1920s he denigrated commercial radio. Edison hated the idea that the radio audience was obliged to hear other people's choices of music and amusement. But the (free) delights of radio soon caused phonograph sales to stall and decline—and Edison would not let his company (now ostensibly run by his sons) design and sell radios until it was too late to compete.

As Edison headed into the final decades of a long and highly productive career as inventor and entrepreneur, he found himself the most admired man in America, a genuine national hero whose huge accomplishments were made all the more endearing by his folksy persona. The press still loved him, and he rarely disappointed, regaling reporters with forthright and humorous remarks. Who but Edison, after a huge and devastating fire at his West Orange laboratory in 1914, could say, "Oh shucks, it's all right. We've just got rid of a lot of old rubbish."[29] His birthday became the occasion of annual admiring articles. Edison spent more and more time enjoying the tropical pleasures of his waterfront home, Seminole Lodge, in the Florida Everglades. There, underwritten by his late-life friends Henry Ford and Harvey Firestone, the aging Edison passed his four final years happily pursuing the holy grail of finding a wartime source of American rubber for industrial purposes.

The public viewed Edison as the lovable man who had brought them many wonderful things that profoundly changed their world—certainly the phonograph was an astonishing addition to daily life, while movies had become a national passion. Honors and medals in profusion were his. Edison was as devoted as any workaholic could be to his younger second wife. His six children by his two marriages saw little of him as they grew up and found him a disappointing father. Some were estranged. Two sons ran his companies. In the end, the companies, which he owned, were worth about $12 million, a fraction of the value of Westinghouse's industrial corporations.

To this day, Thomas Edison remains the American inventor with the most patents: 1,093. His Menlo Park workshop served as the early prototype of the highly productive industrial laboratory. Above all, Edison in the 1920s was seen as the patron saint of the invisible energy that made life so much easier, so much better for so many—electric light. And though he had long been removed from the electric power industry, Edison's name was still attached to many local lighting companies and he was happy to bask in that role. When he died in 1931 at age eighty-four, *The New York Times* lauded the inventor whose "genius so magically transformed the everyday world.... No one in the long roll of those who have benefitted humanity has done more to make existence easy and comfortable."[30]

NIKOLA TESLA

Nikola Tesla outlived both his original champion, George Westinghouse, and his early rival, Edison. After the triumph of the Niagara

Falls Power Company plant, Tesla concentrated his electrical research and experimentation on producing extremely high voltages and frequencies, all with the ultimate aim of wireless transmission of energy, for either communication or power. His dreams were big, far beyond the imaginings of his peers, and very expensive. By the end of 1897 he was already having money troubles, as he confided in an importuning letter to Ernest Heinrichs. The Westinghouse PR man wrote back, "While I am very glad to hear that you are physically and mentally in perfect condition, it grieves me very much to have you inform me that you are ailing with what you choose to call 'financial anemia.' . . . I shall remember your wish about speaking of you to the boys in Pittsburgh and trust that some of them will feel disposed to send you a christmas rememberance [*sic*]."[31]

Tesla's idealistic and generous renunciation of his AC royalties was beginning to haunt him. With both Westinghouse and GE now sharing the Tesla patents and building induction motors, this income would have easily supported his most profligate laboratory research and still left him the very rich man he deserved to be. His first biographer and longtime friend, John J. O'Neill, estimated that by 1905, when Tesla's AC motor patent expired, American induction motors alone were generating 7 *million* horsepower. At $2.50 per horsepower, one can quickly calculate that Tesla had nobly forfeited a princely and heartbreaking $17.5 million in royalties.[32] Why didn't George Westinghouse, now that his company was prospering, reward Tesla in some way for this sacrifice? We simply do not know. The investments of men like Edward Dean Adams, who had bankrolled Tesla to a certain point, were small change compared to this lost royalty income.

Fortunately, Tesla's post-Niagara status as an electrical genius extraordinaire was sufficient to persuade several of his rich New York society friends and admirers to ante up tens of thousands to perfect his latest amazing invention. By early May 1898, with America's newly declared war with Spain reaching fever pitch, Tesla was anxious to display his new work before a select audience of invited guests—wealthy potential investors—during an electrical exposition at the vast Moorish-style Madison Square Garden. One always expected something unusual, dazzling, and uncanny at a Tesla lecture, strange gadgets and electrical gizmos capable of throwing out great bolts of light and lightning, but the top-hatted millionaires who walked into the private auditorium this time found only a very large tank of water and what looked like every boy's dream, a gigantic toy boat five feet long by three feet wide. Nikola Tesla explained that they were looking at his perfected remote-controlled robotic boat, what he called his "teleautomaton." He used a handheld transmitter to direct his boat to sail forward, change directions, and turn its lights on and off. While cementing his reputation as a weird and wonderful wizard, this display thoroughly obscured the two major scientific advances—a multichannel broadcasting system and remote-control electronics—embodied in this seemingly oversize toy. Tesla declared, "You see there the first of a race of robots, mechanical men which will do the laborious work of the human race."[33] The millionaires were not at all convinced and left unwilling to open their checkbooks.

With little commercial interest in his automatons, Tesla returned to his previous love—the infinite generation of wireless electrical power and transmission. Of course, this would also take large sums of money to move forward. He targeted his close acquaintance the fabulously wealthy John Jacob Astor IV, exhorting him to become a backer of his manifold grandiose electrical dreams. "You will see how many enterprises can be built up on that novel principle, Colonel. It is for a reason that I am often and viciously attacked, because my inventions threaten a number of established industries." Astor brought Tesla smartly back to earth, writing, "Let us stick to oscillators and cold lights. Let me see some success in the marketplace with these two enterprises, before you go off saving the world with an invention of an entirely different order, and then I will commit more than my good wishes." Receiving even that nibble, Tesla maintained a persistent pursuit, tantalizing Astor by describing how "I can run on a wire sufficient for one incandescent lamp more than 1000 of my own lamps, giving fully 5000 times as much light. Let me ask you, Colonel, how much is this alone worth when you consider that there are hundreds of millions invested today in electric lights?"[34] By late 1898, Astor capitulated and bought $30,000 worth of stock in Tesla's company.

Living in his new patron's famous hotel seemed to be part of this deal, and Nikola Tesla left behind the upper-middle-class comforts of the Gerlach Hotel for the prodigal opulence and society cachet of the luxurious Waldorf-Astoria, an eleven-story German-Renaissance palace at Fifth Avenue and 34th Street. With its numerous elegant public rooms, the Waldorf-Astoria by the last turn of the century was *the* fashionable watering hole for the New York elite, as well as the preferred hostelry for visiting royalty. Whether it was gala balls or intimate tête-à-têtes where one could watch swells and society ladies passing through Peacock Alley, the Waldorf-Astoria radiated luxury, privilege, and ostentation. Nikola Tesla settled in happily.

Despite Tesla's practical talk of his new and better light bulb sweeping the world, as soon as he had Colonel Astor's money safely in hand, the elegant inventor immediately set aside such prosaic concerns to return to the electrical frontier he so loved to explore. As he built bigger and more powerful oscillators and generators that threw out ever greater streamers of electricity, Tesla realized he could not safely continue in his Houston Street laboratory. Leonard Curtis, a friend and patent attorney, had retreated from the stress and strain of New York legal life to the frontier simplicities of Colorado Springs, a pretty resort at the foot of the pristine Rocky Mountains. There he ran the electric power company, a now essential part of the local mining industry. He offered Tesla free land for a makeshift laboratory outside of town with a majestic view of Pike's Peak, and free electricity. Tesla began sending ahead his electrical equipment—copper bars, great rolls of electrical wire, huge generators and motors, while he himself followed that spring.

When the tall, slender Tesla stepped off the train into the dry, cool air of Colorado Springs on May 19, 1899, he was as far west as he had ever been. Before him rose the Rocky Mountains in all their snow-capped violet glory, while the crisp blue sky at this high altitude seemed to go on forever. The lovely town, set among such magnificent scenery and ringed by meadows of wildflowers, was also famous for its healthful mineral springs. Tesla's old friend Curtis and the town's leading dignitaries hospitably welcomed the New York wizard, delighted to have such an eminent personage grace their small community. They escorted him to the Alta Vista Hotel and then formally feted him at a great banquet. Nikola Tesla stayed in room 207 (divisible by three) at the Alta Vista, his window looking out on the mountains, and had eighteen fresh towels delivered each morning. Tesla told the local reporters he planned "to send a message from Pike's Peak to Paris [for the Paris Exposition]. . . . I will investigate electrical disturbances in the earth. There are great laws, which I want to discover, and principles to command."[35]

While Colonel Astor, off on a European tour, no doubt assumed Tesla was attending to the practical details of conquering the world with his "cold" light, the great inventor was instead contemplating far more exalted concerns as he began to construct his Colorado Springs laboratory at six thousand feet above sea level. The first concern was how to design and build the most powerful electrical transmitter ever. Then once that was up and running, Tesla was seeking to individualize and isolate the energy that his supertransmitter sent forth into the atmosphere. Finally, he wanted to determine whether the earth and its atmosphere were—as he suspected—resonating at specific frequencies. How did such energy waves travel?

To answer these questions, Tesla had constructed, far beyond the edge of town on a prairie pasture, a huge wooden barnlike structure with a retractable roof, braced on three sides with wooden supports. Jutting up was a wooden tower holding a single two-hundred-foot-tall copper pole topped with a yard-wide copper ball. The copper pole rose from the largest Tesla coil yet, which the inventor referred to as his "magnifying transformer," capable of generating 100 million volts. The inside of the barnlike structure appeared to be some kind of riding ring, a wide circular space hemmed in on all sides by a tall wooden fence. The plain plank floor was filled with electrical equipment that turned out to be Tesla's transmitter, built to operate on the local AC from the Colorado Springs power station. It was then sent through a transformer to hugely increase voltage. Outside the wooden arena sat Tesla's receiving stations, which to the untrained eye mainly looked like gigantic cans. Tesla had one assistant he had brought from New York, a young man sworn to deepest secrecy. To keep away the inevitable electrical sightseers, Tesla put up a tall fence and posted many signs warning "Keep Out, Great Danger." One reporter who ventured into the compound to peer through the windows found Tesla's assistant next to him saying, "Your life is in peril and you would be a great deal safer if you would remove yourself from the vicinity."[36]

To Nikola Tesla's delight, Colorado Springs had frequent, gigantic, crashing thunderstorms. On July 3, 1899, Tesla was working on getting his laboratory installed when he noticed dark clouds massing in the west, boiling up until they erupted into a violent electrical storm that passed over the plain and receded far into the distance. It was then "I obtained the first decisive experimental evidence of a truth of overwhelming importance for the advancement of humanity," he would later write. On his instruments Tesla saw recorded "heavy and long persistent [electrical] arcs" as the storm passed and then became fainter and fainter. The storm came and went, but the equipment continued to record electrical activity waxing and waning at a regular rate. These were standing electrical waves. "No doubt whatever remained: I was observing stationary waves. . . . The tremendous significance of this fact in the transmission of energy by my system had already become quite clear to me. Not only was it practicable to send telegraphic messages to any distance without wires, as I recognized long ago, but also to impress upon the entire globe the faint modulations of the human voice, far more still, to transmit power, in unlimited amounts to any terrestrial distance and almost without loss."[37] Tesla now firmly believed energy could be transmitted without wires.

He continued to construct his powerful magnifying transformer, and finally the evening came when he was ready to test it. As the darkness settled its cool velvet mantle over the Rockies, Tesla stationed himself where he could watch the copper pole. Dressed for the occasion in his best cutaway, derby hat, and thick rubber-soled shoes, he signaled his helper inside to turn on the switch. He saw a ten-foot blue spark spurt from the copper ball, then another, and another, each longer, bluer, until they crackled forth in huge bolts forty, fifty, sixty feet long, then a hundred feet, each setting off a great crack of thunder. Inside the barn laboratory, the whole atmosphere was alive with a blue aura and dancing sparks. Joyously, Tesla looked back up at the copper ball—lightning 130 feet long was spurting out, a brilliant flash that brought on a gigantic crash of thunder. Then all went dead and the smell of ozone was overpowering. It became so quiet, the wind could be heard rustling through the mountain grasses. Tesla rushed in to scold his assistant, but all their power was gone. When they looked over at Colorado Springs, it, too, was dark. Tesla had blown out the town's powerhouse.

All summer, fall, and winter, Tesla happily explored this unmapped terrain of high voltages, the earth's resonance, and electrical waves. He had spectacular photographs taken showing him sitting in the laboratory reading while great bolts of lightning played about him. These were, in fact, double exposures, but when published subsequently, they only reinforced his reputation as an incredible electrical wizard, a man capable of making lightning to rival the angry gods. Finally, in late January of 1900, Tesla had to pull himself away from these fascinating experiments and return to New York.

When Nikola Tesla debarked in effulgent glory at Gotham, he found the pulsating city in a greater frenzy of moneymaking than ever. At long, long last, the American economy had recovered from the searing depression of the 1890s. Manhattan's streets were still jammed with horse-drawn vehicles, but now they sparred for space with the electric trolleys and the new plaything of the wealthy, the motorcar. Tesla settled happily back into the opulence of the Waldorf-Astoria, enjoying his own celebrity as he strolled through Peacock Alley, resplendent in his beautifully tailored Prince Albert coat, four-in-hand tie, white silk shirt, favorite green suede high-top boots, silver-topped cane, kidskin gloves, and those famously flashing eyes. Nikola Tesla reclaimed his place in New York society's whirl of lavish dinners, opera soirees, and dinners at Delmonico's or Sherry's, always keeping an eye out for possible wealthy investors. His laboratory in Colorado Springs had been an exhilarating, expensive foray, and once again his coffers were empty. Moreover, he had no major commercial venture for generating new and future income. Of course, he approached Colonel Astor again, only to find him now politely indifferent to the man who had put aside "cold" light to pursue electrical chimeras.

To keep his profile high and entice new backers, Tesla offered to write an article for his dearest friend, Robert Underwood Johnson, an editor at *Century* magazine. Johnson, who eagerly commissioned the piece, was appalled when Tesla submitted a long, tendentious mélange of philosophy and science titled "The Problem of Increasing Human Energy." Tesla refused all editorial suggestions, so *Century* went ahead and published the whole rambling thing in June 1900. The article was most memorable for the amazing photos of Tesla taken at his Colorado Springs laboratory. It all caused a sensation, much of it negative. For some years now, Tesla had kept himself arrogantly aloof from his fellow engineers, disparaging the work of others, blithely describing feats and projects of his own that then never materialized. A decade earlier, he had begun criticizing Edison's incandescent light bulb as expensive and wasteful and prophesying its imminent demise at the hands of his, Tesla's, far superior "cold" light. Yet where was this much ballyhooed light? Tesla had done too much self-aggrandizing and expressed too much disdain for other electricians. His electrical peers were more than ready to pounce upon his strange and boastful writings. His subsequent assertion that he was in touch with Martians, based on radio signals received from outer space while in Colorado, did nothing to lessen the scoffing. Broke as ever, in early December 1900 Tesla asked Westinghouse if he could extend an earlier $3,000 loan he could not afford to repay.

Just at this discouraging moment, Nikola Tesla was rescued from financial despair by none other than J. Pierpont Morgan, now Wall Street's most fearsome, feared, and powerful financier. Tesla, circulating as he did in New York's richest social circles, had become something of a favorite with Morgan's grown daughter, Anne, twenty-eight. One evening after dinner at the family's Italianate mansion, bedecked with a rich mix of Morgan's fast-growing collections of European paintings and precious antiques and curiosities, Tesla had managed to persuade the great banker to consider backing his new venture, for Tesla was fervent in his desire to build his great electrical dream, his "World System" of power transmission. However, just as Colonel Astor wanted to invest in "cold" light, so J. Pierpont Morgan had one focused interest—world telegraphy. He knew Tesla's AC patents had trumped all else at Niagara. What was to say his wireless patents would not prevail again in this latest race, the race to span the ocean skies with invisible, electrical words?

Morgan seems to have also alleviated Tesla's immediate pressing shortage of money, for on December 12 Tesla penned a scrawled, almost pathetic letter of gratitude: "How can I begin to thank you. . . . My work will proclaim loudly your name to the world!"[38] Over the next few months, Tesla, who had initially imagined himself selling $150,000 worth of shares in a company to Morgan, found himself thoroughly cornered, with the ruthless financier at the last minute adding (as a condition to closing the deal) 51 percent ownership of Tesla's patents not just for wireless, but also for the same "cold" light that Colonel Astor imagined he had a major interest in. Tesla, desperate to silence the rising clamor of his sneering critics with his "World System" of worldwide power and instant communication, acquiesced. By early March 1901, he finally had the prospect of funds. Yet could either Morgan or Tesla possibly have imagined this would be sufficient to launch any major electrical venture? Lighting the World's Fair had been a $500,000 contract. Niagara alone had cost $6 million before ever a penny was earned.

Presumably, all Tesla could think about was at least getting launched. He purchased two hundred acres out on a rural tract at Shoreham, Long Island, and by July 1901 was building his great dream, the Wardenclyffe tower, a strange, giant erector-set structure that rose 187 feet and was topped with a large globelike dome. Below the tower a shaft traveled 120 feet into the ground, while sixteen iron pipes pushed into the bowels yet another 300 feet, "gripping the earth." Across the fields stood an attractive laboratory building designed by his friend Stanford White. But just as Tesla was busy creating his mysterious tower, Marconi stunned the world on December 12, 1901, by successfully transmitting the letter *S* across the Atlantic. Tesla had believed for some time that Marconi was infringing on many of his patents, but Tesla, like Edison in the early days of the light bulb, could not be bothered at first by inferior "patent pirates." And by the time he saw Marconi for the true threat he was, he could not possibly afford—as Edison and Westinghouse had—to send in battalions of lawyers. He could barely pay for his expensive life at the Waldorf-Astoria.

By January 2, 1902, Tesla needed more money if he was to complete his electrical tower and was pressing Morgan accordingly: "Now, Mr. Morgan, am I backed by the greatest financier of all times? And shall I lose great triumphs and an immense fortune because I need a sum of money!!"[39] But Morgan could not see why he should pour more money into such a huge and expensive enterprise for wireless communication when Guglielmo Marconi had already succeeded with much less. Tesla explained the far more grandiose plans he had—the almost finished tower was not for mere prosaic transatlantic telegraphy but was a gigantic transmitter that would be capable of generating 10 *million* horsepower to straddle the globe with wireless communication and cheap electric power. "Will you help me or let my great work—almost complete—go to pot?" he asked.

Morgan was not impressed, just deeply irritated that he had been snookered into squandering his money on visionary schemes. Not only was he adamant that "I should not feel disposed at present to make any further advances," he was also unwilling to return to Tesla his patents, thereby preventing him from finding others to back him commercially.[40] When Tesla humbly presented himself at Morgan's Wall Street office in January of 1904 to "show you that I have done the best that could be done, you fire me out like an office boy and roar so that you are heard six blocks away; not a cent. It is spread all over town, I am discredited, the laughing stock of my enemies."[41] (We don't know if Tesla even approached Westinghouse for money for Wardenclyffe, but it seems likely that he did and was rebuffed there, also.)

Tesla's biographers have all wondered whether J. P. Morgan, with his big investment in GE, deliberately sabotaged Tesla in his dream of developing a superior system of wireless energy. Morgan certainly had huge stakes in still expanding electrical power systems all over America (and elsewhere in the world). He drove a hard bargain to obtain control of Tesla's patents and then would not relinquish them for development by someone else. Morgan might have had compelling reasons to thwart a new, untried, and possibly revolutionary technology, one that would make obsolete Morgan's own efforts and giant investments. But by this time the ever idealistic, ever naive Tesla also had a terrible track record—with the notable exception of his Westinghouse AC collaboration—of turning his brilliant ideas and inventions into commercially viable products. So the hard-nosed Morgan may simply have felt that he had paid good money and had nothing to show for it but the possibly valuable patents.

The desperate Tesla wrote Morgan long, lovelorn letters, piteous pleas that enumerated for page after page the reasons the giant of Wall Street should recommit to their brief, unfinished relationship. Tesla just did not seem to grasp that a tough-hearted titan like Morgan was not likely to be moved by such flowery confessions as "There has been hardly a night when my pillow was not bathed in tears, but you must not think me weak for that, I am perfectly sure to finish my task, come what may."[42] This elicited a reply from Morgan's private secretary that "Mr. J. P. Morgan wishes me to inform you that it will be impossible for him to do anything more in the matter."[43] Tesla consoled himself by writing to the indifferent Morgan that the financier's work was "wrought in passing form. Mine is immortal."

To this day, Wardenclyffe remains a scientific mystery. How exactly did Tesla plan to carry out his incredibly ambitious plan? Did he possess technology that would really work and do what he claimed? There had been an enormous amount of practical work and gigantic financial investments required to make Tesla's AC motors work in the real world, necessary development work that he often airily overlooked. Not only could Tesla not afford to move forward on his almost finished tower, he could not even afford the lawyers he desperately needed to defend his many patents, depriving him of all sorts of royalties. A loner by nature, unattached to the power and prestige of a great university or a major corporation, Tesla was at a total disadvantage. By 1905, he could no longer afford the laboratory out at Wardenclyffe. Over the next decade, he poured his inventive energy into turbines, but lack of money and the skepticism of his peers meant that yet again these machines never became fully developed commercial products. After J. P. Morgan died in 1913, Tesla soon began appealing to his financier son. Just before Christmas of 1913, he wrote, "I need money badly and I cannot get it in these dreadful times."[44] The younger Morgan took pity and sent Tesla $5,000 at the Waldorf-Astoria.

Four years later, when the American Institute of Electrical Engineers bestowed its 1917 Edison Medal upon Tesla, the main speaker pointed out, "Were we to seize and eliminate from our industrial world the results of Mr. Tesla's work, the wheels of industry would cease to turn, our electric cars and trains would stop, our towns would be dark, our mills would be dead and idle. Yes, so far reaching is this work, that it has become the warp and woof of industry."[45] Yet in the previous year, 1916, the man whose alternating current inventions literally powered the modern world had had to declare bankruptcy. Unable to pay his bills, Tesla eventually had to relinquish the mortgage for his beloved (and never completed) Wardenclyffe tower to the Waldorf-Astoria, where he had lived since 1898 and was $20,000 in arrears on his bill. In an effort to make the land easier to sell, the hotel blew up the tower.

George Westinghouse had died in 1914, but Tesla continued to ply the electric company with ideas in the hopes that they would once again develop his inventions, especially when they entered the new field of radio. In 1920, Tesla was still offering his services, "provided your company is willing to come to an understanding with me on terms decidedly more generous than those under which they acquired my system of power transmission thirty years ago."[46] Again and again, the company politely said no. In 1930, Tesla became convinced that the Westinghouse people had pirated some of his early transmission patents and threatened in a letter, "It would be painful to me to have to resort to legal proceedings against a great corporation whose business is largely founded on my inventions."[47] Sadly broke and understandably bitter, Tesla wrote a letter to the *New York World* complaining, "Had the Edison companies not finally adopted my invention they would have been wiped out of existence, and yet not the slightest acknowledgment of my labors has ever been made by any of them, a remarkable instance of the proverbial unfairness and ingratitude of corporations."[48] This proud, brilliant eccentric continued to earn fees occasionally as a consultant, but he did not do well in corporate settings. He wrote articles from time to time, often nostalgic pieces about his role in the rise of electricity.

Yet so great were Nikola Tesla's early contributions and so prescient did his many "fantastical" predictions prove as the years and decades unfolded, the aging inventor always had a steady and loyal coterie of admirers who sought him out, young science writers like John J. O'Neill, up-and-coming scientists inspired by his work and writings, and old comrades from the Westinghouse days like Charles Scott, eminent professor of engineering. They made sure he received some honors, organizing on the occasion of his seventy-fifth birthday in 1931 a Festschrift that hailed his great contributions. *Time* magazine featured Tesla on the cover and celebrated him as an eccentric genius for the ages. "I have been leading a secluded life, one of continuous, concentrated thought and deep meditation," he was quoted as saying. "Naturally enough I have accumulated a great number of ideas. The question is whether my physical powers will be adequate to working them out and giving them to the world."[49]

The Westinghouse Electric & Manufacturing Company—worried about the bad publicity of an elderly, impoverished Tesla—decided in 1934 to begin paying his monthly bill at his new residence, room 3327 (divisible by three) on the thirty-third floor of the new skyscraper Hotel New Yorker. It was a piddling and insulting sum—$125 a month—for one who had given up so many millions to save the company long ago. In contrast, GE paid the inventor William Stanley, another elderly electrical pioneer on hard times, a stipend of $1,000 a month. The Yugoslav government also began to contribute a small pension. Tesla continued to invent, and when he announced patenting a new "death ray" weapon at the start of World War II, the press gave it great play. All told, Tesla had 111 American patents, and there were apparently many other inventions he never bothered to register. But lacking any major backer, most of his inventions remained either completely theoretical or were never fully developed for real commercial use.

More and more, Tesla lived in his own world, as big a romantic as ever, and as eccentric. He had his vegetarian meals specially cooked by the hotel chef and insisted that the help not get closer to him than a few feet, part of his phobia of germs. He had finally abandoned wearing his old-fashioned frock coats and his kid gloves for a regular well-tailored business suit. Nikola Tesla in his later years developed a strange passion for pigeons. Never having married and with all his relatives back in Europe, he found solace and a certain familial companionship with these cooing, waddling birds. He had long outlived most of his old friends, Robert Johnson, Stanford White, Mark Twain. So he took to whiling away the hours by feeding and talking to the pigeons outside the New York Public Library and St. Patrick's Cathedral, often late at night. If he found a bird that was sick or injured, he rescued it and smuggled it into his hotel room to nurse it back to health. A particularly elegant white pigeon was deeply beloved.

One evening the elderly, almost cadaverous Nikola Tesla told John J. O'Neill about the white pigeon while they were sitting and visiting in the lobby of the Hotel New Yorker. Tesla said, "I loved that pigeon. Yes, I loved her as a man loves a woman, and she loved me. . . . As long as I had her, there was a purpose in my life. Then one night as I was lying in my bed in the dark, solving problems, as usual, she flew in through the open window and stood on my desk. . . . As I looked at her I knew she wanted to tell me—she was dying. . . . When that pigeon died, something went out of my life. Up to that time I knew with a certainty that I would complete my work, no matter how ambitious my program, but when that something went out of my life I knew my life's work was finished."[50]

On January 7, 1943, as snow tumbled past his room on the thirty-third floor of the Hotel New Yorker and World War II raged across the globe, Nikola Tesla died in his bed, age eighty-six, alone and impecunious. Though Tesla had long been an American citizen, the Yugoslav government made sure his funeral was a grand occasion at the magnificent Cathedral of St. John the Divine up in Morningside Heights. Two thousand mourners came on January 12 to pay their respects to this electrical genius. The U.S. government secretly confiscated some of his papers, concerned that they contained potentially important scientific material.

Sadly, Tesla died too soon to savor one last wonderful vindication. The whole world believed Marconi to be the father of radio. Yet when Marconi brought suit against the U.S. government for infringing on his radio patents, the U.S. Supreme Court ruled that the record showed that Marconi's patents infringed on those of Tesla! Though here, as with most of his inventions, Tesla had not done the hard labor of creating a commercially viable product.

To this day, Nikola Tesla remains a brilliant but enigmatic figure, a scientist, inventor, dreamer, and visionary whose post-Niagara scientific contributions are the source of great debate. Did he degenerate into a kook, or was he just decades ahead of his time? Electricity had created many, many millionaires. But Tesla, who made possible the electric age, was never one of them. Still, he did live to see his AC system straddle the globe, illuminating nation after nation and powering millions of motors. Almost sixty years after he had stepped ashore in New York, dreaming a big dream of electrifying the world, that dream had more than come to pass.

ELECTRICITY

And what of electricity itself, that "subtle, vivifying fluid"? As electric power became more versatile, more reliable, more prosaic, the most far-fetched electrical dreams of these three Promethean creators—Edison, Westinghouse, and Tesla—were fulfilled. The Niagara powerhouses added generator after generator and even in 1902 were providing a fifth of all electricity in the United States. As described in historian David E. Nye's book *Electrifying America,* by 1910 American business and industry were eagerly incorporating electricity into their daily operations, dramatically raising productivity. By 1940, when electricity had spread throughout society, American productivity had risen 300 percent. Edison biographer Matthew Josephson calculated that assembly-line plants such as those run by automaker Henry Ford saw efficiency rise 50 percent. While electricity was by no means the only explanation, it played an important part in the tremendous ensuing rise in living standards. The cost of electricity declined steadily as it was generated far more efficiently. In 1902, 7.3 pounds of coal were needed to generate 1 kilowatt-hour in a central station. By 1932, that figure was down to 1.5 pounds of coal.[51]

Residential electrical service, however, spread far more slowly. This was partly because power companies concentrated first on wooing and supplying more profitable business customers. But it was also because in the early years electricity was still a luxury, more expensive than gas, and not as reliable. Nationwide, in 1907, only 8 percent of Americans lived in homes served by electricity. By 1920, that figure had risen, but only to 35 percent. As electric service improved and costs dropped, Americans signed up for electricity in their homes as eagerly as had business. In major cities like Chicago, 95 percent of homes had electricity by the 1920s.

To appreciate the universality of electricity in American *urban* homes by 1930, one has only to consider that even in a midsize midwestern city like Muncie, Indiana, 95 percent of families had electricity. This was true even though more than a third still had no bath and a fifth still used outhouses. Yet virtually every Muncie household opted for the electricity that allowed one to sit for hours during the darkest evenings in a bright room and suffer no strained eyes or gas-induced headaches. Equally persuasive were such labor-saving conveniences as electric irons, vacuums (750,000 sold in 1919), washing machines, toasters, and hot-water heaters. Just thirty years after Edison figured out how to build the first power grid, electricity was changing the daily tenor of life all over urban America. Many a wife happily dispensed with the maid and did the now easier job of housekeeping herself.

For all the convenience of lamps and washing machines, probably no early electrical appliance brought such change (and pleasure) into the American home as the radio in the early 1920s. For the first time, just as Tesla had once predicted, people could tune in to the world far distant, receiving new voices and sounds from a big wooden box. Now, when their ball team played out of town, or the president gave a speech, or a famous soprano sang in New York, Americans could follow these events. The soap operas kept them enthralled, comedy hours lightened their days. Was it surprising that within a decade almost every American family (with electricity) had a radio? It would probably be hard to overstate the influence of radio in both knitting together the nation and inspiring many a small-town boy or girl to more exciting ambitions. Electrical power made possible the important and the frivolous, the noble and the idiotic.

Because electricity was viewed strictly as a commercial commodity in these early decades, access to electricity came far more slowly for citizens out in the hinterlands, where a third of Americans still lived in the 1930s. As late as 1934, only a tenth of the nation's farms had electricity, for service was determined by profitability. It was President Franklin Delano Roosevelt who first proposed the radical notion that isolated farmers were as entitled to the liberating benefits of electricity as Americans living in cities and suburbs. The private utilities, which had long seen electricity strictly as an economic proposition, balked. The New Deal stepped forward with government-backed rural electrification projects. Even so, it would take twenty-five years for the high-tension wires carrying electricity to straddle the hills and prairies into even the most remote farms, finally ending their long dependence on tallow candles and kerosene lamps. During these decades, as electricity became the invisible lifeblood of our modern civilization, its price dropped almost by half. Farmers were as quick as the businessmen and housewives to put electricity to work doing all kinds of drudgery.

For all the immense gains and almost magical gifts that electricity bestowed, there were, of course, some small losses, felt only by a few. The world, now powered by machines, became far noisier. The hum, rumble, and roar of motors and engines, the persistent sound of electrically amplified noises, all came to fill the air. Natural sounds, just plain silence, were drowned out by man-made din. The once ubiquitous horse largely disappeared, no longer needed for mobility or sheer horsepower. The American night sky, once truly black and blazing with billions of glistening stars, decade by decade became steadily more permeated by man-made electric light, until now, when the sun sets and true darkness descends, it is not inky black, but an orange gray color, especially just above the horizon. In consequence, the stars, especially the Milky Way, are difficult and often impossible to see. In *Electrifying America,* one man laments, "We of the age of machines, having delivered ourselves of nocturnal enemies, now have a dislike of night itself. . . . Today's civilization is full of people who have not the slightest notion of the character or the poetry of the night, who have never even seen night."[52] Electricity allowed a greater regimentation of life, ripping away the natural rhythms of time and season. The quieting of work and home as the natural light disappeared no longer existed, nor did families gather about the hearth for heat and light. Men, women, and children retreated more and more into their comfortable and convenient homes or became more and more obliged to toil on in their well-lighted offices and factories.

The rise of such electrical machines as radio, movies, television, videos, computers, and the Internet also meant the rise of life at a remove, life experienced passively through a screen, the watching and observing of activities organized and presented by others. Before electricity, men, women, and children who wanted to enjoy music, theater, dance, politics, speeches, or sports were present as active participants or spectators. Nineteenth-century human experience was firsthand, *in person,* intimate, authentic, with the notable exception of that described by the written word and still images, whether paintings or photographs. With electricity came a veritable cornucopia of possible new human experiences, far more than most modern people can begin to absorb. But many of them are experienced secondhand.

Nikola Tesla lived to see his great invention, his great gift to mankind, spread across the landscape, brightening homes, enlivening communities, and enriching the whole country, just as he had hoped. Despite the travails and frustrations of his later life, Tesla, ever the idealist, said: "I continually experience an inexpressible satisfaction from the knowledge that my polyphase system is used throughout the world to lighten the burdens of mankind and increase comfort and happiness." The rise of the railroads and the telegraph changed forever age-old notions of distance and time; the steam engine had already hinted at the potential of machine-created energy. Electricity unleashed the Second Industrial Revolution, bestowing on man incredible gifts: the untold hours once lost to simple darkness, the even greater hours lost to drudging human labor, and the consequent freeing and flourishing of the human mind and imagination. Even acknowledging a loss of nineteenth-century intimacy and authenticity, the intrusion of perpetual din, and the mechanized rush of modern life, the coming of electricity ultimately expanded the whole human sense of time, energy, and possibility. Great, indeed, has been the power of electricity.

Acknowledgments

In the course of researching and writing this book, I have been helped by many people. First, I would like to thank my agent, Eric Simonoff at Janklow & Nesbit, who offered excellent advice every step of the way, and Katie Hall, my editor at Random House, who provided superlative editing and all-around energy and intelligent enthusiasm.

Paul Israel at the Thomas Edison Papers at Rutgers University responded to many questions, helping me and Sarah McKinney of Random House navigate the vast and wondrous Edison on-line archive. He also carefully read and commented upon my manuscript. Both Israel's excellent biography, *Edison: A Life of Invention,* and his fascinating work with Robert Friedel and Bernard S. Finn, *Edison's Electric Light: Biography of an Invention,* were indispensable to this book. Over at the Edison National Historic Site in West Orange, New Jersey, Doug Tarr was most helpful with photographs. At the Smithsonian Institution, Bernard S. Finn, and Harold Wallace at the Museum of American History answered numerous queries and kindly demonstrated some of the early electrical apparatus in their collections.

Robert F. Dischner, Ph.D., of USA Service Company, Inc., stands out for his wonderful enthusiasm for this project and his constant support. He came from Syracuse to serve as a personal guide around

Buffalo and Niagara Falls, introducing me to Maureen Fennie, head of the local history room at the Niagara Falls Public Library. The librarians at the Buffalo & Erie County Historical Society were also efficient and most helpful during my visit to that institution. Dischner provided invaluable access to the original Niagara Power archives and the papers of Edward Dean Adams. National Grid USA archivist Joseph Santore did a yeoman's job of locating files, letters, and photographs.

Out in Pittsburgh, Edward Reis, head of the George Westinghouse Museum in Wilmerding, arranged for me to spend two days in their archives there, which yielded invaluable material. My daughter Hilary was my first-rate assistant there. Professor Tim Ziaukas of the University of Pittsburgh's Bradford campus smoothed the way for that trip and then shared what he had found in his own work in the Westinghouse files. He then kindly reviewed my manuscript. Richard Price at the Historical Society of Western Pennsylvania performed useful research there on my behalf. In Boston, Debbie Funkhouser carefully combed the historical files of the Harvard Business School's Baker Library and the Massachusetts Historical Society, looking for Westinghouse material. In New York, Sarah McKinney of Random House spent many hours gathering old newspaper coverage about the electric chair battle and mining the on-line Edison archives for all manner of important documents. Also in New York, Christine Nelson at the Morgan Library provided needed manuscripts from their archives. In Philadelphia, Valerie-Anne Lutz of the American Philosophical Society quickly sent along the Niagara Falls–related papers of Coleman Sellers. Christopher Baer of the Hagley Museum and Library outside Wilmington, Delaware, rapidly answered all queries having to do with the Pennsylvania Railroad. At Random House, Danielle Durkin was very helpful in the final stages of the book.

In Washington, D.C., the librarians at the Library of Congress's Division of Manuscripts were most friendly when assisting me with the Tesla papers. Filmmaker Robert Uth, whose two-part documentary on Tesla ran on PBS, was very helpful in putting me in touch with the Nikola Tesla Museum in Belgrade and other Tesla groups when I was looking for photos. Professor W. Bernard Carlson, who has been working on a new Tesla biography, generously shared an article in draft and had a number of useful discussions as I was getting started. Here in Baltimore, I must once again thank the librarians at Johns Hopkins University's Milton S. Eisenhower Library for all their stellar efforts, especially those in the Inter-Library Loan Office. I could not write my books without their good-natured, informed help. Also here in Baltimore, I was fortunate to visit the incomparable Light Bulb Museum, where one can see original Edison bulbs and Westinghouse stopper lights.

I would also like to offer heartfelt appreciation to the numerous people who took the time to read and critique my manuscript. They were Paul Israel; Robert Dischner; Tim Ziaukas; my husband, Christopher Ross; my uncle, inventor and entrepreneur Nelson Jonnes; IBM executive (and onetime science teacher) Victor Romita; and New York City history buff Robert Sarlin. All had many useful comments that helped improve the manuscript. I am grateful to Christopher Buck, whose early enthusiasm for this story fired my own, and to Deborah Buck for her delightful hospitality. Finally, special thanks to my father-in-law, artist John Ross, who drew such excellent and useful diagrams to explain the mysteries of electricity.

Bibliography

I am deeply indebted to the numerous historians and biographers who have written so well on early electricity, whether their subject was the scientists and inventors who blazed those first pioneering paths or the electrical industry itself.

Many biographies have been written about Thomas A. Edison, but the most recent (1998) and most useful for me was Paul Israel's *Edison: A Life of Invention,* which concentrates on the science and business. Paul Israel also coauthored with Robert Friedel and Bernard S. Finn *Edison's Electric Light: Biography of an Invention* (1986), which provided wonderful specifics about the momentous and arduous invention of the light bulb. Francis Jehl's three-volume *Menlo Park Diaries* (1937) serves up an incredible wealth of engaging detail but is not always reliable. Matthew Josephson's *Edison: A Biography* (1959), the first to draw on the large and amazing Edison archives, is a good read and gives a nice flavor of the era. About the time I began work on this book, almost half of the vast Rutgers University Edison archives, a cornucopia of invaluable primary source material, became available on-line.

For Nikola Tesla, John J. O'Neill's biography, *Prodigal Genius,* was a great introduction, authored as it was by a science writer who knew Tesla. It was published in 1944 shortly after Tesla's death. Marc Seifer's helpful 1996 biography, *Wizard: The Life and Times of Nikola*

Tesla, was able to draw on much material unavailable to O'Neill. Other important sources were Tesla's letters in the Library of Congress and the Niagara Mohawk archives. There are numerous useful published collections of Tesla's lectures, writings, and patent applications, as well as articles about him, often labors of love by Tesla's many dedicated fans. Margaret Cheney and filmmaker Robert Uth wrote *Tesla: Master of Lightning* (1999), an excellent heavily illustrated coffee table biography that is the companion book to Uth's PBS documentary of the same title.

Unfortunately, no new biography has been written of George Westinghouse since Henry G. Prout's *A Life of George Westinghouse* (1926) was commissioned by the Westinghouse Corporation, which apparently felt Francis E. Leupp's *George Westinghouse: His Life and Achievements* (1918) was inadequate. Anyone contemplating a new biography of the great Pittsburgh magnate would have quickly been discouraged by the incredible paucity of primary sources, a problem even Prout and Leupp complained about. Westinghouse did not, like Edison and Tesla, preserve his personal letters and papers, nor did he make himself available to the press. However, when business historian Harold C. Passer wrote his magnificent classic work, *The Electrical Manufacturers, 1875–1900* (1953), he was able to elaborate on many fascinating Westinghouse episodes using what he called the "Historic Westinghouse Archives" at the company's Pittsburgh headquarters. Tellingly, Passer, a Harvard Business School scholar, appears to be the only historian ever given unfettered access to those archives. Thereafter, all access was heavily vetted. Sadly, when CBS dismantled the Westinghouse Corporation in the late 1990s, the historic archives seemed to disappear. However, fragments remain in a small archive at the George Westinghouse Museum in Wilmerding outside Pittsburgh, where one can find a large and well-organized trove of reminiscences about Westinghouse gathered and organized in the 1930s.

David E. Nye's *Electrifying America: Social Meanings of a New Technology* (1990) was most illuminating on the national experience as light and power spread, while Andreas Bluhm and Louise Lippincott's *Light! The Industrial Age 1750–1900* (2000), a sumptuous art catalog to a 2001 show by the same name in Pittsburgh, provided a fascinating look at how societies and artists absorbed these new technologies.

When writing about New York City, I found Edwin G. Burrows and Mike Wallace's wonderful *Gotham* (1999) and Kenneth T. Jackson's incredible *Encyclopedia of New York City* (1995) invaluable. For Pittsburgh, I was grateful to find Stefan Lorant's great work, *Pittsburgh* (1964). For Chicago, I turned to Donald L. Miller's excellent *City of the Century* (1996), and for Niagara Falls, to Pierre Berton's amusing *Niagara* (1992) and the *Anthology and Bibliography of Niagara Falls* compiled by Charles Mason Dow (1921). Other important Niagara sources were Robert Belfield's 1981 Ph.D. thesis, "Niagara Frontier," and the archives Edward Dean Adams carefully assembled to document the building of the Niagara hydroelectric power plants. These archives, preserved by the Niagara Mohawk Power Corporation in Syracuse (since acquired by National Grid USA Company, Inc.), are a necessary antidote to Edward Dean Adams's official two-volume history, *Niagara Power* (1927), which for all its majesty is still often confusing and sanitizes or simply ignores all troubles.

Notes

CHAPTER 1

"Morgan's House Was Lighted Up Last Night"

1. James D. McCabe, *New York by Sunlight and Gaslight* (Philadelphia: Douglass Bros., 1882), p. 332.
2. Letter from Major Sherbourne Eaton to Thomas A. Edison dated June 8, 1882. Thomas A. Edison Archives website, Rutgers University.
3. George William Sheldon, *Artistic Houses* (New York: D. Appleton and Company, 1883), pp. 75–80. Many of the photos reprinted by Dover Books as *Opulent Interiors of the Gilded Age.*
4. Herbert Satterlee, *J. Pierpont Morgan: An Intimate Portrait* (New York: Macmillan, 1940), p. 208.
5. Letter from J. P. Morgan to Mr. James Brown dated December 1, 1882. Morgan Library Archives, New York, New York.
6. Letter from J. P. Morgan to Sherbourne Eaton dated December 27, 1882. Morgan Library Archives, New York, New York.
7. Satterlee, *J. Pierpont Morgan,* p. 208.
8. Edward H. Johnson, "Personal Recollections of Mr. Morgan's Contribution to the Modern Electrical Era," November 1914. Morgan Library Archives, New York, New York.
9. Satterlee, *J. Pierpont Morgan,* p. 214.
10. Ibid., p. 216.
11. "The Doom of Gas," *St. Louis Post-Dispatch,* May 1, 1882. Item #2192 in Charles Batchelor Scrapbook (1881–1882), Thomas A. Edison Archives website, Rutgers University.

12. Frank L. Dyer and T. Commerford Martin, *Edison: His Life and Inventions* (New York: Harper & Bros., 1910), p. 374.
13. "The Doom of Gas."

CHAPTER 2
"Endeavor to Make It Useful"

1. J. L. Heilbron, *Electricity in the 17th and 18th Centuries* (Berkeley: University of California Press, 1979), p. 171.
2. Ibid., p. 244.
3. Ibid., p. 314.
4. Ibid., p. 318.
5. Carl Van Doren, *Benjamin Franklin* (New York: Viking Press, 1938), p. 158.
6. Marcello Pera, *The Ambiguous Frog* (Princeton, N.J.: Princeton University Press, 1992), p. 15.
7. Van Doren, *Benjamin Franklin,* p. 158.
8. Ibid., p. 159.
9. Ibid., p. 159.
10. Ibid., p. 161.
11. Pera, *The Ambiguous Frog,* p. 14.
12. Ibid., p. 18.
13. Sanford P. Bordeau, *Volts to Hertz* (Minneapolis: Burgess Pub. Co., 1982), p. 47.
14. Ibid.
15. Hal Hellman, *Great Feuds in Medicine* (New York: Wiley, 2001), p. 25.
16. Ibid., p. 27.
17. Bordeau, *Volts to Hertz,* p. 53.
18. John M. Thomas, *Michael Faraday and the Royal Institution* (New York: Adam Hilger, 1991), p. 5.
19. Bordeau, *Volts to Hertz,* p. 64.
20. Thomas, *Michael Faraday and the Royal Institution,* p. 16.
21. Ibid., p. 40.
22. John Tyndall, *Faraday as a Discoverer* (New York: Thomas Crowell, 1961), pp. 22–23.
23. Thomas, *Michael Faraday and the Royal Institution,* p. 43.
24. Herbert W. Meyer, *A History of Electricity and Magnetism* (Cambridge: MIT Press, 1971), p. 61.
25. Thomas, *Michael Faraday and the Royal Institution,* p. 96.
26. David Gooding and Frank A. J. L. James, *Faraday Rediscovered* (London: Stockton Press, 1985), p. 63.
27. Ibid.
28. Thomas, *Michael Faraday and the Royal Institution,* p. 1.
29. Ibid., p. 121.
30. Robert Louis Stevenson, "A Plea for Gas Lamps," in Mike Jay and Michael Neve, *1900* (New York: Penguin, 1999), p. 58.

31. Wolfgang Schivelbusch, *Disenchanted Night: The Industrialization of Light in the Nineteenth Century* (Berkeley: University of California Press, 1988), p. 221.
32. Bordeau, *Volts to Hertz,* p. 147.
33. "Electric Lamps," *St.-Louis Daily Globe Democrat* (n.d.), sent on August 23, 1878, to Thomas Edison by Professor George Barker. Thomas A. Edison Archives website, Rutgers University.
34. Stevenson, "A Plea for Gas Lamps," p. 59.
35. Paul Israel, *Edison: A Life of Invention* (New York: Wiley, 2000), p. 165.
36. Matthew Josephson, *Edison: A Biography* (New York: McGraw-Hill, 1959), p. 178.
37. Israel, *Edison: A Life of Invention,* p. 166.

CHAPTER 3

Thomas Edison: "The Wizard of Menlo Park"

1. Paul Israel, *Edison: A Life of Invention* (New York: Wiley, 2000), p. 120.
2. Matthew Josephson, *Edison: A Biography* (New York: McGraw-Hill, 1959), p. 134.
3. Israel, *Edison: A Life of Invention,* p. 11.
4. Ibid., p. 17.
5. Charles D. Lanier, "Two Giants of the Electric Age," *Review of Reviews* 8 (1893), p. 48.
6. *New York Daily Graphic,* April 2, 1878.
7. Israel, *Edison: A Life of Invention,* p. 165.
8. "Edison's Newest Marvel," *New York Sun,* September 16, 1878.
9. Israel, *Edison: A Life of Invention,* pp. 169–70.
10. Editorial Page, *New York Sun,* September 16, 1878.
11. Robert Friedel and Paul Israel with Bernard S. Finn, *Edison's Electric Light: Biography of an Invention* (New Brunswick, N.J.: Rutgers University Press, 1986), p. 22.
12. Ibid., p. 26.
13. Francis Jehl, *Menlo Park Reminiscences,* vol. 1 (Dearborn, Mich.: Edison Institute, 1937), p. 232.
14. Ibid., p. 197.
15. Josephson, *Edison: A Biography,* p. 196.
16. "Edison's Electric Light," *New York Herald,* March 27, 1879, Thomas A. Edison Archives website, Rutgers University.
17. Friedel and Israel, *Edison's Electric Light,* p. 71.
18. Israel, *Edison: A Life of Invention,* p. 182.
19. Friedel and Israel, *Edison's Electric Light,* p. 101.
20. Jehl, *Menlo Park Reminiscences,* vol. 1, p. 338.
21. Friedel and Israel, *Edison's Electric Light,* p. 109.
22. Ibid., p. 111.
23. Israel, *Edison: A Life of Invention,* p. 167.

24. Ibid., p. 196.
25. Friedel and Israel, *Edison's Electric Light,* p. 179.
26. Josephson, *Edison: A Biography,* p. 231.
27. Friedel and Israel, *Edison's Electric Light,* p. 180.
28. Ibid., p. 181.
29. George S. Bryan, *Edison: The Man and His Work* (New York: Knopf, 1926), p. 152.
30. Josephson, *Edison: A Biography,* pp. 244–45.
31. "The Aldermen Visit Edison," *The New York Times,* December 22, 1880, p. 2.
32. Friedel and Israel, *Edison's Electric Light,* p. 182.
33. "The Aldermen Visit Edison."
34. "Moonlight on Broadway," *New York Evening Post,* December 21, 1880.
35. "Lights for a Great City," *The New York Times,* December 22, 1880, p. 2.
36. Josephson, *Edison: A Biography,* p. 248.
37. George T. Ferris, ed., *Our Native Land* (New York: Appleton, 1886), p. 551.
38. Josephson, *Edison: A Biography,* p. 252.
39. Ibid., pp. 263–64.
40. "Summary of Events for 1881," *Index to the New-York Daily Tribune* (New York: Daily Tribune, 1882), p. v.
41. Iza Duffus Hardy, *Between Two Oceans* (London: Hurst & Blackwood, 1884), pp. 94–95.
42. "Summary of Events for 1882," *Index to the New-York Daily Tribune* (New York: Daily Tribune, 1883), p. vii.

CHAPTER 4

Nikola Tesla: "Our Parisian"

1. Nikola Tesla, *My Inventions: The Autobiography of Nikola Tesla* (Williston, Vt.: Hart Bros., 1982), p. 66. Originally a series in *Electrical Experimenter* magazine, 1919.
2. Ibid.
3. Neil Baldwin, *Edison: Inventing the Century* (New York: Hyperion, 1995), p. 132.
4. Tesla, *My Inventions,* p. 36.
5. Ibid., p. 29.
6. Ibid., p. 31.
7. John T. Ratzlaff, ed., *Tesla Said* (Millbrae, Calif.: Tesla Book Company, 1984), p. 284 ("A Story of Youth Told by Age").
8. Tesla, *My Inventions,* p. 54.
9. Ibid., p. 57.
10. Ibid., pp. 59–60.
11. Ibid., p. 61.
12. John J. O'Neill, *Prodigal Genius: The Life of Nikola Tesla* (New York: McKay, 1944), p. 49.

13. Tesla, *My Inventions,* p. 65.
14. Andreas Bluhm and Louise Lippincott, *Light! The Industrial Age 1750–1900* (New York: Thames & Hudson, 2000), p. 31.
15. Paul Israel, *Edison: A Life of Invention* (New York: Wiley, 1998), p. 214.
16. Tesla, *My Inventions,* p. 66.
17. Ibid., p. 67.
18. W. Bernard Carlson, *Innovation as a Social Process: Elihu Thomson and the Rise of General Electric, 1870–1900* (New York: Cambridge University Press, 1991), p. 206.
19. Tesla, *My Inventions,* p. 67.
20. Ibid.
21. Ibid., p. 70.
22. Ibid., p. 34.
23. Ibid., p. 70.
24. George T. Ferris, ed., *Our Native Land* (New York: Appleton, 1886), p. 554.
25. Tesla, *My Inventions,* p. 71.
26. Ratzlaff, *Tesla Said,* p. 280 ("Letter to the Institute of Immigrant Welfare" dated May 12, 1938).
27. Tesla, *My Inventions,* p. 72.
28. O'Neill, *Prodigal Genius,* p. 62.
29. Edwin G. Burrow and Mike Wallace, *Gotham: A History of New York City to 1898* (New York: Oxford University Press, 1999), p. 1152.
30. O'Neill, *Prodigal Genius,* p. 64.
31. Alfred O. Tate, *Edison's Open Door* (New York: E. P. Dutton, 1938), pp. 146–47.
32. "Tesla Says Edison Was an Empiricist," *The New York Times,* October 19, 1931, p. 25.
33. Ibid.
34. Marc Seifer, *Wizard: The Life and Times of Nikola Tesla* (Secaucus, N.J.: Birch Lane Press, 1996), p. 41; Tesla ad, *Electrical Review,* September 14, 1886, p. 14.
35. Ratzlaff, *Tesla Said,* p. 280.
36. Ibid.
37. Tesla, *My Inventions,* p. 72.
38. "Summary of Events for 1886," *Index to the New-York Daily Tribune for 1886* (New York: Daily Tribune, 1887), pp. iv–v.
39. Ratzlaff, *Tesla Said,* p. 280.
40. Seifer, *Wizard,* p. 40.
41. Matthew Josephson, *Edison: A Biography* (New York: McGraw-Hill, 1959), p. 340.
42. Israel, *Edison: A Life of Invention,* p. 254.
43. Ibid.
44. O'Neill, *Prodigal Genius,* p. 65.
45. H. Gernback, "Tesla's Egg of Columbus," *Electrical Experimenter,* March 1919, p. 775.
46. Ibid.

CHAPTER 5

George Westinghouse: "He Is Ubiquitous"

1. Steven W. Usselman, "From Novelty to Utility: George Westinghouse and the Business of Innovation during the Age of Edison," *Business History Review* 66, no. 2 (1992), p. 287.
2. Ibid., p. 289.
3. Guido Pantaleoni, "The Real Character of the Man as I Saw Him," April 1939, p. 5. George Westinghouse: Anecdotes and Reminiscences, vol. 3, box 1, file folder 8, George Westinghouse Museum Archives, Wilmerding, Pennsylvania.
4. George Wise, "William Stanley's Search for Immortality," *Invention & Technology* (summer–spring 1988), p. 43.
5. "The Stanley and Thomson Incandescent Lamp," *Electrical World,* September 27, 1884, p. 118.
6. Henry G. Prout, *A Life of George Westinghouse* (New York: Scribner's, 1926), p. 5.
7. Francis E. Leupp, *George Westinghouse: His Life and Achievements* (Boston: Little, Brown, 1918), p. 287.
8. Prout, *A Life of George Westinghouse,* p. 293.
9. C. A. Smith, "A Lesson Without Words," June 1939, p. 2. George Westinghouse: Anecdotes and Reminiscences, vol. 4, box 1, file folder 9, George Westinghouse Museum Archives, Wilmerding, Pennsylvania.
10. Leupp, *George Westinghouse,* p. 294.
11. Thomas P. Hughes, *Networks of Power: Electrification in Western Society, 1880–1930* (Baltimore: Johns Hopkins University Press, 1983), p. 94.
12. Harold C. Passer, *The Electrical Manufacturers, 1875–1900: A Study in Competition, Entrepreneurship, Technical Change, and Economic Growth* (Cambridge: Harvard University Press, 1953), p. 132.
13. "A Nation at a Tomb," *The New York Times,* August 9, 1885, p. 1:1.
14. Reginald Belfield, "Westinghouse and the Alternating Current," 1935–1937. George Westinghouse: Anecdotes and Reminiscences, vol. 1, box 1, file folder 3, George Westinghouse Museum Archives, Wilmerding, Pennsylvania.
15. Stefan Lorant, *Pittsburgh* (New York: Doubleday & Co., 1964), p. 168.
16. Joseph Frazier Wall, *Andrew Carnegie* (Pittsburgh: University of Pittsburgh Press, 1989), p. 386.
17. George T. Ferris, ed., *Our Native Land* (New York: Appleton, 1886), pp. 517–18.
18. Prout, *A Life of George Westinghouse,* p. 302.
19. Adelaide Nevin, *The Social Mirror* (Pittsburgh: T. W. Nevin, 1888), p. 93.
20. Pantaleoni, "The Real Character of the Man as I Saw Him."
21. Usselman, "From Novelty to Utility," p. 272.
22. Wise, "William Stanley's Search for Immortality," p. 44.
23. Ibid., p. 43.
24. Ibid., p. 44.
25. Belfield, "Westinghouse and the Alternating Current."

26. Bernard A. Drew and Gerard Chapman, "William Stanley Lighted a Town and Powered an Industry," *Berkshire History* 6, no. 1 (fall 1985), p. 8.
27. Ibid., p. 10.
28. Passer, *The Electrical Manufacturers,* p. 133.
29. Belfield, "Westinghouse and the Alternating Current."
30. Passer, *The Electrical Manufacturers,* p. 132.
31. Belfield, "Westinghouse and the Alternating Current."
32. Passer, *The Electrical Manufacturers,* p. 136.
33. Drew and Chapman, "William Stanley Lighted a Town and Powered an Industry," p. 11.
34. Ibid., p. 1.
35. Passer, *The Electrical Manufacturers,* p. 137.
36. Drew and Chapman, "William Stanley Lighted a Town and Powered an Industry," p. 12.
37. "Adam, Meldrum & Anderson, a Brilliant Illumination," *Buffalo Commercial Advertiser,* November 27, 1886, p. 1.
38. Matthew Josephson, *Edison: A Biography* (New York: McGraw-Hill, 1959), p. 346.

CHAPTER 6

Edison Declares War

1. "In a Blizzard's Grasp," *The New York Times,* March 13, 1888, p. 1.
2. *New-York Daily Tribune,* April 17, 1888.
3. Francis E. Leupp, *George Westinghouse: His Life and Achievements* (Boston: Little, Brown, 1918), pp. 143–44.
4. "Wireman's Recklessness," *The New York Times,* May 12, 1888, p. 8.
5. Matthew Josephson, *Edison: A Biography* (New York: McGraw-Hill, 1959), p. 346.
6. Edison Electric Light Company Report of the Board of Trustees to the Stockholders at their Annual Meeting, October 25, 1887, p. 13. Thomas A. Edison Archives website, Rutgers University.
7. Ibid., p. 18.
8. Robert Conot, *A Streak of Luck* (New York: Seaview Books, 1979), p. 255.
9. Francis Jehl, *Menlo Park Reminiscences,* vol. II (Dearborn Park, Mich.: Edison Institute, 1938), pp. 832–33.
10. W. Bernard Carlson and A. J. Millard, "Defining Risk within a Business Context: Thomas Edison, Elihu Thomson, and the a.c.-d.c. Controversy, 1885–1900," in Branden B. Johnson and Vincent T. Covello, eds., *The Social and Cultural Construction of Risk* (Boston: Reidel Publishing, 1987), p. 279.
11. Kenneth R. Toole, "The Anaconda Copper Mining Company, a Price War and a Copper Corner," *Pacific Northwest Quarterly* 41 (1950), p. 322.
12. "Copper," *Electrical Engineer* 7 (February 18, 1888), p. 42.

13. Toole, "The Anaconda Copper Mining Company," p. 324.
14. Thom Metzger, *Blood and Volts: Edison, Tesla and the Electric Chair* (Brooklyn: Autonomedia, 1996), pp. 28–29.
15. Terry S. Reynolds and Theodore Bernstein, "Edison and 'the Chair,' " *IEEE Technology & Society,* March 1989, p. 20.
16. Ibid., p. 21.
17. Harold Passer, *The Electrical Manufacturers, 1875–1900* (Cambridge: Harvard University Press, 1953), p. 153.
18. Ibid., p. 154.
19. Conot, *A Streak of Luck,* p. 253.
20. "A Warning from the Edison Electric Light Co.," February 1888, p. 31. Thomas Edison Archives website, Rutgers University.
21. Ibid., p. 25.
22. Ibid., p. 26.
23. Paul Israel, *Edison: A Life of Invention* (New York: Wiley, 1998), p. 326.
24. "A Warning from the Edison Electric Light Co.," p. 72.
25. John J. O'Neill, *Prodigal Genius: The Life of Nikola Tesla* (New York: McKay, 1944), p. 67.
26. Kenneth M. Swezey, "Nikola Tesla," *Science* 127, no. 3307 (May 16, 1958), p. 1149.
27. T. Commerford Martin, *The Inventions, Researches and Writings of Nikola Tesla,* 2nd ed. (New York: Barnes & Noble Books, 1995; first publication 1893), p. 9.
28. Ibid., p. 10.
29. Ibid.
30. Robert Lomas, *The Man Who Invented the Twentieth Century* (London: Headline, 1999), pp. 24–25.
31. "Discussion," *Electrical Engineer,* June 1888, p. 276.
32. Ibid.
33. Passer, *The Electrical Manufacturers,* p. 277.
34. Ibid., p. 278.
35. Marc Seifer, *Wizard: The Life and Times of Nikola Tesla* (New York: Citadel Press, 1998), p. 50.
36. Passer, *The Electrical Manufacturers,* p. 278.
37. Nikola Tesla, "George Westinghouse," *Electrical World* 26, no. 12 (March 21, 1914), p. 637.
38. John T. Ratzlaff, ed., *Tesla Said* (Millbrae, Calif.: Tesla Book Co., 1984), p. 281 ("Letter to the Institute of Immigrant Welfare" dated May 12, 1938).

CHAPTER 7

"Constant Danger from Sudden Death"

1. Letter from George Westinghouse to Thomas A. Edison dated June 7, 1888. Thomas A. Edison Archives website, Rutgers University.

2. Letter from Thomas A. Edison to George Westinghouse dated June 12, 1888. Thomas A. Edison Archives website, Rutgers University.
3. "High Potential Systems Before the Board of Electrical Control of New York City," *Electrical Engineer* 7 (August 1888), p. 369.
4. *People of the State of N.Y. Ex. Rel. Wm Kemmler* vs. *Charles F. Durston, as Warden of the State Prison at Auburn, N.Y.* In Proceedings under Writ of Habeas Corpus commencing July 8, 1889. See Exhibit A, p. xiii.
5. Matthew Josephson, *Edison: A Biography* (New York: McGraw-Hill, 1959), p. 315.
6. Thomas P. Hughes, "Harold P. Brown and the Executioner's Current: An Incident in the AC-DC Controversy," *Business History Review* 32 (June 1958), p. 148.
7. "Died for Science's Sake," *The New York Times,* July 31, 1888, p. 8.
8. "Mr. Brown's Rejoinder, Electrical Dog Killing," *Electrical Engineer* 7 (August 1888), p. 369.
9. "Died for Science's Sake," p. 8.
10. Josephson, *Edison: A Biography,* p. 347.
11. *People of the State of N.Y. Ex. Rel. Wm Kemmler* vs. *Charles F. Durston.* Exhibit A, p. xvi.
12. Letter from Harold P. Brown to Arthur Kennelly dated August 4, 1888. Thomas A. Edison Archives website, Rutgers University.
13. Frank Friedel, *The Presidents of the United States* (Washington, D.C.: White House Historical Association, 1989), p. 52.
14. Terry S. Reynolds and Theodore Bernstein, "Edison and 'the Chair,' " *IEEE Technology & Society,* March 1989, p. 22.
15. Ibid.
16. "Surer Than the Rope," *The New York Times,* December 6, 1888, p. 5.
17. "Electricity on Animals," *The New York Times,* December 13, 1888.
18. George Westinghouse, "No Special Danger," *The New York Times,* December 13, 1888, p. 5.
19. Harold Brown, "Electric Currents," *The New York Times,* December 13, 1888, p. 5.
20. Kenneth Ross Toole, "The Anaconda Copper Mining Company: A Price War and a Copper Corner," *Pacific Northwest Quarterly* 41 (October 1950), pp. 326–27.
21. Josephson, *Edison: A Biography,* pp. 355–56.
22. Nikola Tesla, "George Westinghouse," *Electrical World* 63, no. 12 (March 21, 1914), p. 637.
23. Marc Seifer, *Wizard: The Life and Times of Nikola Tesla* (New York: Citadel Press, 1998), p. 54.
24. Charles F. Scott, "Early Days in the Westinghouse Shops," *Electrical World* 84 (September 20, 1924), p. 587.
25. John T. Ratzlaff, ed., *Tesla Said* (Millbrae, Calif.: Tesla Book Co., 1984), p. 272 ("Press Statement" dated July 10, 1937).
26. John J. O'Neill, *Prodigal Genius: The Life of Nikola Tesla* (New York: McKay, 1944), pp. 76–77.

CHAPTER 8

"The Horrible Experiment"

1. "Jealousy," *The Buffalo Evening News,* April 2, 1889, p. 1.
2. "Electric Death," *The Buffalo Evening News,* April 4, 1889, p. 1.
3. "For Shame, Brown!," *New York Sun,* August 25, 1889, p. 6.
4. Letter from F. S. Hastings to Thomas Edison dated March 8, 1889. Thomas A. Edison Archives website, Rutgers University.
5. "For Shame, Brown!," *New York Sun,* August 25, 1889, p. 6. Letter from Harold Brown to Thomas Edison dated March 27, 1889.
6. Ibid.
7. Ibid.
8. Thom Metzger, *Blood and Volts: Edison, Tesla and the Electric Chair* (Brooklyn: Autonomedia, 1996), p. 119.
9. "For Shame, Brown!," p. 6. Letter from Arthur Kennelly to Harold Brown dated June 19, 1889.
10. "Expert Brown's Views," *The New York Times,* July 11, 1889, p. 6.
11. "Power of Electricity," *The New York Times,* July 16, 1889, p. 8.
12. Ibid.
13. "Edison Says It Will Kill," *New-York Daily Tribune,* July 24, 1889; and "Testimony of the Wizard," *The New York Times,* July 24, 1889, p. 2.
14. Terry S. Reynolds and Theodore Bernstein, "Edison and 'the Chair,' " *IEEE Technology and Society,* March 1989, p. 24.
15. Neil Baldwin, *Edison: Inventing the Century* (New York: Hyperion, 1995), pp. 204–05.
16. Ibid., p. 206.
17. *Proceedings of the National Electric Light Association,* August 6, 7, 8, 1889 (New York: J. Kempster Printing, 1890), p. 139.
18. "For Shame, Brown!," p. 6.
19. "His Desk Robbed," *New York Journal,* September 4, 1889.
20. "Met Death in the Wires," *The New York Times,* October 12, 1889, p. 1.
21. "Like a City in Mourning," *The New York Times,* October 16, 1889, p. 1.
22. Thomas A. Edison, "The Danger of Electric Lighting," *North American Review* 149 (November 1889), pp. 625–33.
23. E. H. Heinrichs, "Anecdotes and Reminiscences of George Westinghouse," October 1931, p. 14–15. George Westinghouse: Anecdotes and Reminiscences, vol. 2, box 1, file folder 7, George Westinghouse Museum Archives, Wilmerding, Pennsylvania.
24. Ibid.
25. George Westinghouse Jr., "A Reply to Mr. Edison," *North American Review* 149 (November 1889), pp. 653–64.
26. Lewis B. Stillwell, "Alternating Current Versus Direct Current," *Electrical Engineering* 53 (May 1934), p. 710.

27. Metzger, *Blood and Volts,* p. 131.
28. "The Law If Constitutional," *The New York Times,* December 31, 1889, p. 2.
29. Harold P. Brown, "The New Instrument of Execution," *North American Review* 149 (November 1889), pp. 592–93.
30. Craig Brandon, *The Electric Chair: An Unnatural American History* (Jefferson, N.C.: MacFarland & Co., 1999), p. 145.
31. Ibid., p. 142.
32. Metzger, *Blood and Volts,* p. 136.
33. Baldwin, *Edison: Inventing the Century,* p. 202.
34. "Far Worse than Hanging," *The New York Times,* August 7, 1890, p. 1.
35. Ibid.
36. "Inhuman!," *The Buffalo Evening News,* August 7, 1890, p. 1.
37. Ibid.
38. "Far Worse than Hanging," p. 2.
39. Ibid.
40. Brandon, *The Electric Chair,* p. 187.
41. "Far Worse than Hanging," p. 2.
42. Letter to *New York World,* November 29, 1929, p. 10.

CHAPTER 9

1891: "Fear Everywhere of Worse to Come"

1. Francis E. Leupp, *George Westinghouse: His Life and Achievements* (Boston: Little, Brown & Company, 1918), p. 157.
2. A. G. Uptegraff, "The Home Life of George Westinghouse," July 1936, p. 2. George Westinghouse: Anecdotes and Reminiscences, vol. 4, box 1, file folder 9, George Westinghouse Museum Archives, Wilmerding, Pennsylvania.
3. "Activity of the Westinghouse Electric & Manufacturing Co.," *Electrical Engineer,* October 8, 1890, p. 404.
4. Paul Israel, *Edison: A Life of Invention* (New York: Wiley, 1998), p. 335.
5. Henry G. Prout, *A Life of George Westinghouse* (New York: Scribner's, 1926), p. 275.
6. Leupp, *George Westinghouse,* p. 158.
7. Ibid.
8. "Changes in Westinghouse Management," *Electrical Engineer,* December 24, 1890, p. 710.
9. Leupp, *George Westinghouse,* p. 159.
10. Prout, *A Life of George Westinghouse,* p. 275.
11. Matthew Josephson, *Edison: A Biography* (New York: McGraw-Hill, 1959), p. 354.
12. Jean Strouse, *Morgan: American Financier* (New York: Random House, 1999), p. 312.
13. Marc Seifer, *Wizard: The Life and Times of Nikola Tesla* (New York: Citadel Press, 1998), p. 31.

14. John J. O'Neill, *Prodigal Genius: The Life of Nikola Tesla* (New York: McKay, 1944), pp. 82–83.
15. Joseph Wetzler, "Electric Lamps," *Harper's Weekly,* July 11, 1891, p. 524.
16. T. Commerford Martin, ed., *The Inventions, Researches and Writings of Nikola Tesla* (New York: Barnes & Noble, 1995, 2nd ed.; 1st ed. published 1893.
17. "High Frequency Experiments," *Electrical World,* May 30, 1891, p. 385.
18. Israel, *Edison: A Life of Invention,* p. 334.
19. Josephson, *Edison: A Biography,* p. 360.
20. Bernard W. Carlson, *Innovation as a Social Process* (New York: Cambridge University Press, 1991), p. 292.
21. Ibid., p. 293.
22. Clarence W. Barron, *More They Told Barron* (New York: Harper Bros., 1931), pp. 38–39.
23. "An Inventor at Sixteen," *The New York Times,* January 29, 1891, p. 6.
24. "Boston Takes Many Shares," *The New York Times,* February 5, 1891, p. 1.
25. Harold Passer, *The Electrical Manufacturers, 1875–1900* (Cambridge: Harvard University Press, 1953), p. 152.
26. "Edison's Patent Upheld," *The New York Times,* July 15, 1891, p. 3.
27. "The Edison Lamp Decision," *Electrical Engineer,* July 22, 1891, pp. 90–91.
28. American Society of Mechanical Engineers, *George Westinghouse Commemoration* (New York: ASME, 1937), pp. 57–58.
29. "Rumors about Villard," *The New York Times,* December 16, 1891, p. 8.
30. Alfred O. Tate, *Edison's Open Door* (New York: E. P. Dutton, 1938), pp. 260–61.
31. Carlson, *Innovation as a Social Process,* p. 294.
32. Josephson, *Edison: A Biography,* p. 363.
33. Ibid.
34. Ibid.
35. "Edison Makes Objection," *New-York Daily Tribune,* February 20, 1892, p. 2.
36. "Mr. Edison Is Satisfied," *The New York Times,* February 21, 1892, p. 2.
37. "Mr. Edison's Mistake," *Electrical Engineer,* February 17, 1892, p. 162.
38. Tate, *Edison's Open Door,* pp. 278–79.

CHAPTER 10

The World's Fair: "The Electrician's Ideal City"

1. Donald L. Miller, *City of the Century: The Epic of Chicago and the Making of America* (New York: Simon & Schuster, 1996), p. 181.
2. Bessie Louise Pierce, *As Others See Chicago* (Chicago: University of Chicago Press, 1933). pp. 395–96.
3. Francis E. Leupp, *George Westinghouse: His Life and Achievements* (Boston: Little, Brown, 1918), p. 163.
4. E. S. McClelland, "Notes on My Career with Westinghouse," April 1939, George Westinghouse: Anecdotes and Reminiscences, vol. 3, box 1, file folder 8, George Westinghouse Museum Archives, Wilmerding, Pennsylvania.

5. E. H. Heinrichs, "Anecdotes and Reminiscences of Westinghouse," October 1931, pp. 19–20. George Westinghouse: Anecdotes and Reminiscences, vol. 2, box 1, file folder 7, George Westinghouse Museum Archives, Wilmerding, Pennsylvania.
6. "Will Underbid the Trust," *Chicago Times*, April 26, 1892.
7. David F. Burg, *Chicago's White City of 1893* (Lexington: University Press of Kentucky, 1976), p. 91.
8. Charles H. Baker, *Life and Character of William Taylor Baker* (New York: Premier Press, 1908), p. 159.
9. "World's Fair Doings," *Daily Interocean*, May 17, 1892, p. 5.
10. Ibid., May 18, 1892, p. 5.
11. Ibid., May 24, 1892, p. 5.
12. *Chicago Tribune*, May 24, 1892, p. 6.
13. "Westinghouse," *Electrical Engineer*, June 1, 1892, p. 555.
14. E. S. McClelland, "Notes on My Career with Westinghouse," pp. 5–6.
15. Henry G. Prout, *A Life of George Westinghouse* (New York: Scribner's, 1926), p. 136.
16. E. E. Keller, "Geo. Westinghouse Memories," April 1936, p. 4. George Westinghouse: Anecdotes and Reminiscences, vol. 3, box 1, file folder 8, George Westinghouse Museum Archives, Wilmerding, Pennsylvania.
17. "The Edison Light Bulb," *The New York Times*, October 5, 1892, p. 9.
18. "A Most Dangerous Trust," *The New York Times*, November 19, 1892, p. 5.
19. Leupp, *George Westinghouse*, pp. 167–69.
20. Benjamin Lamme, *Benjamin Garver Lamme* (New York: Putnam's, 1926), p. 61.
21. "The Westinghouse World's Fair Exhibit," *Electrical Engineer*, January 25, 1893, p. 100.
22. F. Herbert Stead, "An Englishman's Impressions at the Fair," *Review of Reviews* 8 (July 1893), pp. 30–31.
23. Ibid., p. 32.
24. J. P. Barrett, "Electricity," in G. R. Davis, *World's Columbian Exposition* (Chicago: Elliott Beezley, 1893), p. 301.
25. J. R. Cravath, "Electricity at the World's Fair," *Review of Reviews* 8 July 1893, p. 35.
26. Burg, *Chicago's White City of 1893*, p. 232.
27. Rossiter Johnson, ed., *History of the World's Columbian Exposition*, vol. 1 (New York: Appleton's, 1897, 4 vols.), pp. 481–82.
28. "The Progress of the World," *Review of Reviews* 8 July 1893, p. 1.
29. Johnson, *History of the World's Columbian Exposition*, vol. 1, p. 188.
30. Leupp, *George Westinghouse*, p. 169.
31. "Dazzles Ben's Eyes," *Chicago Tribune*, June 2, 1893, p. 1.
32. "Westinghouse Work at the Fair," *Electrical Engineer*, August 16, 1893, p. 153.
33. J. P. Barrett, *Electricity at the Columbian Exposition* (Chicago: R. R. Donnelly, 1894), pp. 168–69.
34. "Colored People's Day," *Chicago Record*, August 26, 1893.

35. Marc Seifer, *Wizard: The Life and Times of Nikola Tesla* (New York: Citadel Press, 1998), p. 117.
36. T. Commerford Martin, "Nikola Tesla," *Century,* February 1894, p. 584.
37. Leupp, *George Westinghouse,* p. 170.
38. Miller, *City of the Century,* p. 534.

CHAPTER 11

Niagara Power: "What a Fall of Bright-Green Water!"

1. Pierre Berton, *Niagara: A History of the Falls* (New York: Penguin, 1992), p. 51.
2. Ibid., p. 111.
3. Ibid., p. 151.
4. Charles F. Scott, "Personality of the Pioneers of Niagara Power," March 31, 1938. Niagara Mohawk Archives, Syracuse, New York.
5. Robert Belfield, "Niagara Frontier: The Evolution of Electric Power Systems in New York and Ontario, 1880–1935," Ph.D. Thesis, University of Pennsylvania, 1981, p. 8.
6. Letter from Coleman Sellers to Edward Dean Adams dated October 5, 1889. Thomas A. Edison Archives website, Rutgers University.
7. Letter from Coleman Sellers to Edward Dean Adams dated December 17, 1889. Thomas A. Edison Archives website, Rutgers University.
8. George Forbes, "The Electrical Transmission of Power from Niagara Falls," *Journal of the Institution of Electrical Engineers* 22, no. 108 (November 9, 1893), p. 485.
9. Stillwell, quoted in Edward Dean Adams, *Niagara Power,* vol. 1 (Niagara Falls: Niagara Falls Power Co., 1927), p. 363.
10. Adams, *Niagara Power,* vol. 1, pp. 191 ff.
11. Ibid., vol. 2, p. 178.
12. Letter from Coleman Sellers to Edward Dean Adams dated March 17, 1893. Niagara Mohawk Archives, Syracuse, New York.
13. Adams, *Niagara Power,* vol. 2, p. 174.
14. Steven Lubar, "Transmitting the Power of Niagara: Scientific, Technological, and Cultural Contexts of an Engineering Decision," *IEEE Technology & Society,* March 1989, p. 14.
15. Belfield, "Niagara Frontier," p. 18.
16. William Stanley, "Notes on the Distribution of Power by AC," *Electrical World,* February 6, 1892, p. 88.
17. Adams, *Niagara Power,* vol. 2, p. 173.
18. Charles F. Scott, "Long Distance Transmission for Lighting and Power," *Electrical Engineer,* June 15, 1892, p. 601.
19. Edwin G. Burrows and Mike Wallace, *Gotham: A History of New York City to 1898* (New York: Oxford, 1999), pp. 1167–69.
20. Charles F. Scott, "My Own Story of AC and Electrical Power Development, 1887–1895," February 1938. Niagara Mohawk Archives, Syracuse, New York.

21. ——, "Nikola Tesla's Achievements in the Electrical Art," *Electrical Engineering,* August 1943.
22. Letter from Nikola Tesla to the Westinghouse Company dated September 27, 1892. Library of Congress, Manuscript Division, Tesla Papers.
23. George Forbes, "The Utilization of Niagara," *Electrical Engineer,* January 18, 1893, p. 65.
24. "Winter Wonders at Niagara," *The New York Times,* January 6, 1893, p. 9.
25. George Forbes, "Harnessing Niagara," *Blackwood's,* September 1895, pp. 431–32.
26. Coleman Sellers, "Report on Dynamos," March 17, 1893, p. 14. Box 11.8, Niagara Mohawk Archives, Syracuse, New York.
27. Henry Rowland's March 1, 1893, Final Report, p. 75. Rowland Papers, Special Collections and Archives, The Johns Hopkins University, ms. 6, box 50, series 7.
28. Coleman Sellers, "Report on Dynamos," p. 25.
29. Ibid., p. 7.
30. Nikola Tesla, *My Inventions: The Autobiography of Nikola Tesla* (Williston, Vt.: Hart Bros., 1982), p. 48.
31. Letters from Nikola Tesla to Edward Dean Adams dated March–May 1893. Niagara Mohawk Archives, Syracuse, New York.
32. Adams, *Niagara Power,* vol. 2, p. 256.
33. Ibid., p. 182.
34. Ibid., p. 193.
35. "The Alleged Theft of Westinghouse Blueprints," *Electrical Engineer,* June 14, 1893, p. 587; "Theft of the Westinghouse Blue Prints," *Electrical Engineer,* September 13, 1893, p. 251.
36. Letter from Charles A. Coffin to H. McK. Twombly dated May 9, 1893. Henry Lee Higginson Collection 1870–1919, mss. 783, box XII-30, folder 3-102, C. A. Coffin, 1893, Baker Library, Harvard Business School.
37. Letter from Edward Dean Adams to GE dated May 11, 1893. Thomas A. Edison Archives website, Rutgers University.
38. Silvanus P. Thompson, "Utilizing Niagara," *Saturday Review,* August 3, 1895, p. 34.
39. All these letters are in the Niagara Mohawk Archives, Syracuse, New York.

CHAPTER 12

"Yoked to the Cataract!"

1. George Forbes, "Harnessing Niagara," *Blackwood's,* September 1895, pp. 431–32.
2. Ibid., p. 441.
3. Daniel M. Dumych, "William Birch Rankine" (pamphlet published by Drumdow Press, N. Tonawanda, N.Y., 1991), p. 19.
4. Jean Strouse, *Morgan: American Financier* (New York: Random House, 1999), p. 324.

5. Harold Passer, *The Electrical Manufacturers, 1875–1900* (Cambridge: Harvard University Press, 1953), p. 289.
6. Coleman Sellers, "Memo on Visits of Westinghouse Representatives to Niagara Before Presentation of Final Dynamo Design," February 2, 1894, box 11.8. Niagara Mohawk Archives, Syracuse, New York.
7. Passer, *The Electrical Manufacturers,* p. 289.
8. Sellers, "Memo on Visits of Westinghouse Representatives to Niagara Before Presentation of Final Dynamo Design."
9. Edward Dean Adams, *Niagara Power,* vol. 2 (Niagara Falls: Niagara Falls Power Company, 1927), p. 410.
10. Confidential memo from Coleman Sellers to Edward Dean Adams dated December 27, 1893. Niagara Mohawk Archives, Syracuse, New York.
11. Letter from Edward Wickes to Edward Dean Adams dated February 6, 1894, box 26.7. Niagara Mohawk Archives, Syracuse, New York.
12. "The Niagara Dynamo Controversy," *Electrical Engineer,* April 3, 1895, p. 308.
13. Forbes, "Harnessing Niagara," pp. 431–32.
14. Henry G. Prout, *A Life of George Westinghouse* (New York: Scribner's, 1926), pp. 152–53.
15. Page Smith, *The Rise of Industrial America* (New York: Penguin, 1984), p. 525.
16. Passer, *The Electrical Manufacturers,* p. 291.
17. Thomas W. Lawson, *Frenzied Finance,* vol. 1 (New York: Ridgway-Thayer Co., 1905), p. 90.
18. Marc Seifer, *Wizard: The Life and Times of Nikola Tesla* (New York: Citadel Press, 1998), p. 130.
19. Robert Underwood Johnson, *Remembered Yesterdays* (Boston: Little, Brown, 1923), p. 400.
20. Seifer, *Wizard,* p. 161.
21. T. Commerford Martin, "Nikola Tesla," *Century,* February 1894, p. 582.
22. Arthur Brisbane, *Sunday World,* July 22, 1894, p. 5.
23. "The Nikola Tesla Company," *Electrical Engineer,* February 13, 1895, p. 149.
24. Editorial, *New York Sun,* March 14, 1895, p. 6.
25. "Mr. Tesla's Great Loss," *The New York Times,* March 14, 1895, p. 9.
26. Margaret Cheney, *Tesla, Man Out of Time* (Englewood Cliffs, N.J.: Prentice-Hall, 1981), p. 107.
27. Letter from Nikola Tesla to Albert Schmid dated March 22, 1895; letter from Nikola Tesla to Charles Scott dated May 9, 1895. Library of Congress, Manuscript Division, Tesla Papers.
28. Pierre Berton, *Niagara: A History of the Falls* (New York: Penguin, 1992), p. 167.
29. Ibid., p. 170.
30. Adams, *Niagara Power,* vol. 2, p. 417.
31. "Niagara Is Finally Harnessed," *The New York Times,* August 27, 1895, p. 9.
32. Ron Chernow, *The House of Morgan* (New York: Atlantic Monthly Press, 1990), p. 76.

33. H. G. Wells, "The End of Niagara," *Harper's Weekly,* July 21, 1906, pp. 1018–20.
34. Adams, *Niagara Power,* vol. 2, p. 336.
35. Orrin E. Dunlap, "Nikola Tesla at Niagara Falls," *Western Electrician,* August 1, 1896.
36. "Power for Buffalo," *Daily Cataract* (Niagara Falls), July 20, 1896, p. 1.
37. "Yoked to the Cataract!," *Buffalo Enquirer,* November 16, 1896, p. 1.
38. "A Few Cold Facts About Buffalo," *The Buffalo Evening News,* January 8, 1897, p. 8.
39. "Magnificent Power Celebration Banquet at the Ellicott Club," *Buffalo Morning Express,* January 13, 1897, p. 1.
40. Ibid.

CHAPTER 13
Afterward

1. Guido Pantaleoni, "The Real Character of the Man as I Saw Him," April 1939, p. 5. George Westinghouse: Anecdotes and Reminiscences, vol. 3, box 1, file folder 8, George Westinghouse Museum Archives, Wilmerding, Pennsylvania.
2. Henry G. Prout, *A Life of George Westinghouse* (New York: Scribners, 1926), p. 206.
3. Ibid.
4. Maurice Coster, "Personal Reminiscences of George Westinghouse," November 1936, p. 1. George Westinghouse: Anecdotes and Reminiscences, vol. 1, box 1, file folder 1, George Westinghouse Museum Archives, Wilmerding, Pennsylvania.
5. Westinghouse Electric Corporation, "George Westinghouse, 1846–1914," 1946. Box 1, file folder 12, George Westinghouse Museum Archives, Wilmerding, Pennsylvania.
6. Jean Strouse, *Morgan: American Financier* (New York: Random House, 1999), p. 574.
7. Francis E. Leupp, *George Westinghouse: His Life and Achievements* (Boston: Little, Brown, 1918), p. 209.
8. Ibid., p. 210.
9. E. H. Heinrichs, "Anecdotes and Reminiscences of George Westinghouse," October 1931, pp. 31–32. George Westinghouse: Anecdotes and Reminiscences, vol. 2, box 1, file folder 7, George Westinghouse Museum Archives, Wilmerding, Pennsylvania.
10. Leupp, *George Westinghouse,* p. 210.
11. Strouse, *Morgan: American Financier,* p. 595.
12. Ibid., p. 574.
13. Stefan Lorant, *Pittsburgh* (New York: Doubleday & Co., 1964), p. 180.
14. Heinrichs, "Anecdotes and Reminiscences of George Westinghouse."
15. Ibid.

16. Leupp, *George Westinghouse,* pp. 224–25.
17. Alexander Uptegraff, "The Home Life of George Westinghouse," July 1936, p. 4. George Westinghouse: Anecdotes and Reminiscences, vol. 4, box 1, file folder 9, George Westinghouse Museum Archives, Wilmerding, Pennsylvania.
18. "George Westinghouse," *The New York Times,* October 24, 1907, p. 10.
19. Heinrichs, "Anecdotes and Reminiscences of George Westinghouse."
20. Alfred O. Tate, *Edison's Open Door* (New York: E. P. Dutton, 1938), p. 278.
21. Paul Israel, *Edison: A Life of Invention* (New York: Wiley, 1998), p. 347.
22. Matthew Josephson, *Edison: A Biography* (New York: McGraw-Hill, 1959), p. 372.
23. Israel, *Edison: A Life of Invention,* p. 361.
24. osephson, *Edison: A Biography,* p. 378.
25. Ibid., p. 379.
26. Ibid., p. 399.
27. Israel, *Edison: A Life of Invention,* p. 415.
28. Josephson, *Edison: A Biography,* p. 429.
29. Ibid., p. 430.
30. "World Made Over by Edison's Magic," *The New York Times,* October 18, 1931, section II, p. 1.
31. Letter from E. H. Heinrichs to Nikola Tesla dated December 8, 1897. Library of Congress, Manuscript Division, Tesla Papers.
32. John J. O'Neill, *Prodigal Genius: The Life of Nikola Tesla* (New York: David McKay, 1944), p. 81.
33. Ibid., p. 167.
34. Marc Seifer, *Wizard: The Life and Times of Nikola Tesla* (New York: Citadel Press, 2000), p. 210.
35. Harry L. Goldman, "Nikola Tesla's Bold Adventure," *American West* 8, no. 2 (March 1971), p. 5.
36. Ibid., p. 7.
37. Nikola Tesla, "The Transmission of Electric Energy Without Wires," *Electrical World and Engineer,* March 5, 1904.
38. Letter from Nikola Tesla to J. P. Morgan dated December 12, 1900. Library of Congress, Manuscript Division, Tesla Papers.
39. Ibid., January 9, 1901.
40. Letter from J. P. Morgan to Nikola Tesla dated July 17, 1903. Library of Congress, Manuscript Division, Tesla Papers.
41. Letter from Nikola Tesla to J. P. Morgan dated January 14, 1904. Library of Congress, Manuscript Division, Tesla Papers.
42. Ibid., October 13, 1904.
43. Letter from C. W. King to Nikola Tesla dated October 15, 1904. Library of Congress, Manuscript Division, Tesla Papers.
44. Letter from Nikola Tesla to J. P. Morgan dated December 23, 1913. Library of Congress, Manuscript Division, Tesla Papers.

45. Seifer, *Wizard,* p. 384.
46. Letter from Nikola Tesla to E. M. Herr dated October 19, 1920. Library of Congress, Manuscript Division, Tesla Papers.
47. Letter from Nikola Tesla to Westinghouse Company dated January 29, 1930. Library of Congress, Manuscript Division, Tesla Papers.
48. "Mr. Tesla Speaks Out," *New York World,* November 29, 1929, p. 10.
49. "Tesla at 75," *Time,* July 30, 1931.
50. O'Neill, *Prodigal Genius,* p. 317.
51. Arthur G. Woolf, "Electricity, Productivity, and Labor-Saving: American Manufacturing, 1900–1929," *Explorations in American Economic History* 21 (1984), p. 179.
52. David E. Nye, *Electrifying America* (Cambridge: MIT Press, 1997), p. 389.

Photograph Credits

CHAPTER 1: Photograph by George William Sheldon (attributed), from *The Opulent Interiors of the Gilded Age*, published by D. Appleton and Company, New York, 1883–84. Reprinted by Dover Publishers, 1987, p. 146.

CHAPTER 2: Courtesy of the Library of Congress

CHAPTER 3: Photograph by R. F. Outcault. Courtesy of the U.S. Department of the Interior, National Park Service, Edison National Historic Site

CHAPTER 4: Courtesy of the Nikola Tesla Museum

CHAPTER 5: Copyright 2002, George Westinghouse Museum

CHAPTER 6: Courtesy of the New York Historical Society

CHAPTER 7: Courtesy of the U.S. Department of the Interior, National Park Service, Edison National Historic Site

CHAPTER 8: Courtesy of the Cayuga Museum of History and Art

CHAPTER 9: Courtesy of the George Westinghouse Museum

CHAPTER 10: Photograph by G. Hunter Bartlett. Courtesy of the Chicago Historical Society

CHAPTER 11: Courtesy of the Niagara Falls Public Library

CHAPTER 12: Courtesy of Niagara Mohawk, National Grid USA Service Company, Inc.

CHAPTER 13: George Westinghouse, Courtesy of George Westinghouse Museum; Thomas Edison, courtesy of the Library of Congress; Nikola Tesla, courtesy of the Nikola Tesla Museum

Index

EMPIRES *of* LIGHT

Jill Jonnes

A Conversation with the Author

Q: Electricity is something most of us take for granted. How did you become interested in electricity and its history?

Jill Jonnes: I was reading an old biography of George Westinghouse and it included the amazing episode of the first electric chair. We all know Thomas Edison as a boyish and brilliant American hero, so I was quite astonished to find he could be so cutthroat and ruthless. He promoted the development and first use of the electric chair purely to discredit his business rival's product—alternating current electricity. That drama completely drew me in, and I quickly found that the rest of the characters and the story were incredibly compelling. We take electricity utterly for granted, so it was wonderful to relive the period when it was such an exotic technology that only a few visionaries could understand its potential. After all, electricity is invisible and thus quite mysterious. Very few people imagined at the time how completely electricity would remake the world.

Q: J. P. Morgan wanted to be the first person ever to have electricity run through his house. What happened when Edison wired his mansion?

JJ: When J. P. Morgan, one of Edison's earliest financial backers, renovated his large Madison Avenue brownstone mansion, he daringly decided to install this luxurious, cutting-edge technology—electricity—for his lighting, rather than prosaic gas lighting. Edison's men came and dug up Morgan's stable, installed a steam engine and DC electric generator, and ran the electric wires across the garden and into the house, where they were snaked through the existing pipes for gas lighting. Imagine many clusters of tiny light bulbs dangling baldly from the ceiling on wires and you have the idea. Every day an engineer came to Morgan's house before dusk, started the generator out in the stable, and ran the electrical system until 11:00 P.M. or so.

Nonetheless, there were many problems, and eventually an Edison executive came back to rewire the whole house. This time, he thought he'd make a special arrangement so Morgan could have a light on the desk in his magnificent library. The light shorted, and the whole fabulous room went up in flames while the Morgans were at the opera. Morgan was very understanding, allowing the wiring to be fixed, and he enjoyed impressing his wealthy friends with his electrical lighting.

Q: How did the general public react to the lighting of Manhattan?

JJ: The public loved having the streets and avenues of Manhattan lit up, and it all looked wonderfully romantic, especially as shop windows and theaters came ablaze with lights. The first electric streetlights were not Edison's, but intensely powerful and glaring "arc" lights run by other electrical companies. These arc lights had to be perched very high up on tall poles because the light was so brilliant and harsh to the eye. Broadway was one of the first avenues to be lit with arc lights—hence, the Great White Way—and everyone just marveled at how well they could see, and what an interesting effect the light had. People came out in droves in the evenings to enjoy it all. However, arc lights operated on very high voltages and people also began to discover that the wires could be quite deadly.

Q: How did the War of the Electric Currents begin, and what were the original motives behind it?

JJ: The War of the Electric Currents began on December 9, 1887, when Thomas Edison, who had declared that he was against the death penalty and wanted no part of an electric chair, changed his mind and wrote the New York State Commission that the quickest and cleanest way of applying the death penalty would be to use an alternating current machine produced by George Westinghouse. Edison was furious that Westinghouse, a hugely successful inventor and Pittsburgh corporate magnate, was invading his field with what seemed a better product. Edison was determined to convince the public that Westinghouse's kind of electricity was deadly. Edison was a great man, and this was certainly not his finest hour. Seeing the dark side of Edison was most fascinating.

Q: And yet in public, Thomas Edison pretended to be an "unschooled hick."

JJ: Edison, by the time he was inventing the light bulb and launching his electric company, was quite rich by anyone's standards. But he deliberately shunned the sartorial conventions of the day—suit, top hat—preferring to wear a shabby workman's smock and little cap. I think Edison enjoyed this "playing the rube" because he liked to operate outside the rules and because he led a very hands-on work life. He was already so famous that none of this prevented him from finding plenty of financial backers at first. But when electricity turned out to be a far more complex technology than, say, the telegraph, his backers got cold feet. Edison risked much of his own fortune on creating his electrical empire.

Q: The quirky inventor Nikola Tesla discovered electricity as we know it today—alternating current, or AC—and yet he is virtually forgotten. What happened?

JJ: This brilliant but highly eccentric inventor—among his many phobias was a horror of women wearing pearls—viewed his AC work as just a crude start to understanding and mastering the mysteries of electricity. Tesla's research was always visionary, focusing on wireless transmission of energy for communication or power. Tesla should have been rich just from the royalties on AC, but, ever the idealist, he gave them up to save the Westinghouse Electric Company in a time of dire financial peril. Without his AC royalties, which would have been $17 million, Tesla never had enough capital to launch his expensive and ambitious projects. J. P. Morgan briefly financed Tesla's "World System of Power" transmission, a huge tower on Long Island, but then refused further help. This proud genius was reduced to writing pathetic letters begging for help, living out his last years at the Hotel New Yorker, his great love being a beautiful white pigeon.

Q: George Westinghouse stated that his goal, besides helping the world, was to give "many persons an opportunity to earn money by their own efforts." How did this ambition shape the way he approached producing electricity and the War of the Electric Currents?

JJ: The Gilded Age of the 1880s and 1890s was an incredibly rocky economic period, and there was no such thing as a social safety net—just hard times such as few Americans today can imagine. When George Westinghouse, a self-made man and idealist, realized he had a talent for taking inventions (his own or those of others) and creating big and successful companies, he strived to create vast enterprises spanning the globe with as many decent jobs as possible. Eventually, he employed 50,000 people. In turn, when Westinghouse faced losing his companies because of money woes, his employees always offered help to their beleaguered boss. Westinghouse was certainly tough, but his essential decency shone through during the War of the Electric Currents. He knew he had a sterling reputation in an era of robber barons and he firmly believed that the superiority of his technology would win the day—as it did.

Q: Why was the safety of Tesla and Westinghouse's version of electricity, AC, so contested?

JJ: When Thomas Edison first learned about the AC system invented by Nikola Tesla and developed by George Westinghouse, he darkly warned that, "Just as certain as death, Westinghouse will kill a customer within six months." Safety was a paramount issue for Edison: He buried all his electrical conduits under the streets and prided himself that any part of his system gave only a mild electric shock. AC traveled long distances at high voltages, and Edison insisted this electrical current could not be safely and reliably brought down before the electricity entered houses and offices. Even as time and experience showed he was wrong, Edison used his enormous prestige and power to try and thwart Westinghouse at every turn.

Q: The ax murderer William Kemmler became the first person ever sentenced to die by electricity, in what the papers called "the horrible experiment" of the electric chair. How did the Kemmler case change how people thought about electricity?

JJ: On August 6, 1890, when William Kemmler became the first person to die in an electric chair, in Auburn State Penitentiary, electricity was still a very exotic and cutting-edge technology. Even as late as 1907, only 8 percent of American households had electricity because gaslight and kerosene were either cheaper or the only choice. So in 1890, very few Americans had ever seen or experienced manufactured electricity. Edison won the first big battle of the War of the Electric Currents when William Kemmler was electrocuted with Westinghouse generators. But because Kemmler did not die instantaneously or easily, it certainly cast grave doubt on Edison's assertion that AC electricity was a stone killer. One suspects the whole episode of the electric chair simply confirmed electricity as a deeply mysterious but powerful form of energy.

Q: There was a dramatic fight over who would get to electrify the Chicago World's Fair of 1893. How did this change the circumstances of the War of the Electric Currents?

JJ: The second great clash in the War of the Electric Currents came in Chicago over the World's Fair of 1893. Who would win the all-important lighting contract for this international exposition that could showcase electricity to millions who had never seen it? J. P. Morgan had ousted Edison and merged his company with others to become General Electric. Westinghouse was just reorganizing *his* electric company to pacify his creditors. And so GE, known as the "electrical trust," felt free to make an extortionate bid to light the fair. The Chicago businessmen were outraged. Westinghouse then swooped in to Chicago and saved the day, underbidding the trust by a cool million dollars. The Chicago World's Fair was an electrical wonder (all operating on Tesla's AC system) that dazzled all who came.

Q: Explain the controversy around the Cataract Company's decision to have Professor George Forbes design the Niagara Falls dynamo.

JJ: The third and final battleground of the War of the Electric Currents was Niagara Falls. Here the Cataract Company proposed to build the world's first great hydroelectric power plant to generate electricity and transmit it twenty-six miles to booming Buffalo. Once again, George Westinghouse faced off against General Electric to win a huge and groundbreaking electric contract. When the Cataract Company abruptly announced that their own expert, the supercilious Scotsman George Forbes, would design the all-important dynamos, Westinghouse—who had shared all his firm's secrets—was enraged. Nikola Tesla quietly informed the Cataract Company that no one could make such a design without infringing his AC patents. After months of wrangling, the Cataract Company concluded this was true. They awarded the contract on October 27, 1893, to a triumphant George Westinghouse. Three years later, Westinghouse engineers transmitted electricity to Buffalo. With Nikola Tesla's brilliant AC system perfected, electricity would now power the world.

About the Type

This book was set in FF Celeste, a digital font that its designer, Chris Burke, classifies as a modern humanistic typeface. Celeste was influenced by Bodoni and Waldman, but the strokeweight contrast is less pronounced, making it more suitable for current digital typesetting and offset-printing techniques. The serifs tend to the triangular, and the italics harmonize well with the roman in tone and width. It is a robust and readable text face that is less stark and modular than many of the modern fonts and has many of the friendlier old-face features.